Gender and Social Computing:

Interactions, Differences and Relationships

Celia Romm Livermore
Wayne State University, USA

Senior Editorial Director:	Kristin Klinger
Director of Book Publications:	Julia Mosemann
Editorial Director:	Lindsay Johnston
Acquisitions Editor:	Erika Carter
Development Editor:	Joel Gamon
Production Editor:	Sean Woznicki
Typesetters:	Mike Brehm, Jennifer Romanchak
Print Coordinator:	Jamie Snavely
Cover Design:	Nick Newcomer

Published in the United States of America by
Information Science Reference (an imprint of IGI Global)
701 E. Chocolate Avenue
Hershey PA 17033
Tel: 717-533-8845
Fax: 717-533-8661
E-mail: cust@igi-global.com
Web site: http://www.igi-global.com

Copyright © 2012 by IGI Global. All rights reserved. No part of this publication may be reproduced, stored or distributed in any form or by any means, electronic or mechanical, including photocopying, without written permission from the publisher. Product or company names used in this set are for identification purposes only. Inclusion of the names of the products or companies does not indicate a claim of ownership by IGI Global of the trademark or registered trademark.

Library of Congress Cataloging-in-Publication Data

Gender and social computing: interactions, differences, and relationships / Celia Romm, editor.
   p. cm.
  Includes bibliographical references and index.
  Summary: "This book provides an overview of the major questions that researchers and practitioners are addressing, outlining possible future directions for theory development and empirical research on gender and computing"--Provided by publisher.
  ISBN 978-1-60960-759-3 (hardcover) -- ISBN 978-1-60960-760-9 (ebook) -- ISBN 978-1-60960-761-6 (print & perpetual access) 1. Computers and women. 2. Women in computer science. 3. Information technology--Social aspects. I. Romm-Livermore, Celia, 1954-
  QA76.9.W65G46 2012
  303.48'33--dc23
                             2011022934

British Cataloguing in Publication Data
A Cataloguing in Publication record for this book is available from the British Library.

All work contributed to this book is new, previously-unpublished material. The views expressed in this book are those of the authors, but not necessarily of the publisher.

Table of Contents

Preface xiii

Acknowledgment xvii

Section 1
Gender and Computing in the Work Arena

Chapter 1
Gender and Anonymity in Virtual Teams: An Exploratory Study 1
    Elizabeth Koh, National University of Singapore, Singapore
    Na Liu, National University of Singapore, Singapore
    John Lim, National University of Singapore, Singapore

Chapter 2
Adopting ICT in the Mompreneurs Business: A Strategy for Growth? 17
    Yvonne Costin, University of Limerick, Ireland

Chapter 3
Women in IT Careers: Investigating Support for Women in the Information Technology Workforce 35
    Elaine K. Yakura, Michigan State University, USA
    Louise Soe, California State Polytechnic University, USA
    Ruth Guthrie, California State Polytechnic University, USA

Chapter 4
Gender Segregation and ICT: An Indo-British Comparison 50
    Sunrita Dhar-Bhattacharjee, University of Salford, UK
    Haifa Takruri-Rizk, University of Salford, UK

Chapter 5
Gender and E-Marketing: The Role of Gender Differences in Online Purchasing Behaviors 72
    Erkan Özdemir, Uludag University, Turkey

Chapter 6
Women in Information Communication Technologies 87
    Olca Surgevil, Dokuz Eylul University, Turkey
    Mustafa F. Özbilgin, Brunel University, UK & Université Paris-Dauphine, France

Chapter 7
The Impact of Gender in ICT Usage, Education and Career: Comparisons between Greece and
Germany 98
    Bernhard Ertl, Universität der Bundeswehr München, Germany
    Kathrin Helling, Leopold-Franzens-Universität Innsbruck, Germany
    Kathy Kikis-Papadakis, Institute for Applied Computational Mathematics, Greece

Chapter 8
The Effect of Gender on Associations between Driving Forces to Adopt ICT and Benefits Derived
from that Adoption in Medical Practices in Australia.. 120
    Rob Macgregor, University of Wollongong, Australia
    Peter Hyland, University of Wollongong, Australia
    Charles Harvey, University of Wollongong, Australia

Section 2
Gender and Computing in Cyberspace

Chapter 9
The Not So Level Playing Field: Disability Identity and Gender Representation in Second Life.......... 144
    Abbe E. Forman, Temple University, USA
    Paul M. A. Baker, Georgia Institute of Technology, USA
    Jessica Pater, Georgia Institute of Technology, USA
    Kel Smith, Anikto LLC, USA

Chapter 10
Overcoming the Segregation/Stereotyping Dilemma: Computer Mediated Communication for
Business Women and Professionals... 162
    Natalie Anne Sappleton, Manchester Metropolitan University, UK

Chapter 11
Would Elizabeth Cady Stanton Blog? Women Bloggers, Politics, and Political Participation 183
    Antoinette Pole, Montclair State University, USA

Chapter 12
Gender Differences in Social Networking Presence Effects on Web-Based Impression Formation 200
    Leslie Jordan Albert, San Jose State University, USA
    Timothy R. Hill, San Jose State University, USA
    Shailaja Venkatsubramanyan, San Jose State University, USA

Section 3
Gender and eDating

Chapter 13
E-Dating: The Five Phases on Online Dating 222
   Monica T. Whitty, Nottingham Trent University, UK

Chapter 14
How E-Daters Behave Online: Theory and Empirical Observations 236
   Celia Romm Livermore, Wayne State University, USA
   Toni Somers, Wayne State University, USA
   Kristina Setzekorn, Smith Barney, Inc., USA
   Ashley Lynn-Grace King, Wayne State University, USA

Chapter 15
A Trination Analysis of Social Exchange Relationships in E-Dating 257
   Sudhir H. Kale, Bond University, Australia
   Mark T. Spence, Bond University, Australia

Chapter 16
Online Matrimonial Sites and the Transformation of Arranged Marriage in India 272
   Nainika Seth, University of Alabama in Huntsville, USA
   Ravi Patnayakuni, University of Alabama in Huntsville, USA

Compilation of References 296

About the Contributors 332

Index 340

Detailed Table of Contents

Preface xiii

Acknowledgment xvii

Section 1
Gender and Computing in the Work Arena

Chapter 1
Gender and Anonymity in Virtual Teams: An Exploratory Study 1
    Elizabeth Koh, National University of Singapore, Singapore
    Na Liu, National University of Singapore, Singapore
    John Lim, National University of Singapore, Singapore

With the advancement of information and communication technology, virtual teams are becoming ever more popular as geographical constraints in collaboration have become non-issue. Features of technology and characteristics of group influence interaction processes and outcomes. Two elements are the focus of the current work. The first is anonymity, which has been made feasible by technology. The other element concerns gender. In and of itself, gender is an important research target; its role in groupwork must not be overlooked. Both of these elements have aroused much interest across multiple research fields. The existing literature shows their potential in influencing team collaboration processes, satisfaction and performance. The current study presents a process-based interpretation of virtual team collaboration incorporating the anonymity of technology feature and the gender difference of team members. Using a multiple case study approach, the chapter identifies a key set of process variables that help shape team performance. Over and above these individual variables, the study examines the interdependencies among the processes. Task-related activity that occurred during team discussion is found to be affected by gender anonymity, and this in turn influences group performance and members' satisfaction towards the collaboration process. Group dynamics including member awareness, leader emergence and member's conformity are found to be salient process variables that affect the virtual team performance as well.

Chapter 2
Adopting ICT in the Mompreneurs Business: A Strategy for Growth? 17
    Yvonne Costin, University of Limerick, Ireland

It is advocated that the role of technology is instrumental in determining the effectiveness and efficiency of where, when and how business transactions are undertaken to meet stakeholder requirements in a competitive manner. However, research by the Small Business Forum (2006) suggest the use and application of ICT in small firms overall is poor where entrepreneurs do not capitalise sufficiently on the benefits of ICT. To succeed and grow, mompreneurs' businesses should be using ICT as a backbone for the business in an integrated manner. The objective of this chapter is to examine the adoption and application of ICT in the mompreneur business and challenges encountered in its effective use. A specific emphasis is placed on the issue of ICT and its use by the mompreneur in undertaking business transactions and as a means of facilitating small firm growth.

Chapter 3
Women in IT Careers: Investigating Support for Women in the Information Technology Workforce 35
    Elaine K. Yakura, Michigan State University, USA
    Louise Soe, California State Polytechnic University, USA
    Ruth Guthrie, California State Polytechnic University, USA

This chapter examines issues of support for women with Information Technology (IT) careers. Interviews with 38 women lasted about 90 minutes. Questions were open-ended regarding aspects of their careers and career paths. Women represented a wide variety of experience and nine different industry sectors and at varying organizational levels. Research on the lack of women in STEM disciplines focuses mainly on undergraduate education and attracting women to STEM disciplines, focusing on 'filling the pipeline.' This research seeks to understand what it takes to have a successful, satisfying career; this manuscript highlights areas of support for women that may influence their success in IT careers. Knowing this may give women help in planning their careers and navigating a male dominated industry.

Chapter 4
Gender Segregation and ICT: An Indo-British Comparison 50
    Sunrita Dhar-Bhattacharjee, University of Salford, UK
    Haifa Takruri-Rizk, University of Salford, UK

Gender segregation in science, engineering, construction, technology (SECT) is a common persistent feature, both in India and U.K. Even though culturally the two countries differ in various ways, under-representation of women in SECT is widespread and a cause for general apprehension and in recent years this has attracted centre stage in the study of gender, work and family. This chapter discusses the authors' research findings of a comparative study undertaken between India and Britain in the ICT sector. With twenty seven interviews with ICT professionals in the two countries, the authors discuss their views on ICT education, recruitment and employment practices, work-life balance, changing gender relations, opportunities for progression and retention in the two countries taking into consideration women's role in power and politics in the both countries; how 'public' and 'private' patriarchy shapes women's position in the labour market, with an essential backdrop of 'patrifocality' in the Indian context.

Chapter 5
Gender and E-Marketing: The Role of Gender Differences in Online Purchasing Behaviors 72
    Erkan Özdemir, Uludag University, Turkey

The increasingly widespread use of the Internet and the increasing participation of females in business have been significant changes leading to society today. As females have begun to have a more important role in business life, in addition to their crucial role within the family in the decision-making process of making purchases, they have had more financial independence. Additionally, the gap in Internet use in favor of males has also begun to narrow. This makes it necessary for e-marketers that carry out some or all of their transactions online to be fully aware of the effects of gender differences in online purchasing behaviors. This chapter consists of a literature review on the subject of Internet usage and online purchasing behaviors with a focus on gender-based differences. Accordingly, the aim of this chapter is to explore gender-based differences in Internet usage and online purchasing behavior and to suggest some e-marketing strategies for e-marketers. Additionally, this chapter provides a foundation on which to build future studies.

Chapter 6
Women in Information Communication Technologies 87
    Olca Surgevil, Dokuz Eylul University, Turkey
    Mustafa F. Özbilgin, Brunel University, UK & Université Paris-Dauphine, France

Employment in the Information Communication Technology (ICT) sectors around the world has been affected by gender discrimination and stereotyping practices. In the light of this, in this study it is aimed to analyze women's attendance to ICT field and discuss the qualifications (fundamental capabilities and training) which they need to get into the sector. Drawing on interdisciplinary insights, this study questions the implications of numerical feminization in the context of gendered cultures and processes of work in the information technology (IT) sector, and proposes some directions for future research.

Chapter 7
The Impact of Gender in ICT Usage, Education and Career: Comparisons between Greece and
Germany 98
    Bernhard Ertl, Universität der Bundeswehr München, Germany
    Kathrin Helling, Leopold-Franzens-Universität Innsbruck, Germany
    Kathy Kikis-Papadakis, Institute for Applied Computational Mathematics, Greece

Gender is an important issue in the context of information and communication technologies (ICT). Studies show that ICT use is subject to gender bias, e.g. in relation to ICT use and interests. This contribution describes the current situation of gender and ICT professions in Germany and Greece. Based on an empirical study, it shows particular areas in ICT education that suffer from gender inequalities in both countries. Furthermore, the chapter elaborates how gender inequalities develop from secondary to professional ICT careers based on statistics from Germany and Greece.

Chapter 8
The Effect of Gender on Associations between Driving Forces to Adopt ICT and Benefits Derived
from that Adoption in Medical Practices in Australia 120
    Rob Macgregor, University of Wollongong, Australia
    Peter Hyland, University of Wollongong, Australia
    Charles Harvey, University of Wollongong, Australia

Like other Small to Medium Enterprises (SMEs), medical practices can gain a great deal by adopting and using Information and Communication Technologies (ICTs). Unlike other SMEs, little is known about General Practitioners' (GPs) perceptions of the benefits of ICT use or about the differences between these perceptions by male and female GPs. This chapter reports a survey of these perceptions of the drivers for and benefits of ICT use by male and female GPs in Australia.

Section 2
Gender and Computing in Cyberspace

Chapter 9
The Not So Level Playing Field: Disability Identity and Gender Representation in Second Life.......... 144
    Abbe E. Forman, Temple University, USA
    Paul M. A. Baker, Georgia Institute of Technology, USA
    Jessica Pater, Georgia Institute of Technology, USA
    Kel Smith, Anikto LLC, USA

This chapter examines the portrayal of disability, gender and identity in virtual communities where representation is a matter of convenience, style or whim. A survey was conducted of groups, identifying themselves as disabled or having a disability, with a focus on gender, in the virtual space, Second Life. Four distinctive categories were analyzed in this study: groups associated with disabilities or being disabled, race/ethnicity, gender, aging, and sexuality. In the "real world", the visual cues that activate schemas can serve as an explanation for the stigmas and ensuing isolation often felt by people with disabilities. However, in Second Life where the visual cues are removed, users with disabilities are still associating with others who identify as having disabilities. Additionally, gender appears to play a role in the group (i.e. "communities") found in Second Life. Regardless of binary gender framework, the differences between the groups that are externally classified as having some degree of disability, and those who choose to self identify, or affiliate with disability related groups, have rich import for the sociology of online communities as well as for the design and characteristics of games.

Chapter 10
Overcoming the Segregation/Stereotyping Dilemma: Computer Mediated Communication for
Business Women and Professionals 162
    Natalie Anne Sappleton, Manchester Metropolitan University, UK

Since the 1970s, there has been a sizeable uptake of business ownership among women in the United States. Moreover, the fastest recent growth in women-owned firms has been in traditionally male industries such as telecommunications and construction. This development has increased the importance of cross-sex networking to women professionals and entrepreneurs. Setting up a business is no easy task, and network members may be able to provide direct or indirect access to the required assistance, support, information and tangible resources. Categories of persons with whom associations might be fruitful include: individuals in the same industry who have experienced the process of starting and developing a business, potential clients, buyers of the product or service, suppliers, previous colleagues, connections from relevant trade bodies or professional associations, or experts in business matters (lawyers, accountants, financiers, marketers and so on) – all roles that remain overwhelmingly male-dominated.

However, women business owners (particularly those working in traditionally male sectors) may be hindered in their efforts to build collaborative relations with male ties because of gender stereotyping. As a response, women may elect to join all-female networks, but because women in male-dominated sectors are relatively scarce, this strategy may reduce access to resources. In this chapter, it is suggested that virtual networking via web pages, email, chat rooms and networking sites can provide a solution for women in male dominated contexts stuck in a 'segregation/stereotyping bind'. As Sproull and Kiseler (1991: 13) predicted almost twenty years ago, social computing allows individuals to "cross barriers of space, time and social category to share expertise, opinions and ideas". Additionally, virtual communities can provide access to a very large number of diverse others and network maintenance is substantially easier than face-to-face interaction. A review of the literature provides evidence to support these propositions. The chapter concludes with suggestions for the direction of future research.

Chapter 11
Would Elizabeth Cady Stanton Blog? Women Bloggers, Politics, and Political Participation 183
Antoinette Pole, Montclair State University, USA

This study examines the role of women political bloggers and how they use their blogs for purposes related to politics, public policy and current events. Based on a combined purposive snowball sample, in-depth interviews were conducted with 20 women political bloggers in October 2006. Findings show respondents blog about a range of topics, not necessarily unique to women. Generally, women use their blogs to inform their readers, check the media, engage in advocacy efforts, and solicit charitable contributions from their readers and more specifically, women ask their readers to vote and contact elected officials. Data show women deal with a range of challenges blogging the political, most notably discrimination. Though a majority of women political bloggers reported they did not face discrimination, interviewees qualified their responses saying they witnessed discrimination and discriminatory attitudes, suggesting the political blogosphere is somewhat inhospitable to women.

Chapter 12
Gender Differences in Social Networking Presence Effects on Web-Based Impression Formation 200
Leslie Jordan Albert, San Jose State University, USA
Timothy R. Hill, San Jose State University, USA
Shailaja Venkatsubramanyan, San Jose State University, USA

As the Web has expanded in its use and utility it has fundamentally changed the way in which individuals gather and use information. The authors suggest that those changes give rise to tangible and significant effects in the impressions we form of others using Web-based information. This study explores the impacts of perceiver gender, target gender and social networking presence on subjects' perceptions of potential teammates otherwise unknown to them as revealed by ratings they assign based only on search engine results. Experiments reveal differences in how male and female perceivers view others' social networking activity in general and suggest that how the perceiver gender matches, or differs, from the gender of the target affects how social networking presence plays into impression formation. Findings hold implications for professionals, academics and individuals concerned with the role that Web-based information plays in impression formation and how inherent gender-based biases may affect power and politics in the workplace and beyond.

Section 3
Gender and eDating

Chapter 13
E-Dating: The Five Phases on Online Dating 222
  Monica T. Whitty, Nottingham Trent University, UK

Online dating continues to grow in popularity as a way for individuals to locate a potential romantic partner. Researchers have examined how people present themselves on these sites, which presentations are more likely to lead to success, the effectiveness of the matchmaking tools that some companies employ, the stigma attached to using these sites and the types of people who are drawn to online dating. However, there is an absence of scholarly work on how these relationships progress compared to traditional models of courtship. This chapter sets out a model for the phases of online dating and compares this model with Givens' (1979) work on a traditional model of courtship. It is argued here the phases of online dating are very different to other courtship models. These differences pose new challenges and create new benefits to those who elect to find a partner via one of these sites.

Chapter 14
How E-Daters Behave Online: Theory and Empirical Observations 236
  Celia Romm Livermore, Wayne State University, USA
  Toni Somers, Wayne State University, USA
  Kristina Setzekorn, Smith Barney, Inc., USA
  Ashley Lynn-Grace King, Wayne State University, USA

Following a review of the literature on eDating, this chapter introduces the eDating development model and discusses a number of hypotheses that can be derived from it. Also presented in the chapter are some findings from a preliminary empirical research that explored the hypotheses. The findings supported all the hypotheses, indicating that: (1) male and female eDaters follow different stages in their eDating evolution; (2) that the behaviors that males and females exhibit as eDaters are different; and (3) that the feedback that male and female eDaters receive from the environment is different too. The chapter is concluded with a discussion of the implications from this research to eDating theory development and empirical research.

Chapter 15
A Trination Analysis of Social Exchange Relationships in E-Dating.. 257
  Sudhir H. Kale, Bond University, Australia
  Mark T. Spence, Bond University, Australia

More than half a billion users across the globe have availed themselves of eDating services. This chapter looks at the marketing and cross-cultural aspects of mate seeking behavior in eDating. The authors content analyzed 238 advertisements from online matrimonial sites in three countries: India (n=79), Hong Kong (n=80), and Australia (n=79). Frequencies of mention of the following ten attribute categories in the advertiser's self-description were established using post hoc quantitative analysis: love, physical status, educational status, intellectual status, occupational status, entertainment services, money, demographic information, ethnic information, and personality traits. Past research on mate selection using personal ads and the three countries' positions on Hofstede's dimensions of culture were used in hypotheses generation. The results support several culture-based differences in people's self description in online personal ads; however, some anticipated differences were not realized, suggesting that some cultural differences may not be as strong as Hofstede (2001) suggests.

Chapter 16
Online Matrimonial Sites and the Transformation of Arranged Marriage in India 272
    Nainika Seth, University of Alabama in Huntsville, USA
    Ravi Patnayakuni, University of Alabama in Huntsville, USA

Online personals have been a remarkably successful in the western world and have been emulated in other cultural contexts. The introduction of the Internet can have vastly different implications on traditional societies and practices such as arranged marriages in India. This chapter seeks to investigate using an ethnographic approach the role of matrimonial web sites in the process of arranging marriages in India. It seeks to explore how these web sites have been appropriated by key stakeholders in arranging marriage and how such appropriation is changing the process and traditions associated with arranged marriage. The key contributions of this study are in that it is an investigation of complex social processes in a societal context different from traditional western research contexts and an exploration of how modern technologies confront societal traditions and long standing ways of doing things. This investigation suggests that the use of matrimonial web sites have implications for family disintermediation, cultural convergence, continuous information flows ease of disengagement, virtual dating and reduced stigma in arranged marriages in India.

Compilation of References 296

About the Contributors 332

Index 340

Preface

This book, Gender and Social Computing: Interactions, Differences and Relationships, provides an overview of the major questions that researchers and practitioners are addressing at this time, outlining possible future directions for theory development and empirical research on gender and computing. In doing so, this volume contributes toward closing the gap between the public growing interest in gender and computing and the relative scarcity of texts on this topic.

The book focuses on three areas of research on gender and computing, (1) gender and computing in the work arena, (2) gender and computing in cyberspace, and, (3) gender in eDating.

1. GENDER AND COMPUTING IN THE WORK ARENA

In Chapter 1, Elizabeth Koh, Liu Na, and John Lim focus on the interaction between anonymity and gender and their impact on groupwork, particularly in terms of team collaboration processes, satisfaction and performance. The study that the authors report on in the chapter presents a process-based interpretation of virtual team collaboration, incorporating the anonymity of technology feature and the gender difference of team members. Using a multiple case study approach, the authors identify a key set of process variables that help shape team performance, examining the interdependencies among the processes. The major findings from the research are that task-related activities that occurred during team discussion are affected by gender anonymity, and that this in turn influences group performance and members' satisfaction towards the collaboration process. Group dynamics including member awareness, leader emergence and member's conformity are also salient process variables that affect the virtual team performance.

In Chapter 2, Yvonne Costin discusses why women start new business, how they select the industry for their business, and where they place their business. The focus of the chapter is on the home worker/entrepreneur or the mompreneur, namely, mothers who decide primarily for family reasons, to leave paid employment and establish a business operated from home. The chapter examines the adoption and application of ICT in the mompreneur business and discusses challenges encountered in this process. A specific emphasis is placed on the issue of ICT and its use by the mompreneur in undertaking business transactions and as a means of facilitating small firm growth.

In Chapter 3, Elaine Yakura, Louise Soe, and Ruth Guthrie examine issues of support for women in Information Technology (IT) careers. Based on open ended interviews with 38 women from nine different industry sectors and at varying organizational levels, the authors highlight areas of support for women that may influence their success in IT careers, pointing out that knowing these areas of support may give women help in planning their careers in a male dominated industry.

In Chapter 4, Sunrita Dhar-Bhattacharjee and Haifa Takruri-Rizk investigate the dynamics of gender segregation in science, engineering, construction, and technology (SECT) in India and the U.K. the authors point out that even though culturally the two countries differ in various ways, under-representation of women in SECT is widespread in both. Based on in-depth interviews with twenty seven ICT professionals in the two countries, the authors comment on differences between the two countries in education, recruitment and employment practices, work-life balance, changing gender relations, and opportunities for progression and retention, taking into consideration women's role in power and politics in the two countries and how 'public' and 'private' patriarchy shapes women's position in the labour market.

In chapter 5, Erkan Özdemir discusses the role that gender has in e-marketing, particularly as it pertains to gender differences in online purchasing behaviors. The chapter consists of a literature review on the subject of Internet usage and online purchasing behaviors with a focus on gender-based differences. The aim of the chapter is to explore gender-based differences in Internet usage and online purchasing behavior and to suggest e-marketing strategies for e-marketers.

In Chapter 6, Olca Surgevil and Mustafa F. Özbilgin analyze women's participation in the ICT field and the qualifications (fundamental capabilities and training) that women need to get into the sector. Drawing on interdisciplinary insights, this chapter explores the implications of numerical feminization in the context of gendered cultures and processes of work in the information technology (IT) sector, and proposes directions for future research on gender in the ICT field.

In Chapter 7, Bernhard Ertl, Kathrin Helling, and Kathy Kikis-Papadakis describe the situation of gender and ICT in Germany and Greece. Based on an empirical study, the chapter outlines areas in ICT education that suffer from gender inequalities in both countries, explaining how gender inequalities in education affect women's professional in ICT in Germany and Greece.

In Chapter 8, Rob Macgregor, Peter Hyland, and Charles Harvey discuss the adoption of ICT's by male and female General Practitioners' (GPs) in Australia, with particular emphasis on the practitioners' perceptions of the benefits of ICT to them. Based on a survey of GP's perceptions, the authors discuss the differences between male and female understanding of the role that ICT can play in their practices and the benefits that they can derive from ICT's.

2. GENDER AND COMPUTING IN CYBERSPACE

In Chapter 9, Abbe E. Forman, Paul Baker, Jessica Pater, and Kel Smith examine the portrayal of disability, gender, and identity in virtual communities where representation is a matter of convenience, style, or whim. Based on survey of groups that identified themselves as disabled or having a disability, with a focus on gender, four distinctive categories were analyzed in this study: groups associated with disabilities or being disabled, race/ethnicity, gender, aging, and sexual orientation. The findings from this research are that in virtual reality environments (like Second Life) where the visual cues are removed; users with disabilities are still associating with others who identify themselves as having disabilities. Furthermore, gender appears to play a role in the group (i.e. "communities") in Second Life. The authors point out that the differences between the groups that are externally classified as having some degree of disability and those who choose to self identify or affiliate with disability related groups have rich import for the sociology of online communities, as well as, for the design and characteristics of games.

In Chapter 10, Natalie Sappleton suggests that virtual networking via web pages, email, chat rooms and networking sites can provide a solution for women in male dominated contexts stuck in a 'segregation/stereotyping bind'. Social computing allows individuals to cross barriers of space, time, and social category and to share expertise, opinions and ideas with each other. As a result, virtual communities of female business owners in non-traditional areas can provide access to a large number of diverse others and opportunities for network maintenance that may not be available to females in these situations.

In Chapter 11, Antoinette Pole examines the role of women political bloggers and how they use their blogs for purposes related to politics, public policy, and current events. Based on a combined purposive snowball sample, in-depth interviews were conducted with 20 women political bloggers. The findings show that respondents blog about a range of topics, not necessarily unique to women, use their blogs to inform their readers, check the media, engage in advocacy efforts, and solicit charitable contributions from their readers. The data also show that women deal with a range of challenges blogging the political, including discrimination. Though a majority of women political bloggers report that they do not face discrimination, interviewees qualified their responses saying they witnessed discrimination and discriminatory attitudes, suggesting the political blogosphere is somewhat inhospitable to women.

In Chapter 12, Tim Hill, Leslie Albert, and Shai Venkatsubramanyan explore the impacts of perceiver gender, target gender, and social networking presence on subjects' perceptions of potential teammates otherwise unknown to them and revealed to them by ratings based only on search engine results. The results of the study on which the chapter is based reveal differences in how male and female perceivers view others' social networking activity in general and suggest that how the perceiver gender matches, or differs, from the gender of the target affects how social networking presence plays into impression formation. The findings hold implications for professionals, academics and individuals concerned with the role that Web-based information plays in impression formation and how inherent gender-based biases may affect power and politics in the workplace and beyond.

3. GENDER AND eDATING

In Chapter 13, Monica Whitty examines the differences between the development of on-line relationships and the more traditional face-to-face courtships. The chapter presents a model for the phases of online dating and compares this model with Givens' (1979) work on a traditional model of courtship. It argues that eDating follows different "phases" than other courtship models and that these differences pose challenges and create benefits that are different from the challenges and benefits that traditional daters face.

In Chapter 14, Celia Romm-Livermore, Toni Somers, Kristina Setzekorn, and Ashley King introduce the eDating development model. The model focuses on the changes that male and female eDaters undergo during the process of eDating. The discussion in the chapter focuses on findings from a preliminary empirical research undertaken by the authors. The findings supported all of the model's hypotheses, indicating that: (1) male and female eDaters follow different stages in their eDating evolvement; (2) the behaviors that males and females exhibit as eDaters are different; and (3) the feedback that male and female eDaters receive from the environment is different too.

In Chapter 15, Sudhir H. Kale and Mark T. Spence consider the marketing and cross-cultural aspects of mate seeking behavior in eDating. The study that is presented in this chapter is based on a content analysis of 238 advertisements from online matrimonial sites in three countries: India (n=79), Hong Kong (n=80), and Australia (n=79). Frequencies of the following ten attribute categories in the advertiser's self-description were established, including, love, physical status, educational status, intellectual status, occupational status, entertainment services, money, demographic information, ethnic information, and personality traits. The results support several culture-based differences in people's self description in online personal ads.

And, finally, in Chapter 16, Nainika Seth and Ravi Patnayakuni use an ethnographic approach to examine the role of matrimonial web sites in the process of arranging marriages in India. The chapter explores how eDating web sites have been appropriated by key stakeholders in arranging marriages and how such appropriation is changing the process and traditions associated with arranged marriages in India. The investigation undertaken by the authors suggests that the use of matrimonial web sites have implications for family disintermediation, cultural convergence, continuous information flows, ease of disengagement, virtual dating and reduced stigma in arranged marriages in India.

Celia Romm Livermore
Wayne State University, USA

Acknowledgment

This book would never come into being if it weren't for the IGI Global team who contacted me with the suggestion to edit this book and provided me with help throughout the book project. I would like to extend my sincere gratitude in particular to Julia Mosemann, Director of Book Publications and Heather Probst, Director of Journal Publications, who have supported me tirelessly throughout this book project.

This book would not have been written if it were not for the pioneers of research on Gender and Computing whose work has inspired me to accept the IGI Global offer. As a female and a professor of information technology, this book is dedicated to these pioneers with admiration and gratitude.

The list of authors who made important contributions to the emerging body of research on Gender and Computing is long. Many of them are quoted in the book and some of them contributed chapters to the book. Without them, this book project would not become a reality.

Finally, I would like to take this opportunity to thank the members of my family, my husband, the Honorable Bill Callahan, and my children, David, Jonathan and Gail. It was their love and support, as well as, their enthusiasm about all my projects that helped me bring this project to completion.

Celia Romm Livermore
Wayne State University, USA

Section 1
Gender and Computing in the Work Arena

Chapter 1
Gender and Anonymity in Virtual Teams:
An Exploratory Study

Elizabeth Koh
National University of Singapore, Singapore

Na Liu
National University of Singapore, Singapore

John Lim
National University of Singapore, Singapore

ABSTRACT

With the advancement of information and communication technology, virtual teams are becoming more popular as geographical constraints in collaboration have become a non-issue. Features of the technology and characteristics of the group influence interaction processes and outcomes. Two elements are the focus of this paper. The first is anonymity, which has been made feasible by technology. The other concerns gender. Gender is an important research target, and its role in groupwork must not be overlooked. Both elements have aroused much interest across multiple research fields. The existing literature shows their potential in influencing team collaboration processes, satisfaction, and performance. In this paper, the authors present a process-based interpretation of virtual team collaboration, incorporating the anonymity of technology and the gender difference of team members. Using a multiple case study approach, the paper identifies a key set of process variables that shape team performance. The study also examines the interdependencies among the processes. Task-related activity that occurred during team discussion was affected by gender anonymity, and this influenced group performance and members' satisfaction toward the collaboration process. Group dynamics, including member awareness, leader emergence, and member's conformity, are salient process variables that affect the virtual team performance as well.

DOI: 10.4018/978-1-60960-759-3.ch001

1 INTRODUCTION

The distributed work force is becoming more prevalent in many organizations. Increasingly, e-business requires employees to work across office boundaries such as from their homes, outstations, and different countries. Employees often work in temporary groups to fulfill particular organizational tasks. These are known as virtual teams, who due to geographical, organizational or time dispersion, rely heavily on IT such as computer-mediated communications (CMC) to accomplish one or more organizational goals (Powell, Piccoli, & Ives, 2004).

The Internet and other forms of CMC have allowed people to mask their identities. Anonymity, a technology characteristic, has affected how people behave and react. In computer-mediated teams, anonymity has the effect of encouraging participation of all members (Nunamaker et al., 1991). According to the minority influence theory (Nemeth, 1986), the resultant increase of minority participation would lead to more and better ideas generated. Earlier work has shown that anonymous communication is more effective for creative tasks (Tyran et al., 1992). However, whether the effect is consistent for different group compositions remains unknown.

Gender has been considered one of the fundamental personal characteristics that profoundly influence individual perceptions, attitudes, and performance (Christofides, Islam, & Desmarais, 2009; Lind, 1999). Gender differences have been shown to affect collaboration as males and females differ in their collaboration and communication styles. For instance, men tend to be more aggressive and argumentative in communication than women (Herring, 1996). Moreover, in electronic discussions, male participants tend to dominate the conversation and are more agonistic (Guiller & Durndell, 2007; Robertson, Hewitt, & Scardamalia, 2003).

Men and women interact differently with different genders in anonymous and identified virtual teams (Lind, 1999; Thomson, 2006a). In anonymous virtual teams, men and women may perceive genders differently and may even "fake" their own gender to garner certain results (Postmes, Spears, & Lea, 1998). Thus, in this study, we attempt to examine the joint effects of gender and anonymity in virtual teams, specifically comparing between all-female and all-male teams.

Based on theoretical research including task and relationship orientation and the social identity model of deindividuation (Spears & Lea, 1992), gender and anonymity are conceived to affect outcomes in virtual teams (Lind, 1999). The paper also intends to explore the team dynamics and intervening processes that enable the outcomes of performance and satisfaction to occur. The research question is, how does gender and anonymity affect group processes and subsequent outcomes of performance and satisfaction in virtual teams?

The paper will begin with a review of gender differences with regard to computer-mediated communication. The effect of anonymity in virtual teams and the interplay of gender and anonymity will then be explored followed by a review of group processes. The next section describes the research methodology. The case study method is employed and template coding and causal loop diagramming methods were utilized to analyze the data, both within the case, and across the cases. Based on these analyses, a theoretical framework is presented that maps the interdependencies of gender, anonymity, group processes and outcomes. The last section highlights several managerial and research implications drawn from the research findings and the study's limitations.

2 LITERATURE REVIEW

2.1 Gender Effects in CMC

Gender research has revealed several differences in communication styles between males and females (Barrett & Davidson, 2006). Similarly, communication styles over CMC have been found to differ between males and females (Lind, 1999; Thomson, 2006a). Males tend to be more aggressive and argumentative in communication compared to females. On the other hand, females tend to be more encouraging and nurturing compared to males. Guiller and Durndell (2007) investigated messages on an online forum and found that male postings were more likely to feature authoritative language and negative socio-emotional content while female postings contained more attenuated language and positive socio-emotional content.

Other research has highlighted that females had more information requests and interactive messages than males, whereas males provided more explanations and had a higher number of messages (Robertson et al., 2003). Research in same gender interaction in small groups has also revealed differences in communication styles. For instance, most female-female interactions were positive while male-male interactions were more often negative than they were positive (Guiller & Durndell, 2007).

Research has theorized two main perspectives for gender differences - task and relationship orientation (Hahn & Litwin, 1995) and gender-role socialization and stereotypes (Kray, Galinsky, & Thompson, 2002). In short, the task and relationship orientation posits that men are task-oriented and value self-sufficiency as they view relationships in terms of status and dominance, while women are relationship-oriented and nurturing and tend to be more sensitive to others' needs. Other research suggests that gender roles are due to socialization. Men and women learn these roles from young i.e. men are competitive and women are cooperative, as it is socially acceptable. This gender-role socialization reinforces such stereotypes.

The effect of gender differences in the dynamics and outcomes of virtual teams is also significant. Studies show that males in virtual teams were less satisfied and perceived less cohesion than females (Lind, 1999). Moreover, females believed that the group conflict was readily resolved compared to males. Using instant messengers, Christofides and colleagues (2009) found that men rated male online support staff to be more competent compared to females.

Despite the findings on gender differences, some authors downplay the phenomena of differences between gender and communication styles. For example, Robertson et al. (2003) argued that gender differences should not be seen as a problem, as the different styles for each gender are necessary for their individual process of knowledge construction. Other research suggests that CMC facilitates a gender-neutral communication environment where gender differences are non-salient. Cohen and Ellis (2008) found that response postings in a discussion forum were equally split between responses to the same gender and cross-gender.

2.2 Anonymity and Gender in Virtual Teams

Anonymity is a crucial feature in collaborative systems and refers to the inability of team members to identify the origin and destination of messages received and sent (Valacich et al., 1992). Drawing from the social identity model of deindividuation (Spears & Lea, 1992), anonymity within a social group would encourage deindividuation which maximizes the opportunity of group members to give full voice to their collective identities. Deindividuation behaviors decreases people's adherence to norms that emerge with the group. Anonymous communication thus promotes a low-threat environment, reduces evaluation apprehension, and breaks down social barriers and conformance pressures.

Some studies have shown that anonymity improves the performance of teams (Tyran et al., 1992), and in turn promotes participation equality and increases the number of ideas generated and the quality of the decision (Nunamaker et al., 1991). However other research reveals that anonymity did not affect the interaction of users.

For instance, a field study comparing the use of a group support system and traditional collaboration revealed that while parallelism and meeting memory was consistently found to increase the participation of group members, anonymity did not affect user participation (Dennis & Garfield, 2003).

The interplay between gender and anonymity in virtual teams could result in several changes. Team members, who traditionally participate less in face-to-face context, such as the females, are more likely to express themselves (Flannigan et al., 2002). Through the use of anonymity technological features, gender as a status characteristic is eliminated which weakens gender based expectations in conversation (Herschel, 1994).

In anonymous virtual teams, females enjoy interacting anonymously more than males, participate more than males, and perceive higher acceptance of ideas (Flannigan et al., 2002; Weber, Wittchen, & Hertel, 2009). In a computer-mediated experiment, Weber and colleagues (2009) found that anonymous females had higher effort and performance than identified females while identified males had higher performance compared to anonymous males in team tasks. The authors explained that females tend to engage in gender-specific norms under identified conditions and non-congruent gender roles in anonymous conditions i.e. females became more competitive in anonymity. On the other hand, males put in more effort in the identified condition as they want others to have a favorable impression of themselves (Weber et al., 2009).

However, other research has demonstrated that even with such anonymity, genders were accurately guessed. Crowston and Kammerer (1998) demonstrated that males contribute significantly more to discussions than other participants do in an anonymous setting. Researchers explain that different gender linguistic styles, such as females being reassuring and soothing compared to males being aggressive, will reveal the gender of members (Herring, 1996). This could potentially downplay the effects of gender and anonymity in online communication.

In addition, some research has found that it is not anonymity per say that affects gendered group dynamics and outcomes; rather, it is the task or topic of discussion. Thomson (2006b) posited that anonymous participants assimilate to the group norm as cued by the gender stereotype of the discussion topic due to communication accommodation. The study found that for male stereotypical topics (sports, cars), participants had higher frequency of opinions, directives, insults and adjectives while for female stereotypical topics (fashion, shopping), more personal information, questions, agreements, self-derogatory comments were shared. This shows that participants are influenced by the language style used by others as well as their expectations of how the social group would speak.

2.3 Group Processes

Group processes can be viewed from two aspects – group communication (the linguistic acts in the communication process) and the actual group dynamics present (Goldberg & Larson, 1975). While the former focuses on the communication process between group members, the latter examines the many behavior patterns of the interaction and interpersonal relationships between group members. Linguistic acts can be seen as a subset of group dynamics and is useful for tracing the observable patterns of the group. On the other hand, group dynamics analyzes the overall impression of the group.

2.3.1 Linguistic Acts

Social interactions can be mediated through language or more precisely *"linguistic acts"* (Klein & Huynh, 1999). Researchers have suggested that the analysis of language acts and communicative practices involved in collaboration brings a deeper understanding on the collaboration processes (Cecez-Kecmanovic & Webb, 2000).

There are three major linguistic acts occurring in team processes: those addressing subject matter and topic of discussion, addressing norms and rules to organize and direct the process of interaction, and those addressing personal experiences, desires and feelings (Cecez-Kecmanovic & Webb, 2000). These are known as task-related acts, norms and rules acts and socio-emotional acts respectively. Members communicate with one another to explore subject matter. They may express their opinions, seek clarification, or even argue with one another on subject-related issues. When dispute occurs, members may discuss norms and rules guiding their interaction and their work process. Members may also share their personal views, experiences, or attitudes towards the cooperation and relationships during the collaboration process.

2.3.2 Group Dynamics

Many group behavior and patterns could form in a virtual team. Research has observed several patterns of group dynamics. In particular, member awareness, conformity and leader emergence are potential behaviors that could affect the performance and satisfaction of teams.

Member awareness is an understanding of the activities of others, which provides a context for the individual's own activity (Dourish & Bellotti, 1992). Awareness of group members is believed to be a critical building block in the construction of a dynamic team.

Conformity refers to the behavior of individuals who adhere to group norms over time (Galanes & Adams, 2007). Team members conform in order to be accepted and liked, to be perceived as right and not appear a fool, and to help the group function as a whole (Galanes & Adams, 2007). Conformity is linked to group productivity unless in excessive levels, in which case, negative groupthink occurs (Janis, 1972).

Leader emergence is the natural rise of individuals who are perceived by other members as best embodies of the group prototype, that is, the behaviors and norms to which the less prototypical members are attempting to adhere (Hackman, 1990). Studies suggest that leaders strongly influence team performance and individual team members' satisfaction in virtual teams (Hiltz, Turoff, & Johnson, 1991). The leader in virtual teams usually takes the responsibility of organizing, delegating assignments, coordinating information, and supporting the contributions of others. It has also been observed that the role of the leader in groups can emerge naturally through communication with other members (Zander, 1971); when one participates more actively, he is more likely to be perceived as the leader even without being assigned leadership.

3 RESEARCH METHODOLOGY

This study adopts a case study methodology to examine the entire collaboration process of three virtual teams. In the field of Information Systems (IS) research, the case study is one of the frequently adopted research methods, and its felicity is widely acknowledged (Orlikowski, 1993). The case study method is believed to provide rich and insightful data which is helpful to understand otherwise complex issues (Yin, 1994). We adopt the case study methodology for this study based on two reasons. First, the case study approach is particularly helpful to analyze the group processes and outcomes in the virtual team. Second, we aim to study contemporary phenomenon within its real-life context without controls over behavioral events to generate theories from practice.

3.1 Case Background

A multiple case study involving three three-person virtual teams was conducted. The three teams in this study, labeled as teams A, B and C, were made

Table 1. Group characteristics

	Member Composition	Anonymity
Team A	All Male	No
Team B	All Male	Yes
Team C	All Female	No

up of undergraduate students from a Management of IS course teaching virtual team skills. They were all in the 2nd or 3rd year of their studies with an average age of 22.1. To complete the course, students needed to finish two projects in virtual teams. The first project was to list and describe the various aspects of the Hewlett-Packard organizational culture. The other was to identify Google's best competitive advantage and give justifications. Both projects were based on the course curriculum; however, students still had to do further research.

Students were allowed to form the group themselves. At end of the group registration, the instructor helped students without a group to form a group. Students were told to use the collaborative technologies of instant messaging software and email only to accomplish the project. These were popular tools used by students for their day-to-day communication and learning. Students were instructed to discuss and finish the project during their computer laboratory period only. All communication between team members was logged and collated. Finally, three teams were selected that met the criteria of our study as shown in Table1.

3.2 Data Collection

Data was collected from multiple sources to enable triangulation of data (Yin, 1994). The team chat logs were archived and used as the main data for the study. Emails between team members were also requested from students and analyzed. After completion of the project, the team members were interviewed individually. Most interviews were recorded with prior permission and then transcribed. The interviews of students were conducted at various places such as labs and canteens. The interview transcript was used as an additional source of data. The final grades of each project for all teams are recorded as reference too.

4 DATA ANALYSIS AND DISCUSSION

4.1 Data Analysis

During data analysis, we used the rich insights available in the case to help us reconstruct and reveal the collaboration process. Template coding (King, 1998) and causal loop diagram (CLD) mapping were conducted to analyze the data. Template coding was used to generate categories to represent the concepts of interest while CLD mapping helped to interrelate the template codes into a causal relationship.

The original template was generated by synthesizing the literature on gender, anonymity and group processes. Template analysis consisted of perusing the data and modifying the template accordingly until a stable template was obtained. The entire analysis was highly iterative and involved moving back and forth among the data and existing literature. The final template based on the data analysis is presented in Table 2.

Next, the template is further verified with each case separately and the mapping technique of CLD was adapted to interrelate the above-generated templates causally (Klein & Huynh, 1999). A CLD consists of variables connected by arrows denoting the causal influences among the variables. Each causal link is assigned a polarity, either positive or negative to indicate the direction of causal relationship. Subsequently, cross-case analysis was conducted to examine the overall pattern and verify the causal relationships between variables in stage two. A detailed analysis will be presented in the following sections.

Table 2. Data template

Categories	Definition/Description
Structure	
Anonymity	Anonymity refers to the inability of team members to identify the origin and destination. The opposite of anonymity is identified.
Gender	Male or Female
Linguistic Acts	
Socio-emotional acts	See section 2.3.1
Task-related acts	See section 2.3.1
Norms and rules	See section 2.3.1
Group Dynamics	
Member awareness	See section 2.3.2
Leader emergence	See section 2.3.2
Conformity	See section 2.3.2
Outcome	
Satisfaction	Members' satisfaction towards the collaboration process
Performance	Average grade of two projects evaluated by the instructor

4.2 Within Case Analysis

4.2.1 Findings from Team A

Team A consisted of 3 male members - Jack, Don and Vernon (pseudonyms). They formed their group themselves and had previously worked with each other before. During their collaboration they were very much task-focused although at times they engaged in small talk and regulation of norms. Members began with trying to understand the task required of them and they also related with each other interpersonally, being a previous history group. For instance, Don greeted Vernon with, *"Have not seen you for sometime..."* Members also shared their current *"state"* such as just having woken up or their other competing tasks in the day ahead e.g. *"many project meetings today"*.

From the start all group members were active and responsive. One particular group member, Vernon, misunderstood the task requirement and was about to do the task individually. Another alert member, Don, realized his mistake and quickly corrected him and clarified that they had to discuss and do the project together. To compensate for the lack of non-verbal communication present in CMC, members explicitly indicated to each other their non-verbal actions. For instance when Vernon was about to read a website, he typed *"hold on... looking at the website... will buzz when I am done."* Similarly, after waiting for some time for group members to read the task, Don asked, *"read the questions?"*, to signal and direct everyone's attention back to the communication medium. He proceeded to wait for everyone's response before discussing the question.

For the first project, this group started with brainstorming ideas. One after another, members would throw in website URLs and points, competing for the most number of unique points. When they had enough pointers, Don initiated the cessation of the brainstorming and asked Vernon to compile the ideas together. Although Don seemed to have directed the discussion, Vernon emerged as the leader as he was more outspoken. He coordinated the ideas and argued the points logically across. The other members respected his expertise and acquiesced. This can be seen in Don's response, *"so how, Vernon?"* and Jack's reply to Vernon's instructions, *"yup..."* Among the three, Jack was the least participative member although he showed his presence by contributing points and volunteering to elaborate on the compiled points.

For the second task, as the norms of working together was already established earlier, the group took a slightly shorter time to complete the task. The same procedure occurred, brainstorming, then one person (Vernon) compiling. After all members looked at the draft and agreed on it, the project was submitted. Moreover, from the interviews, team members indicated that they had prior knowledge about the subject matter for this task. This allowed them to rapidly brainstorm and discuss ideas. After the virtual team experience, students reflected that they were better able to appreciate time management and organization of ideas. Team members were satisfied with the process of collaboration although in overall grades, they had the lowest score.

4.2.2 Findings from Team B

This team consisted of three male members. They were unable to find a group in the course and the tutor arranged for them to be in the same group together. Using the web-based instant messaging client, the tutor gave students the nicknames - student 1, 2, and 3 respectively. Due to the anonymity of the system, students were unable to ascertain each others' identity. Interviews afterward verified that students had no idea who they were collaborating with although they guessed that the other members were male based on their linguistic styles.

Members in team B were very cordial and polite from the beginning. In the interview, one student expressed anxiety over the anonymous condition as he felt it was like an exam but he soon got over his nervousness as the discussion progressed. Team members were very concerned about the collaboration procedure and were explicit about work norms. For instance, student 1 enquired, *"So what now? We go to the initial web page provided and then look for extra information on the Internet?"* Also, as the task began, student 3 initiated a way of compilation, *"You can give me whatever information you have, and I will sum up after that, OK?."* Later, after they had found information about the project, student 2 asked how they were supposed to pass the information to student 3. In this case, student 3 decided to combine the content of the websites that the others had contributed. Student 1 was impressed by this, praising student 3 as *"efficient"*, although he wanted to see the final compiled work. Student 3 emerged as the leader, as he was the main compiler and the initiator of each phase of the project.

As the team proceeded to the unstructured task, there was no more discussion about work norms. Members understood how they were to function and proceeded to collaborate in the same style as earlier. Basically, they first searched for information, when one member suggested a reference link, the other members would follow with either more links of their own or comments about the viability of the earlier link. After a small discussion about the solution, student 3 would compile the final solution.

Students 1 and 3 were slightly more active participants in both tasks than student 2, contributing points and combining answers while student 2 simply added points. In the interviews, students 1 and 3 were annoyed about the lack of participation by student 2, although this was not expressed in any way in the chat logs. In fact, at the end of their collaboration, all members thanked each other for their work and claimed that it was "nice" talking and working with each other. Student 2 shared in the interview that he was not worried about the opinions of others as it was an anonymous discussion. In contrast, students 1 and 3 were concerned about what others thought of them in the task and wanted recognition and agreement for their ideas. All members also felt rather negative towards the overall collaboration experience. In this team, students perceived that they did not learn much from this experience. However, this team received the highest overall score for the two tasks.

4.2.3 Findings from Team C

Team C consists of three females, Dawn, Winnie and Samantha (pseudonyms) who are friends. The team discussion started very task focused with Dawn initiating the discussion for the first project. Team members also expressed their expectations for the task. For instance, Samantha clarified, *"so we are supposed to search for information and categorize and organized into points"*, with Dawn swiftly responding with agreement. The implicit norm in this group was that brainstorming of ideas was first. Besides sharing URLs during this phase, this group also typed their points into the messaging client, allowing all members to read the points easily. When one member wrote their point or URL, the others would normally respond by concurring or providing supporting comments. As they were brainstorming, Dawn informed the group that she was compiling the points for the group, which was met with agreement by the other members. In this way, Dawn emerged as the leader for the group. Besides compiling, she actively facilitated the discussion, contributed points and responded to discussions by other members.

For the second task, the group was also relatively focused, and aimed to finish the project. There was no discussion about group norms but members understood that brainstorming came first before the compilation. For this question, Winnie had very strong opinions on the answer. Dawn, the emergent leader, agreed to her points while Samantha did not respond. This led to Dawn directly seeking Samantha's opinion. Samantha replied with a hesitant *"I guess so"*. Later, she suggested another point to the team. Winnie then responded that they could only have one point, and sought a way to combine both of their ideas through a broader point. To Winnie, this represented a win-win situation for both Winnie and Samantha. However, Samantha held back on her comments and instead, opened a private chat window with Dawn, who she was closer to, to discuss the points. Samantha felt uncertain about the point that Winnie raised, but after discussion with Dawn, was more able to accept the revised point.

To ensure that all members were satisfied with the project, Dawn compiled a first draft of the project and ensured that both Samantha and Winnie were able to read and edit the draft. All parties were then able to resolve their differences and agree to a set of answers for the project. Towards the end, Winnie's gratefulness for the smooth work coordination resulted in her thanking Dawn for the work.

Due to the differing opinions, the group took a slightly longer time to perform the second task compared to the first. However, from the interviews, members felt that they learnt more from the first structured task than the second, as many of them had prior knowledge of the second task. Still, members enjoyed the second task more than the first as it was more interesting to them. In this team, all members remarked that they all contributed "very equally" to the tasks. Students were satisfied with the collaboration process and the team's overall performance was second to the other teams.

4.3 Cross-Case Analysis

Based on the data analysis invoking the template coding and CLD techniques, we summarize the key variables of interest and its relationships. This is depicted in Table 3. While it seems that the male anonymous team performed the best, members of the identified teams were more satisfied. To understand the outcomes, the paper analyzes the multiple case study and evaluates the relationships between gender and anonymity on linguistic acts, group dynamics and outcomes. Based on the data, the paper has mapped out process flows which resulted in a CLD, illustrated in Figure 1. This shows the interdependencies between gender and anonymity, group processes, and outcome variables in virtual team work. The next few sections elaborate on each interdependency alongside suggested propositions (p).

4.3.1 Influence of Anonymity on Group Process

From the case analysis, the anonymous group (Team B) tended to be very polite, and not go into great lengths in discussion compared to the identified groups (Team A and C). Moreover, in Team B, although two members were frustrated

Table 3. Cross case summary

	Factor	Team A	Team B	Team C	Influenced By
Gender		Male	Male	Female	N.A.
Anonymity		No	Yes	No	N.A.
Linguistic acts	Task-related acts	Moderate	High	Moderate	Anonymity; Leader emergence
	Socio-emotional acts	High	Low	Moderate	Anonymity
	Norms and rules acts	Low	High	Low	Member Awareness, Leader emergence
Group dynamics	Member awareness	High	Low	High	Socio-emotional acts
	Leader emergence	High	High	Moderate	Gender, Conformity
	Member conformity	Moderate	Moderate	High	Gender
Outcome	Satisfaction with collaboration process	High	Low	High	Socio-emotional acts, Conformity
	Group Performance	Low	High	Moderate	Task-related acts, Norms and rules acts

with a member as revealed in the interviews, over CMC they did not express their feelings directly to the member, but said that they had a "nice" time working with each other, parting on cordial terms. Conversely the identified groups shared more personal information with others such as their personal experiences and feelings. This indicates that there is *less socio-emotional discussion in anonymous communication* (p1). Anonymous communications help mask any status differences that could potentially be perceived by members in e-collaboration and depersonalizes the communication.

Figure 1. Process flow in virtual teams

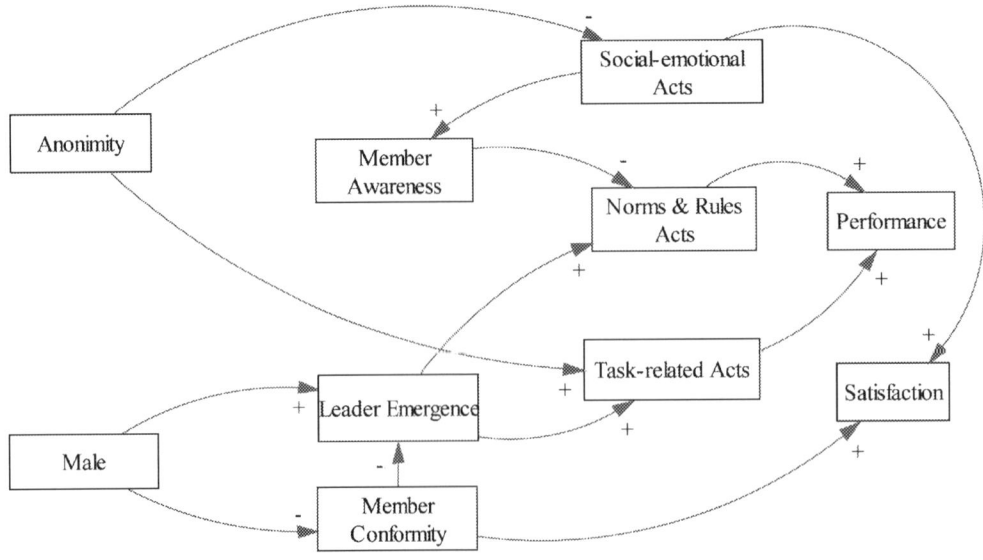

The anonymous mode of communication is shown to be effective in encouraging more effort and participation in the virtual team by overcoming members' inner restraints and evaluation apprehension (Valacich et al., 1992). Anonymity provides members with equal opportunities to express ideas by reducing the cues of social status and thus the fear of disapproval from other members. Research has shown that the reduction of social cues caused by computer-mediated communication encourages greater task-related discussion (Siegel et al., 1986). This is consistent with the findings from the cross-case analysis, i.e., *there are more task-related acts compared to socio-emotional acts in anonymous communication* (p2).

4.3.2 Influence of Gender on Group Process

From our data, gender seems to influence the group process. Generally, females avoided direct conflict in communication as seen from Team C. When women perform a task, they try to foster cooperation and connection within the group. As a result, members of an all-female learning group tend to reserve their opinions and agree readily to group norms to complete the task to maintain a peaceful atmosphere. In contrast, males contended with each other for the better argument as seen in Teams A and B. Males are more likely to elaborate on the pros and cons of alternative points and stick to their own stand unless being persuaded by strong evidence (Herring, 1996). The male groups also tended to question group norms rather than agree to it. Therefore, *males are less likely to conform than females* (p3).

At the same time, we see more competition for leadership in male groups than female groups. In male groups there was a struggle for leadership, especially in Team A. Moreover, in male groups, the leader emerged as the member who was convinced about his points and intent on seeing his point through, besides active participation and being responsible for compiling the project, compared to the female group. In the female group, the "leader" functioned more of a facilitator. Thus, *males are more associated with leader emergence than females* (p4).

4.3.3 Relationship Between Linguistic Acts and Group Dynamics

Socio-emotional acts such as personal feelings and thoughts led to greater member awareness. For instance, in Team A, Vernon's sharing that he found "good stuff" for the task and his assertion for the points, led the other group members to believe that he was enthusiastic about the task and potentially willing to take charge of compiling the project. Subsequently, Don checked with Vernon if he could compile the project which Vernon readily agreed to. Conversely, in Team B, which had very little socio-emotional activity, members in team B felt lost several times during the project discussion process. They were unaware of what other members were doing and frequently had to clarify and ascertain what they should do. Thus, *socio-emotional discussion is positively related to member awareness* (p5).

In a group discussion, when members realized they were not aware of what was going on, they would turn to discuss about the procedure and order of the subtask. This helped them to become clearer about the ongoing progress of the project. Especially in Team B, the lack of member awareness prompted them to discuss about group norms and rules to guide the progress of the discussion. Thus, *member awareness is negatively related to norms and rules acts* (p6).

Norms and rules are also affected by leader emergence. As a leader emerges in the group, the operation rules and procedures are made explicit by the leader. Typically, the leader informs the group members of how things should be done and asserts his or her authority at the same time. The proposition is *leader emergence is positively associated with norms and rules acts* (p7). Leader emergence also coincided with task-related acts. From the cases, we see that the leader demonstrated his leadership by arguing for his point (Team A), or showing his capabilities in summarizing and organizing (Team B and C). Therefore, *leader emergence is positively associated with task-related acts* (p8).

From the case analysis we find that groups that were lower in conformity were more likely to have leader emergence. For instance, members in team B did not follow group norms such as active participation and a leader emerged. Thus we find that *conformity is inversely related to leader emergence* (p9).

4.3.4 Influence of Group Process on Performance and Satisfaction

During the interview, most members agreed that communication related to personal experience and feelings helped them to release the tension in the discussion and resulted in them being more satisfied with the learning process. Thus, *socio-emotional discussion is positively related to the satisfaction with collaboration process* (p10). In the interview, we also find that members who have held back on their opinions in the discussion in favor of overall group harmony expressed negative feelings towards the virtual team process. Thus *conformity is negatively related to satisfaction with collaboration process* (p11).

Moreover, *group performance is positively affected by task-related acts* (p12) *as well as norms and rules acts* (p13). Naturally, the emphasis on the task as demonstrated by task-related discussion and problem-solving should lead to higher group performance. On the other hand, norms and rules acts are important in helping the team function by dividing the task, setting priority as well as solving disputes, and thus contribute to the effectiveness of the team's performance.

5 IMPLICATIONS AND LIMITATIONS

This study is an attempt to contribute to extant research on gender, anonymity and virtual teams. Based on the multiple case study data, the paper has proposed relationships between gender, anonymity, group processes and outcomes throughout the lifespan of a virtual team. These provide specific evidence of the temporal influence of member characteristics and technology characteristics on virtual team outcomes. In general, we found that linguistic acts in the virtual team are primarily affected by anonymity. Anonymity affected socio-emotional and norms and rules acts. Gender as a user characteristic affected the group dynamics of member conformity and leader emergence. In turn, these group processes influenced other process and outcome variables. In particular, group performance was affected by task-related and norms and rules acts while members' satisfaction towards the collaboration process was affected by socio-emotional acts and member conformity.

There are several research implications of this study. First, since a process flow of the impact of gender and anonymity on group processes and outcomes has been mapped, research could further examine each proposition and test their relationship in other settings using quantitative methods.

Second, a major contribution of this paper is the identification of group processes that interplay with gender and anonymity. These salient intervening processes such as linguistic acts and the group dynamics, member awareness, conformity and leader emergence can be useful variables for subsequent research. Gender anonymity does not directly affect outcomes, rather these variables mediate or moderate the degree of performance and satisfaction.

Third, this case study can be complemented with alternate methods of analysis such as content-analysis to investigate the percentage of each pattern of behavior or quantitative methods like surveys and experiments to generalize findings to a wider population.

Table 4. Managerial suggestions

Findings	Managerial Implications
Less socio-emotional activity in anonymous virtual teams (p1). There are more task-related acts in anonymous communication (p2). Group performance is positively affected by task-related acts (p12)	For tasks that require a lot of attention to the content and/or are constrained with minimal face-to-face contact, managers can utilize anonymous virtual teams as team time is spent predominantly on the task at hand. This will ensure higher performance in the team.
Males are less likely to conform than females (p3). Conformity is negatively related to satisfaction with collaboration process (p11).	For tasks requiring creativity, managers should ensure more male members in the virtual team as it will result in more ideas and divergent thinking rather than conformity. Moreover, this will result in more satisfied members.
Males are more associated with leader emergence than females (p4). Conformity is inversely related to leader emergence (p9).	If there are gender quotas for managers of virtual teams, managers should allocate female managers as if the virtual team is not assigned a manager, most likely a male member will emerge as a leader.
Less socio-emotional activity in anonymous virtual teams (p1). Socio-emotional acts are positively related to the satisfaction with collaboration process (p10).	To enhance the satisfaction of work teams, virtual team members should be allowed to identify each other. Managers could include tasks that generate member-support such as ice-breakers that encourage members to share personal experiences about themselves.
Leader emergence is positively associated with norms and rules acts (p7). Leader emergence is positively associated with task-related activities (p8).	In mixed-gender teams, managers need not officially assign leaders to virtual teams as even without official leaders, leaders will emerge who will help regulate the team and steer the team to accomplish the task.
Socio-emotional activities are positively related to member awareness (p5). Member awareness is negatively related to norms and rules acts (p6). Socio-emotional activity is positively related to the satisfaction with collaboration process (p10). Group performance is positively affected by task-related (p12) as well as norms and rules acts (p13).	Generally, successful virtual teams should engage simultaneously in task-related, socio-emotional and norms and rules acts. Managers should monitor these activities closely. If any area is lacking, managers should ensure that it is enhanced. However, if managers are tight-pressed and busy, socio-emotional acts are not as important as the other performing activities in accomplishing the task. Still, managers must note that this low socio-emotional activity will decrease the overall satisfaction of the team.

Besides theoretical implications, this study has implications for practice. The evidence from the case data suggests several strategies in the management of virtual teams. This is illustrated in Table 4.

Nevertheless, this study is not without its limitations. First, the case study methodology while allowing researchers to understand "how" and "why" phenomena occur, can only apply to the specific context. Second, the current study used undergraduate students as subjects instead of actual management teams and findings may not be generalized to the real world. However, the study evaluated graduating students who were taking the MIS course, who could potentially be working in virtual teams.

Secondly, the teams chosen were not zero-history teams and their previous work experience could have dictated norms and rules such as in Team C. However, the researchers did check with the participants about their previous work history and working norms during the interview. For the students that were friends before the formation of the virtual team, they replied that this was the first time they were working in a virtual team and they did not know of any predefined or implicit norms of working in such CMC teams. They previous worked with their teammate face-to-face for only one other project.

Finally, the study could have examined an anonymous, all-female team to complement the other cases. However, the researchers were unable to triangulate enough data for such a group. Still, based on the existing cases and the process model mapped, the paper already suggests group processes and outcomes for this case.

Gender and anonymity are indeed crucial in virtual teams, affecting group processes and performance and satisfaction. Employing a multiple case study, this paper has illuminated several group processes such as team awareness and leader emergence that affected team outcomes. It has shown the salience of gender allocation and technological configurations (anonymity vs. identified) for effective team performance in virtual teams.

REFERENCES

Barrett, M., & Davidson, M. J. (2006). *Gender and Communication at Work*. Aldershot, UK: Ashgate.

Cecez-Kecmanovic, D., & Webb, C. (2008). Towards a communicative model of collaborative web-mediated learning. *Australian Journal of Educational Technology*, *16*(1), 73–85. Retrieved from http://cleo.murdoch.edu.au/ajet/ajet16/cecez-kecmanovic.html.

Christofides, E., Islam, T., & Desmarais, S. (2009). Gender stereotyping over instant messenger: The effects of gender and context. *Computers in Human Behavior*, *25*(4), 897–901. doi:10.1016/j.chb.2009.03.004

Cohen, M. S., & Ellis, T. J. (2008). The Asynchronous Learning Environment (ALN) as a Gender-Neutral Communication Environment. In *Proceedings of the 28th American Society for Engineering Education/IEEE*, Saratoga Springs, NY.

Crowston, K., & Kammerer, E. (1998). Communicative style and gender differences in computer-mediated communications. In Ebo, B. (Ed.), *Cyberghetto or cybertopia? Race, class, and gender on the Internet* (pp. 185–203). Westport, CT: Praeger.

Dennis, A. R., & Garfield, M. J. (2003). The adoption and use of GSS in project teams: Toward more participative processes and outcomes. *Management Information Systems Quarterly*, *27*(2), 289–323.

Dourish, P., & Bellotti, V. (1992). Awareness and Coordination in Shared Workspaces. *ACM CSCW*, 107-114.

Galanes, G. J., & Adams, K. (2007). *Effective group discussion: theory and practice* (12th ed.). New York: McGraw-Hill.

Goldberg, A. A., & Larson, C. E. (1975). *Group communication: Discussion processes and applications*. Englewood Cliffs, NJ: Prentice-Hall.

Guiller, J., & Durndell, A. (2007). Students' linguistic behaviour in online discussion groups: Does gender matter? *Computers in Human Behavior*, *23*(5), 2240–2255. doi:10.1016/j.chb.2006.03.004

Hackman, R. (1990). *Groups that work (and those that don't): Creating conditions for effective teamwork*. San Francisco: Jossey-Bass.

Hahn, S., & Litwin, A. H. (1995). Women and men: Understanding and respecting gender differences in the workplace. In *Managing in the age of change* (pp. 188–198). New York: Irwin.

Herring, S. (1996). Posting in a Different Voice: Gender and Ethics in Computer-Mediated Communication. In *Philosophical Perspectives on Computer Mediated Communication*. Albany, NY: State University of New York Press.

Herschel, R. T. (1994). The Impact of Varying Gender Composition on Group Brainstorming Performance in a GSS Environment. *Computers in Human Behavior*, *10*(2), 209–222. doi:10.1016/0747-5632(94)90004-3

Hiltz, S. R., Turoff, M., & Johnson, K. (1991). Group Decision Support: The Effect of Designated Leader and Statistical Feedback in Computerized Conferences. *Journal of Management Information Systems*, *8*(2), 81–108.

Janis, I. (1972). *Victims of Groupthink*. Boston: Houghton Mifflin.

King, N. (1998). Template Analysis. In Symon, G., & Cassell, C. (Eds.), *Qualitative Methods and Analysis in Organizational Research: A Practical Guide*. Thousand Oaks, CA: Sage.

Klein, H. K., & Huynh, M. Q. (1999). *The potential of the language action perspective in ethnographic analysis*. Binghamton, NY: School of Management, SUNY Binghamton.

Kray, L., Galinsky, A., & Thompson, L. (2002). Reversing the gender gap in negotiations: an exploration of stereotype regeneration. *Organizational Behavior and Human Decision Processes*, 87, 386–409. doi:10.1006/obhd.2001.2979

Lind, M. R. (1999). The gender impact of temporary virtual work groups. *IEEE Transactions on Professional Communication*, 42(4), 276–285. doi:10.1109/47.807966

Nemeth, C. J. (1986). Differential Contributions of Majority and Minority Influence. *Psychological Review*, 93(1), 23–32. doi:10.1037/0033-295X.93.1.23

Nunamaker, J. F., Dennis, A. R., Valacich, J. S., Vogel, D. R., & George, J. F. (1991). Electronic meeting systems to support group work. *Communications of the ACM*, 34(7), 40–61. doi:10.1145/105783.105793

Orlikowski, W. J. (1993). CASE tools as organizational change: Investigating incremental and radical changes in systems development. *Management Information Systems Quarterly*, 17(3), 309–340. doi:10.2307/249774

Postmes, T., Spears, R., & Lea, M. (1998). Breaching or building social boundaries? SIDE effects of computer-mediated communication. *Communication Research*, 25, 689–715. doi:10.1177/009365098025006006

Powell, A., Piccoli, G., & Ives, B. (2004). Virtual teams: a review of current literature and directions for future research. *SIGMIS Database*, 35(1), 6–36. doi:10.1145/968464.968467

Robertson, O., Hewitt, J., & Scardamalia, M. (2003). *Gender Participation Patterns in Knowledge Forum: an Analysis of Two Graduate-Level Classes*. Paper presented at the IKIT Summer Institute, Toronto, ON, Canada.

Siegel, J., Dubrovsky, V. J., Kiesler, S., & McGuire, T. W. (1986). Group processes in computer-mediated communication. *Organizational Behavior and Human Decision Processes*, 3, 157–187. doi:10.1016/0749-5978(86)90050-6

Spears, R., & Lea, M. (1992). Social influence and the influence of the "social" in computer-mediated communication. In M. Lea (Ed.), *Contexts of computer-mediated communication* (pp. 30-65). London: Harvester-Wheatsheaf.

Thomson, R. (2006a). Gender and Electronic Discourse in the Workplace. In Barrett, M., & Davidson, M. J. (Eds.), *Gender and Communication at Work* (pp. 239–249). Aldershot, UK: Ashgate.

Thomson, R. (2006b). The Effect of Topic of Dicussion on Gendered Language in Computer-Mediated Discussion. *Journal of Language and Social Psychology*, 25(2), 167–178. doi:10.1177/0261927X06286452

Tyran, C. K., Dennis, A. R., Vogel, D. R., & Nunamaker, J. F. (1992). The application of electronic meeting technology to support strategic management. *Management Information Systems Quarterly*, 16(3), 313–334. doi:10.2307/249531

Valacich, J. S., Jessup, L. M., Dennis, A. R., & Nunamaker, J. F. (1992). A conceptual framework of anonymity in group support systems. *Group Decision and Negotiation*, 1, 219–241. doi:10.1007/BF00126264

Weber, B., Wittchen, M., & Hertel, G. (2009). Gendered Ways to Motivation Gains in Groups. *Sex Roles*, *60*, 731–774. doi:10.1007/s11199-008-9574-4

Yin, R. (1994). *Case study research: Design and methods*. Beverly Hills, CA: Sage.

Zander, A. (1971). *Motives and Goals in Groups*. New York: Academic Press.

This work was previously published in International Journal of E-Politics (IJEP), Volume 2, Issue 1, edited by Celia Romm Livermore, pp. 1-16, copyright 2011 by IGI Publishing (an imprint of IGI Global).

Chapter 2
Adopting ICT in the Mompreneurs Business:
A Strategy for Growth?[1]

Yvonne Costin
University of Limerick, Ireland

ABSTRACT

The advancing pace of women's entrepreneurial activity across the globe represents a promising trend to fuel economic development and social progress (Fitzsimons & O'Gorman, 2005). Research has shown that women-led businesses have strong growth aspirations, are customer-oriented, value human capital and cultural aspects of business and are geared towards financial performance (McGowan & Henry, 2004). Analysis of the reasons why women start a new business, the choice of product/service and industry sector has highlighted the home-based female entrepreneur. This home worker/entrepreneur now constitutes a segment referred to as the mompreneurs - the mothers who decide primarily for family reasons to leave paid employment and establish a business operated from home. Thus, while mompreneurs gain benefits of operating their business from home, they also experience additional challenges not encountered by more traditional female entrepreneurs, one such issue is the central dependence of the business on ICT (Information Communications Technology).

It is advocated that the role of technology is instrumental in determining the effectiveness and efficiency of where, when and how business transactions are undertaken to meet stakeholder requirements in a competitive manner. However, research by the Small Business Forum (2006) suggest the use and application of ICT in small firms overall is poor where entrepreneurs do not capitalise sufficiently on the benefits of ICT. To succeed and grow, mompreneurs' businesses should be using ICT as a backbone for the business in an integrated manner. The objective of this chapter is to examine the adoption and application of ICT in the mompreneurs business and challenges encountered in its effective use. A specific emphasis is placed on the issue of ICT and its use by the mompreneur in undertaking business transactions and as a means of facilitating small firm growth.

DOI: 10.4018/978-1-60960-759-3.ch002

1. INTRODUCTION

The advancing pace of women's entrepreneurial activity across the globe represents a promising trend to fuel economic development and social progress (Fitzsimons & O' Gorman, 2005). Women now represent more than one-third of all people involved in entrepreneurial activity and are likely to play an even greater role when informal sectors are considered. Female entrepreneurs are increasingly prominent as employers, customers, suppliers and competitors in the global economy. The contribution that female entrepreneurs make to economic development is significant and has increased over time (McGowan & Henry, 2004). Indeed it has been cited there is no greater initiative a country can take to accelerate its pace of entrepreneurial activity than to encourage more of its females to participate. Women entrepreneurs are not only prominent in industries where they are traditionally active, but also in less traditional sectors (e.g. manufacturing, construction and transportation) through the development of a range of services and products (European Commission, 2002). Research has shown that women-led businesses have strong growth aspirations, are customer-oriented, value the human capital and cultural aspects of the business and are geared towards financial performance (McGowan & Henry, 2004).

Analysis of the reasons why women start a new business, choice of product/service and industry sector has highlighted to the author the home-based female entrepreneur, in a somewhat reconfigured context. This home worker/entrepreneur now constitutes a segment referred to as the mompreneurs - the mothers who decide primarily for family reasons to leave paid employment and establish a business operated from home. Thus, while mompreneurs gain the benefits of operating their business from home, they also experience additional challenges not encountered by the more traditional female entrepreneurs, one such issue is the central dependence of the business on ICT (Information Communications Technology). It is advocated that the role of technology is instrumental in determining the effectiveness and efficiency of where, when and how business transactions are undertaken to meet stakeholder requirements in a competitive manner. However, research by the Small Business Forum (2006) suggests that the use and application of ICT in small firms overall is poor where entrepreneurs do not capitalise sufficiently on the benefits of ICT. To succeed and grow, mompreneurs' businesses should be using ICT as a backbone of the business in an integrated manner. The objective of this chapter is to examine the adoption and application of ICT in the mompreneurs business and the challenges encountered in its effective use.

The chapter commences with an overview of the level of enterprising activity and the profile and characteristics of female entrepreneurs in Ireland. An analysis of the reasons why women start a business leads to a discussion on the emergence of the home-based female entrepreneur, in a somewhat reconfigured context - the mompreneur. A specific emphasis is placed on the issue of ICT and its use in undertaking business transactions and as a means of facilitating small firm growth. The information presented provides suggestions on topics which should be addressed in government policy to facilitate greater and more efficient adoption of ICT as an enabler of firm growth for the mompreneur.

2. FEMALE ENTREPRENEURSHIP IN IRELAND

In spite of their growing numbers, women-owned businesses continue to lag behind male owned firms in Ireland both for recently started businesses and for established businesses (Forfas, 2007). In 2006 there were 60,000 male new firm entrepreneurs and 21,000 female new firm entrepreneurs in Ireland where men were 2.9 times more likely than women to be new firm entrepreneurs (4.3% of men compared to 1.5% of women). The imbalance in these trends unfortunately has not improved greatly over the last five years, where the average rate of early stage entrepreneurship for the period 2002 to 2007 was 12.0 per cent for men and 4.8 per cent for women. In addition to the gender difference the number of female established firms decreased from 3.4 per cent in 2004 to 2.8 per cent in 2007 (Fitzsimons & O'Gorman,2007). In Ireland women are underrepresented among all categories of entrepreneurs - only one-third of all entrepreneurs were women – less than one-quarter of high-expectation and high-growth entrepreneurs were women (Fitzsimons & O' Gorman 2007). Moreover in Ireland women are less active as entrepreneurs across all age cohorts, among all income categories and across all educational levels bar one - women with post-graduate qualifications have the same rate of early stage entrepreneurial activity as men. Only one-quarter of the high-expectation and high-growth entrepreneurs are women (Fitzsimons & O' Gorman, 2008). The dearth of females starting high technology businesses was also highlighted by Richardson and Hynes (2007) where it was recommended that educational, social and professional barriers that militate against females participating in engineering and technology entrepreneurial activity could be reduced. Worryingly, these statistics exist in the backdrop where female education standards are higher than ever before, where female participation rates in the labour market have risen significantly and a high proportion of the female population in the age cohort at which entrepreneurial activity is most likely to occur. Therefore, it is argued that greater levels of female activity in starting and growing businesses must be encouraged and supported. This necessitates a more developed understanding of the profile of women who start a business, their aspiration for business growth and the underlying causes of the challenges encountered in achieving this growth. A particular focus is placed on the role of ICT as an enabler of business development for the female entrepreneurs and more so the mompreneur.

3. PROFILING FEMALE ENTREPRENEURS

Female entrepreneurs are a heterogeneous group where their individual motivations for entering self employment and for growing the business differ. Additionally, the type of business and industry sector they establish their business in tends to vary from the male counterparts. In general research would suggest that the motivation for starting a new business, the nature and type of business established by females reflect the wider positions (personally and professionally) they hold in the broader societal and business context (Carter & Shaw, 2006). The heterogeneity of the female entrepreneur population is better understood when analysis is undertaken of factors such as why females start a business, the type of business they establish, the objectives they have for the business growth.

3.1 Motivation to Start the Business

It is a widely held claim that women start a business to have more flexibility and freedom to juggle work, leisure and family commitments. Business growth is not an explicit objective as it would conflict with their purpose of choosing self employment (Arenius & Kovalainen, 2006; Brush, Carter, Gatewood, Greene, & Hart, 2004; Chell & Baines, 1998; De Martino and Barbato, 2003; Kjeldsen & Nielsen, 2000; Orhan & Scott, 2001; Valiulis, Drew, Humbert, & Daverth, 2004). These family issues may be a source of conflict when deciding to grow the business (Babaeva & Chrikova, 1997; Fielden, Davidson, Dawe & Makin, 2003). Family commitments may serve to restrict the time and resources available to the female entrepreneur to grow the business and explains why many female owner/managed firms remain lifestyle businesses. Villanueva and Pavone (2007) suggest that the lower than average rate of growth among women-owned ventures was often a conscious choice by a subset of women entrepreneurs and did not reflect the actual potential of the business itself. Therefore a challenge arises to ensure the reason for starting the business does emerge as a barrier to subsequent firm growth.

3.2 Type of Businesses Established

Female business activity is confined in the main to what are viewed as "feminised occupational industry sectors" (Carter & Bennett, 2006; Henry & Johnston, 2003; Marlow, Carter & Shaw, 2008; Small Business Service, 2003). These include businesses in the catering, personal and business services, training and development and recruitment. According to the Gender Equality Unit (2000) women tend to start businesses in the general services, tourism and financial services sectors, while men tend to start more businesses in manufacturing, construction and technology. For men, these first three industry sectors accounted for half of all early stage entrepreneurship as compared to 77 per cent for women. However, some newer evidence suggests that female entrepreneurs are setting up and running the so-called new economy companies in the high technology and professional services sectors, as well as in other non-traditional businesses (McGowan & Henry, 2004). Moreover, many females operate their business on a part time capacity and use their home as a business base (Small Business Service, 2003; Marlow & Carter, 2004). Technological and telecommunications improvements now more than ever, make it possible for women to operate businesses from home. This in part resolves the child-or-dependent care issue in that women can stay in the workforce and also be with their children. Given this it is questioned if ICT is used to its full potential in an integrated manner in female operated business to facilitate the operations of a business irrespective of the nature of the product/service offering.

3.3 Prior Work Experience

The choice of the industry sector where the business is established is influenced by the prior work experience of females. For example, 88 per cent of employed women are currently working in services compared to 51 per cent of men, while just 9 per cent of women are employed in industry compared to 17 per cent of men (Fitzsimons & O' Gorman, 2007). An examination of the level of the prior work experience or the functional roles assumed by females provide an added dimension on the impact of prior work experience on small firm growth. A number of studies (Arenius & Kovalainen, 2006; Eagly & Karau, 2002; Marlow, 2002; Stevenson, 2003) suggested that this same factor will be a determinant on the subsequent growth of the established small firm. Moreover in the majority of cases the level of experience gained by female owner/managers is not at managerial level (Arenius & Kovalainen, 2006; Eagly & Karau, 2002; Marlow 2002; Stevenson, 2003). According to the SIA Group, the lack of managerial experience is a key determinant on the growth of the enterprise.

"………a serious obstacle for women in business is the need for training be it skill-based, technical, technological or management technique and that such training is an essential contribution to producing an able body of entrepreneurs, who not only survive but thrive and can contribute to both local and global economies" (2001: p17).

The lack of managerial competencies requires the female entrepreneur to employ functional specialists, delegate to existing staff or alternatively acquire the relevant management training themselves. Female entrepreneurs should be encouraged to undertake training and development areas that will provide them with skills and competencies to manage business operations with a focus on people, financial and ICT management. In addition the lack of certain expertise by the female entrepreneur results in a reliance on external sources to compensate their skills deficiencies.

3.4 Networking and Female Businesses

Women are less likely to belong to a business network, to have a business mentor, or to join business associations at business start up (Still & Timms 2000). Forfas (2007) contribute this fact to the personal background and employment experience of Irish females, in addition to the boarder socio-economic and cultural context in which female businesses operate. These factors result in many females not building up sufficient contacts that can assist them in the start up and growth stages of business development (Hampton, Cooper & McGowan, 2007). In particular, female entrepreneurs are less likely to have exposure to successful female entrepreneurs or female role models who they can revert to for advice (Fitzsimons & O'Gorman, 2007). Participation in networks (formal and or informal) provide access to information, support, and a forum for practical advice and sharing of experiences with individuals who are in similar positions and encountering or have encountered the same challenges of business growth (Eastwood 2004; Holden, 2007). Delaney (2007) extends the focus of networks to include the importance of mentoring for female entrepreneurs as it provides inexpensive advice and access to experts who can empathise with the challenges faced by the female entrepreneur (Holden, 2007). Oppedisano (2005, p.248) captures how vital networking and mentoring is when they state "Women want and *need* connections- whether it's through the sewing circle, water well, chat rooms, clubs or associations!" A major element to overcoming this barrier may involve women empowering one another through sharing experiences and assisting each other.

To date, a review of the profile of the female entrepreneur and the type of businesses established had highlighted the need for a balance of work and life resulting in women operating a business from home thus an increased incidence of the home based entrepreneur i.e. the mompreneur.

4. THE MOMPRENEUR

The home worker/entrepreneur now constitutes a segment referred to as the mompreneur - the mother who decides primarily for family reasons to leave paid employment and establish a business operated from the home.

A number of women are not content with choosing between the workplace or being a stay-at-home mum; they want a blend of both. Research conducted by British Telecom (2008) found that for more than a third (38 per cent) of business mums indicated their 'big idea' was based on a career that enabled them to work from home and juggle parenthood. More than a third (36 per cent) of female entrepreneurs were between 26 and 30 years where 74 per cent of respondents had at least one child under two years old. In the main the issues and challenges encountered by mompreneurs mirrored those of the more mainstream entrepreneurs. For instance almost half (49 per cent) found their ongoing family commitments an obstacle when trying to launch their own business; more than a fifth (22 per cent) said they had difficulty in borrowing money because of their 'mum' status; more than two thirds (68.2 per cent) have had to sacrifice 'me time' as a penalty for starting up a business venture, with a further 30 per cent saying it has affected the time they can spend with their partner.

There was a strong sense amongst mompreneurs that making their business a success required round the clock dedication. Furthermore the internet was viewed as instrumental in helping this group of entrepreneurs develop their business. The role of technology facilitated these women to work outside of the traditional 9 to 5 day giving them the freedom to devote time to their businesses as well as their family. Despite the benefits of ICT for the mompreneur, the research study by British Telecom (2008) showed that only a third of business mums access the internet between the hours of 5pm and 9am. Just over 40 per cent admit to accessing the internet more than three hours every day, with more than half of them claiming to use constantly throughout the day. The mompreneurs use the internet for a primarily for finding services, researching competitors, advertising and promotions.

These findings are somewhat surprising given this relatively low application of ICT in this segment of female entrepreneurs where both the nature of the businesses established and the need for flexibility in where and when work is undertaken. In order to identify why this might be the case the research examines if this was characteristic of the broader population of female entrepreneurs. Unfortunately, it is not possible to identify research specific to female entrepreneurs, therefore research on ICT in small firms was used as a reference point. It is envisaged that the issues encountered in the broader population of owner/managers are relevant for the female entrepreneur.

5. INFORMATION COMMUNICATION TECHNOLOGIES: AN ENABLER IN SMALL FIRM DEVELOPMENT

The adoption of ICT is no longer a means of creating a sustainable competitive advantage, but an essential competitive weapon necessary for survival. In ascertaining the role and use of ICT as an enabler or otherwise of small firm growth ICT needs to be defined. A review of the literature and policy documents on this area present a range of terms which are frequently used in an interchangeable manner. It is further contested that this same terminology is a barrier to its adoption in the small firm as it creates confusion in the mind of the non technical entrepreneurs. The eBusiness Strategy Group (2004, p.9) defines eBusiness in the context of SME's as"...*the application of information and communication technologies to business processes in all sectors of the economy to reduce costs, to improve customer value and to find new markets for products and services*".

Building a greater strategic awareness of ICT among female entrepreneurs' will help to integrate eBusiness strategies into the overall business (Department of Enterprise Trade and Employment, 2005). There is a distinction between ICT management skills and ICT user skills. Both are necessary if an owner/manager wants to make informed decisions on how to deploy technology in order to gain competitive advantage. Essentially ICT reflects how Information Technology is integrated into business functions for the exchange of information, buying and selling, promoting of products/services and the adoption of electronic business systems and processes in all stages of the input, process and output firm activities. In summary as described by Forfas (2000), "ICT is not just involving the use of Internet or email but information communication technologies throughout all businesses processes to create sustainable competitive advantage – essentially linking "back office processes to front end customer supports". Research by the Small Business Forum (2006) suggests, the use and application of ICT in small firms overall is poor where entrepreneurs do not capitalise sufficiently on the benefits of ICT. Furthermore, this report endorsed the importance of increased usage of ICT by small firms and indicated that SMEs need to make better decisions about Information Communications Technology investment, to manage the ICT facilities they already have, and to make more and better use of these facilities to increase productivity. The Irish Central Statistics Office (CSO) published a report in 2007 containing statistics on how ICT is being used by small firms in Ireland. In summation, recent statistics by CSO (2007) indicates that the ICT infrastructure in Ireland is improving year on year, with broadband usage in SME's at 68 per cent in 2007 up from 61 per cent in 2006. Further the report highlighted that SMEs in Ireland have good access to the Internet: 92 per cent of enterprises have Internet access and Ireland ranks 7th on this indicator (out of 21 countries). However of note is the fact that SMEs lag behind in terms of access to broadband connectivity in Ireland: 79 per cent of large enterprises have broadband, but this number falls to 40 per cent for medium sized enterprises and to 27 per cent for small enterprises. Within this broader technological infrastructure it is important to distinguish the factors that impact on the implementation of ICT in female operated businesses.

6. ENABLING SMALL BUSINESS DEVELOPMENT: THE ISSUES AND CHALLENGE ENCOUNTERED IN THE ADOPTION OF ICT

A number of common issues emerge as barriers to the adoption and integrated use of ICT in the small firm (Small Business Forum, 2006). The primary reasons relate to a lack of appreciation amongst owners/managers of the contribution that ICT can make to their business, lack of ICT management skills, difficulty in accessing independent advice and the costs associated with acquiring and maintaining Information Technology (IT) systems (Department of Enterprise Trade and Employment, 2006). Similar findings emerged in a study undertaken by CSO (2007) which concluded that the low level of usage of ICT by entrepreneurs was in the main related to a lack of knowledge or education on what ICT entails or indeed how it can specifically benefit their business. The adoption of ICT by entrepreneurs is prevented by a *"lack of understanding of the opportunities available to small businesses"*, by a *"lack of understanding on how to implement these techniques"* and a *"lack of skills amongst the workforce to use them"*. Similar Feuer, Messnarz & Sanchez (2002, p. 2) found that one of the main barriers to eBusiness is a *"lack of technical skills"* in the business resulted in a slow up on ICT in the small firm. These issues are now discussed in more detail below.

- *Lack of perceived need of the benefits of ICT for the small firm*: Owner/managers acts alone in researching the need for and the type of ICT required for their business.
- *Insufficient ICT Education*: The lack of education amongst owner/managers in relation to eBusiness and ICT has lead to uncertainty around its use and implementation (Chapman, Szczygiel & Thompson, 2000; Elsammani, Scown & Hackney, 2001; Lewis & Cockril, 2002; Moussi & Davey, 2000; Nath, Akmnakigil, Hjelm, Sakaguchi & Schultz, 1998; Smith & Webster, 2000; Van Akkeren & Cavaye, 1999). This claim is backed up by findings by the Irish Chamber of Commerce (2002) who found that in many cases *"companies ignore…opportunities offered by ICT"*, *"eBusiness is often confused with having a webpage"*, *"the opportunities offered by ICT are underestimated"* and finally eBusiness is *"not considered a priority and by owner/managers"*. It is advanced that entrepreneurs in the main are not aware of the total range of services which can be exploited via ICT.
- *Lack of Technical skills*: In the small business where there is only the owner/manager or a few staff, then the technical remit of managing ICT falls with the entrepreneur. The Irish Chamber of Commerce (2002) found that a lack of understanding of ICT from the technological viewpoint and an inability to maintain and manage ICT were dominant reasons for the failure of ICT in an integrated manner into their business.
- *Training for ICT*: A key issue relates to the dearth of training and education courses

that address the needs of the non-technological entrepreneur that provide the level of understanding and awareness of the strategic importance of ICT and how it can leverage a competitive advantage for the firm (Department of Enterprise Trade and Employment, 2005; Small Business Forum, 2006). Where such courses are available, their direct relevance, cost, ease of access and time required to undertake them are critical to their success. Owner/managers need to see value in taking time away from immediate business needs to participate in such programmes.

- *Cost of ICT implementation*: The price of technology was identified as a major barrier. Feuer, Messnarz, & Sanchez (2002, p. 2) also alluded to the fact that "*the cost of hardware/computer equipment*" and "*the cost of specialist eBusiness technical skills*" are major barriers for entrepreneurs. Thus, the lack of financial resources in SMEs combined with the perceived cost of eBusiness implementation form a formidable barrier to ICT implementation.

- *Security Fears and Concerns*: Research conducted by Jacobs & Dowsland (2000), Moussi & Davey (2000), Nath, Akmnakigil, Hjelm, Sakaguchi & Schultz, 1998; Strader & Shaw, 2000; Van Akkern which was cited by Jones, Davies & Muir (2003) all indicated that SME's have fears and concerns over security when implementing an eBusiness initiative. However, more recent research conducted by CSO (2007, p. 6) in Ireland found that "*virus checkers or protection software were used by 91% of enterprises in 2007. Firewalls were used by 81% of enterprises while 46% of enterprises indicated that they backed up data offsite*". These statistics indicate that whilst Irish SME's do have concerns over the security of their IT infrastructure, the majority are taking proactive steps to avoid any issues by utilising virus checkers, firewalls and frequent offsite data backups.

- *Knowing where to source advice and information:* Entrepreneurs will have knowledge of where to source finance and or marketing expertise of advice, however this is not the case with advice for ICT. Given their own lack of knowledge in the technical aspects of ICT entrepreneur are often inhibited and do not have the confidence to source advice from technical experts.

Overall the aforementioned issues encountered in the adoption of ICT in the female small firm are manifested in a broader range of challenges encountered by mompreneurs in growing their business.

7. CHALLENGES ENCOUNTERED IN STARTING AND GROWING THE BUSINESS

Goodbody Economic Consultants (2002) found that social conditioning and perceived expectations of females in society; incorrect perceptions of the demands of running a business; lack of female role models; low proportion of women pursuing Science, Engineering and Technology programmes; lack of self confidence and difficulties in reconciling work and family life for females were factors contributing to low levels of female enterprising activity. Additionally, Carter & Rosa, 1998; Carter, 2000; Henry & Johnson, 2003; Marlow & Patton, 2005 assert that a funding gap hinders the growth of women-led businesses. This may be attributed to the fact that females typically start a business with personal savings (Bhide,1993) normally accumulated over time, therefore women are disadvantaged when sourcing funds from commercial financial institutions to start or grow their business (Brush, Carter, Gatewood, Greene & Hart, 2004; Carter & Rosa,

1998; McClelland, 2003; Minniti & Bygrave 2003; Stevenson 2003; Storey 1994). Carter & Shaw (2006) argue that, very often, growth orientated women entrepreneurs will resort to "bootstrapping techniques" in the early years of the business, while they consolidate the company and gain more experience. Barringer & Ireland (2006) define bootstrapping as the use of creativity, ingenuity, and any means possible to obtain resources other than borrowing money or raising finance from traditional sources and may include (re) mortgage, private loan, loan from family and friends or using their family home as a location for the business.

Research suggests another key challenge is the lack of adoption and use of ICT as a means of facilitating small firm growth which will have serious implications for the competitiveness and the potential to grow the business. Moreover, the adoption of ICT amongst female entrepreneurs/mompreneur will also draw on another source of new business creation which remains virtually untapped in Ireland i.e. female entrepreneurs (Henry & Kennedy 2002).

Overall a review of the literature pertaining to the adoption of ICT in female owner/managed firms and the mompreneur business highlighted a vast lacuna of empirical studies. Due to the lack of empirical research studies specific to female entrepreneurs it was deemed important to bridge this information deficit in some form. This lead to the completion of an exploratory study to explore the adoption of ICT in mompreneur businesses. This study will serve to direct much needed attention to this very important function and will bring to the fore topics which are important for inclusion in further more comprehensive quantitative research studies to build evidence based data on this topic. Moreover the findings from this study will serve to identify if common issues emerge in the adoption of ICT with mompreneur and the broader cohort of entrepreneurs.

8. THE EMPIRICAL STUDY

An exploratory study - a series of face to face in-depth interviews with seven female mompreneurs was conducted. The objectives of the research were to obtain an insight into the profile of the entrepreneurs and their objectives for the growth of their business. Detail was obtained on their perception of ICT, its role and use in the operations of their business and the challenges encountered in the adoption of ICT into their business. A summary of the key findings are presented below.

8.1 Respondent Profile

Initially an in-depth profile of the female business owner was developed by investigating their personal characteristics which aligned with the stereotypical profiles documented in previous literature. The seven entrepreneurs had obtained a third level educational award with two respondents holding a Masters Degree, thus reflecting the high levels of education of female owner/managers (Fitzsimons & O' Gorman 2007; Forfas, 2007) The respondents had previous work experience in the same industry sector of their current business (Stevenson, 2003). In keeping with the literature the minority (2) of respondents held managerial positions in previous employment (Arenius & Kovalainen, 2006; Eagly & Karau. 2002; Marlow, 2002; Stevenson, 2003). Push and pull factors (Buttner & Moore 1997; Orhan and Scott, 2001) were evident amongst the participants for starting the business with the majority of respondents (5) establishing their business for positive reasons (the desire 'to be my own boss'. Differing motivations for the female entrepreneur operating from home or running an internet based businesses showed that internal rather than external forces had led them to this route of self-employment with the majority citing that it allowed them "the flexibility to work from home" while others cited that it allowed them the opportunity to "stay at home to raise their family" and "achieve a successful work/life balance". Such findings align with previous literature (Arenius & Kovalainen, 2006; Brush, Carter, Gatewood, Greene & Hart, 2004; Chell & Baines 1998; De Martino & Barbato, 2003; Orhan & Scott 2001; Valiulis, Drew, Humbert & Daverth, 2004) that the motivation for female entrepreneurs starting a business is to allow them "more flexibility and freedom to juggle work and family commitments". All respondents were adamant that the business was a long term investment for their future and not just a stop gap to facilitate child rearing, thus questioning the notion that the majority of females start a business for family reasons only (McClelland, 2003; Stevenson, 2003).

From a firm perspective, the majority of firms operated in "feminised industry sectors" such as retail, professional services, education and training, health and well being sectors (Carter & Bennett 2006; Browne, Farrell, Harris & Sessions, 2007; Henry & Johnston 2003; Marlow, Carter & Shaw, 2008). Results indicate that there was a significant difference in the number of years that these women had been running their business. The number of years in operation ranged from 1-3 years, 8-11 years and 18-25 years. Similarly there were significant differences in the number of employees in each business. This ranged from 1 employee, to 14, to 26. Comparable with research conducted by others (Carter & Bennett 2006; Browne, Farrell, Harris & Sessions, 2007; Henry & Johnston 2003; Marlow, Carter & Shaw, 2008) in the most part the businesses were servicing both industrial and consumer markets, with the consumer market being the most prevalent. The product/service offered to the market varied from actual tangible products, to business consulting services and training services.

8.2 Aspirations for the Growth of their Business

Five respondents indicated that they had specific objectives targeting revenue growth since the outset of their business. The remaining two respondents reported that while they did not devise specific targets for firm growth for the first year of business activities, they did so currently. The specific growth objectives were aligned to the perceptions of firm growth as described by respondents *"Expanding the customer base whilst retaining existing business....ensuring profit margins are never jeopardised"*; *"it means increasing profits year on year with the gaining of new clients each year"*; and *"Increased turnover and profit and more clients"*. These findings concur with the findings that multiple objectives exist for firm growth which are commercially focussed (Barkham, Gudgin, Hart, & Hanvey, 1996; Glancey, 1998; Philipsen & Kemp, 2003; Storey, 1994). Six owner/managers undertook regular business reviews (quarterly basis) to evaluate how well they achieved their growth objectives. All bar one respondent indicated that they had achieved growth in sales and profit for the previous two years. The remaining respondent suggested that firm growth was not achieved as her *"focus had been on product development..... and establishing a solid foundation for growth"* which required short term consolidation of business activities for longer term financial growth.

8.3 The Adoption of ICT in Respondent Firms

All respondents stated that ICT played a major role in the development of their business with 2 of the respondents indicating that their business is solely an internet based business. As indicated in previous studies, female business owners often lack the knowledge and expertise in technology due to many factors, some of which include the subject area studied as part of their education or indeed because of the traditional "feminised" sectors in which they operate in. This study again reflects such findings with all of the respondents indicating that they *"did not have the technical expertise to start the business"* and indeed it was one the main challenges that they faced in starting and running the business, with one respondent indicating that she "*had to learn how to use the Internet*". Other challenges faced by the female entrepreneur included the "*changing pace of technology*" and keeping abreast of technological developments and also "*drastically changing software programmes*". What was interesting to note is that all of the female entrepreneurs were themselves managing the ICT of the business on their own without employing anyone in this position within the business. All of those interviewed were very positive and proactive in recognising their weaknesses in the area of ICT and had received training in order to operate their businesses effectively and efficiently with one of the respondents indicating that she is currently growing her business through partnerships with other internet based businesses. It was cited also by the majority of respondents that the training received should have taken a more practical approach rather than theoretical approach.

Previous studies show that appropriate ICT can help the entrepreneurial firm to create business opportunities (OECD, 2004). It can assist in reducing costs for the business by improving internal processes, improving their product/service through faster communication with their customers, and better promoting and distributing products/services through an online presence (Sakai, 2002). Essentially, ICT has the potential to improve the core business of entrepreneurial businesses in every step of the management of the business. Although all of the respondents recognised the benefits of ICT, the research showed that the adoption of ICT was very much on a piecemeal basis where none of the respondents had fully integrated ICT across or between business functions. Where the use of ICT was most prevalent it was customer lead and centred on selling and promotional activities. Feedback from respondents revealed a strong reliance on the internet and email. The internet was most frequently used and was deemed a critical communicational tool to contact and communicate with customers and suppliers. A reason may be attributed to fact that market and customer base of the firms is local.

Further probing of respondents indicated that ICT was not used for front end activities whereby respondents seemed to be deficient in their comprehension about how ICT could be incorporated into every aspect of the business. They were not leveraging the benefits of ICT for all functions of the business. E-procurement (allowing faster and cheaper communication with suppliers; improved inventory management systems; creating linkages to global supply and outsourcing opportunities) was not being adopted by the respondents as part of the ICT strategy, alongside employing more cost effective accounting and financial management practices through ICT. The opportunity to enhance business planning capabilities and improve customer relationship management through ICT was not recognised either by the respondents. Furthermore the respondents were not familiar with how the adoption of ICT could allow them to manage the Human Resource Management (HRM) function of the business by improving communication with employees through the use of an intranet and also provide an opportunity for e-learning for employee training. In essence, the adoption of ICT by the respondents was to improve comprehension and understanding of business trends and market prices through easier access to information via the internet and e-marketing through websites.

9. CHALLENGES ENCOUNTERED IN MANAGING THEIR BUSINESS

The responses centred on a range of external and internal challenges. From an external perspective increased uncertainty in the economic climate and in the marketplace was a common concern to all respondents. This was particularly the case for firms selling domestic household related services. In a related manner the need for increased marketing and selling was viewed as a weakness and an area the majority of respondents required assistance with. In particular attention was drawn to the need to become more sophisticated in the use of the internet for establishing contacts and a means of selling aligning with previous research which indicates that most entrepreneurial firms appear to be still at a stage where establishing a web site is the main issue (OECD, 2004). Respondents cited challenges in developing websites as they lacked the technological expertise and financial resources to design and implement effective sites for the business. Following on from this obstacle it was found that the respondents *"lacked the time"* to gain the knowledge needed for the effective implementation of a website as they just focus on the day-to-day *"operations of the business"*. Maintenance of the website was another challenge faced by some of the respondents (5), with some indicating that they had invested in the website at the initial stage of the business start up but did not have the internal capabilities to maintain and upgrade the site on a continuous basis. As a result they became dependent on outside partners. ICT systems failure also posed a challenge for some of the respondents with some of them not wanting to *"over rely"* on the website as a way of doing business with risks of computer viruses, systems malfunctions or security risks when making payments through a website. It also emerged from the research that respondents had difficulties integrating external and internal business processes in e-business in order to maximise the use of their web site.

Although challenges were faced by the respondents in terms of the website, the majority of challenges experienced by respondents centred on internal factors of managing a growing business. Increased administration and bureaucracy associated with people management and in cash flow management. Respondents indicated that these tasks consumed an increased management of their time eroded into the time respondents could spend on developing the business. Three respondents indicated the need to become more efficient in the execution of these tasks. Despite this there was a very low usage of ICT in the completion of these administrative tasks. Other common issues related to the ongoing need to monitor and manage the costs of doing business in particular staff and distribution costs which were to a large degree out of the control of the owner/manager. On a personal level problems were encountered with time management and the isolation of working on their own. Overall there was a consensus that *"the issues impacting on firm growth were not different whether it is male or female"* or type of female operated business. In relation to the impact of family life, comments received included *"having children and juggling this with work"* was an issue but not that different from their previous work environment. Comments were received about the need for women to more effectively network beyond *"women in business networks"* and to *"break out of stereotypical sectors"*.

Interestingly it is viewed that despite operating a business from home, the mompreneur faces challenges in childcare, particularity as the firm grows. This presents a critical milestone in the decision of the mompreneur to develop the potential of their business and if so how best to do this or forfeit business growth to ensure they can be at home with their children. It is advanced that ICT can assist the mompreneur to some way achieve growth and still reach a level of balance between business and childcare commitments. Finally, comments suggested that while certain issues were specific to mompreneur firms, it was important to reduce the stereotyping of female business which can in some ways diminish their contribution.

In summary it is argued that female owner/managers are a more heterogeneous group which reflect a range of personal characteristics and variations in the type of business established and how these businesses are operated. The mompreneur is an emerging form of female entrepreneurship.

The feedback from respondents suggests that respondents established their business for positive reasons, develop clear commercial business growth objectives, adopt a focused strategic approach to achieve these objectives and display a very coherent level of awareness of the obstacles and challenges impacting on the achievement of firm growth. The challenges are related in general to internal management issues, typically associated with firm growth irrespective of gender, however gender specific challenges encountered by females included lack of expertise and knowledge in IT and thus in turn ICT.

10. CONCLUDING COMMENTS AND RECOMMENDATIONS

This research has highlighted a number of issues and unanswered questions which require research in order to develop an improved understanding of the role and use of ICT in female operated business, specifically the mompreneur. It was found that respondents did not capture the full benefits of ICT. Despite good access to computers and high levels of connectivity, there is relatively little implementation of ICT-enabled integrated business processes or the adoption of sophisticated online activities in mompreneur businesses. However it is suggested from the review of the general literature that these issues are not specific to the mompreneur but reflect the situation with smaller firms irrespective of gender. Therefore there is a need to develop the level of sophistication of the use of ICT with a focus on of integration within the other business functions both at an intra-firm and inter-firm level. This supports a conclusion that the development agencies will increasingly be required to promote ICT and business development supports as part of an overall package to promote enterprise development.

It is in the interest of policy makers in Ireland to fully ensure the relevant adoption of ICT amongst female entrepreneurs/mompreneurs to both increases their internal efficiency and productivity, as well as facilitate the growth of their firm. Further it is suggested that an enhanced knowledge and skills in ICT will serve as a conduit for starting a new business and as an enabler for business growth and development. Also ICT can aid the mompreneur to participate in professional and social networks to overcome the obstacle of "working in isolation" or alone, so often cited as a challenge faced by female entrepreneurs. In order to develop ICT competencies a number of issues should be considered in small firm supports and interventions.

From a policy perspective there is a need for support agencies to develop a number of support structures and training interventions which address the following:

- Demystify the perceived complexities associated with ICT which are largely founded in the type of language used by ICT suppliers and web designers. It is important that the language used is non-technical and contextualized for a small firm. Moreover it is argued this is not just an issue for the female entrepreneur.
- Inform female entrepreneurs of the role that ICT can play in the start up and growth of a business e.g. support agencies could assist in creating awareness and reducing the psychological barriers to ICT acquisition by showcasing small firm success stories, best practices, and benefits gained through ICT adoption. By clearly demonstrating the modern systems approaches to ICT/business integration, the failure rate of ICT adoption can be reduced.
- Promote "being a mompreneur" as a viable self employment option with the encouragement of value added modern industry sectors. Central to this is the promotion of ICT as a key enabler and catalyst in the establishment and growth of the mompreneur business.

- Strengthen ICT literacy and build capacity in the alignment of business and ICT strategies through appropriate training. Training for the adoption of ICT during all phases of the business life cycle should be available i.e. support agencies and policy makers should target female entrepreneurs/mompreneurs by organising capacity building workshops which focus on ensuring that ICT investments focus on the real needs of the business and also enabling more effective communication between the entrepreneurial firm and ICT functions at a strategic level
- Conduct ICT audits with client companies. This will involve experts conducting a needs analysis of the small firm with regard to ICT and devising an ICT acquisition and implementation plan. Mentoring for the mompreneur is viewed as an important follow up mechanism to assist the owner/manager implement and monitor the ICT function and to develop it as the firm grows.
- Create exposure to and facilitate the building of networks and alliances for female owner/managers to assist in the achievement of small firm growth. The concept of networking should be broadened to include virtual networking (e.g. Twitter). Also social networking should be encouraged for female entrepreneurs or mompreneurs. An internet-based advisory or e-coaching service on advanced ICT solutions, ICT/business integration practices, establishment of virtual offices, and virtual business units could also be established as part of the networking initiative.

Given the lacuna of empirical research specific to the female entrepreneur and the mompreneur and ICT adoption it is suggested that research is warranted in the following areas:

- Investigate long-established female owned firms who have adopted ICT and have grown and maintained a competitive advantage in the market. These will form the basis of important "good practice" case studies to assist female counterparts in the achievement of small firm growth.
- Build a more comprehensive body of knowledge on how internal factors shape the pattern of growth female owner/managed firms. In particular research on how owner/manager characteristics (age, education and work experience), the characteristics of the firm (product/service and industry profile) and the strategic activities of the owner/manager impact on the growth process and whether ICT has had an impact on this.
- Comparative studies of female-owned businesses who have adopted ICT and those that have not.

The findings of this research and the suggestions for further research will serve to progress a research topic that is still in its infancy in and one, which is of importance to academia, government support agencies, policy makers and female owner/managers/mompreneurs alike.

REFERENCES

Arenius, P., & Kovalainen, A. (2006). Similarities and Differences Across the Factors Associated with Women's Self-employment Preference in the Nordic Countries. *International Small Business Journal, 24*(1), 31–57. doi:10.1177/0266242606059778

Babaeva, L., & Chirikova, A. (1997). Women in business. *Russian Social Science Review, 38*(3), 81–92. doi:10.2753/RSS1061-1428380381

Barkham, R., Gudgin, G., Hart, M., & Hanvey, E. (1996). *The Determinants of Small Firm Growth: An Inter - Regional Study in the United Kingdom 1986-1990. Regional Policy and Development Series*. London: Jessica Kingsley Publishers.

Barringer, B., & Ireland, D. (2006). *Entrepreneurship- Successfully Launching New Ventures*. Pearson Education Prentice Hall.

Bhide, A. (1993). Bootstrap Finance: the art of start-ups. *Harvard Business Review*, (Nov-Dec): 109–117.

British Telecom. (2008). *Mum Magnates- A study into entrepreneurial mums and how they keep their new born businesses thriving*. The Red Consultancy-www.bizmums.yell.com

Browne, S., & Farrell, L., Harris & Sessions, J. (2006). Risk Preference and Employment Contract Type. *Journal of the Royal Statistical Society. Series A (General)*, *169*(4), 849–863.

Brush, C. Carter, N M., Gatewood, E.J., Greene, P.G., Hart, M., (2004). *Women Entrepreneurs, Growth, and Implications for the Classroom*. United States Association for Small Business and Entrepreneurship.

Buttner, H., & Moore, D. (1997). Women's organizational exodus to entrepreneurship: self-reported motivations and correlates with success. *Journal of Small Business Management*, *35*(1), 34–47.

Carter, N., & Kolvereid, L. (1997). *Women starting new businesses: The experience in Norway and the US"*. Paper presented at the OECD conference on Women Entrepreneurs in SME's, Paris.

Carter, S. (2000). Improving the numbers and performance of women-owned businesses: some implications for training and advisory services. *Education + Training*, *42*(45), 326–334. doi:10.1108/00400910010373732

Carter, S., & Bennett, D. (2006). *Gender & Entrepreneurship", Enterprise and Small Business- Principles, Practice and Policy* (2nd ed., pp. 176-192). Prentice Hall Financial Times.

Carter, S., & Rosa, P. (1998). The financing of male and female owned businesses. *Entrepreneurship and Regional Development*, *10*(3), 225–241. doi:10.1080/08985629800000013

Carter, S., & Shaw, E. (2006). *Women's Business Ownership- Recent Research and Policy Development*. Report to the Small Business Service.

Chapman, P., James-Moore, M., Szczygiel, M., & Thompson, D. (2000). Building internet capabilities in SMEs. *Journal of Enterprise Information Management*, *13*(6), 353–361.

Chell, E., & Baines, S. (1998). Goes gender affect business performance? A study of micro business services in the UK. *Entrepreneurship and Regional Development*, *10*(2), 117–135. doi:10.1080/08985629800000007

CSO. (2008, March 31). *Information Society and Telecommunications, 2007* [press release]. Retrieved May 21, 2008 from http://www.cso.ie/newsevents/pr_informationsociety2007.htm

Delaney, L. (2007, February). Trading tools- gain an edge by managing global shipments online. *The Entrepreneur Magazine*. Retrieved from www.entrepreneur.com

DeMartino, R., & Barbato, R. (2003). Differences between women and men MBA entrepreneurs: Exploring family flexibility and wealth creation as career motivators. *Journal of Business Venturing*, *18*, 815–832. doi:10.1016/S0883-9026(03)00003-X

Department of Enterprise. Trade and Employment (1999). White Paper on Human Resource Development, Government Publications, Dublin.

Eagly, A., & Karau, S. (2002). Role Congruity Theory of Prejudice Toward Female Leaders. *Psychological Review, 109*, 573–598. doi:10.1037/0033-295X.109.3.573

eBusiness Strategy Group (2004). *eBusiness Strategy: Optimising usage of ICTs by Irish SME's and Microenterprises*. Department of Enterprise, Trade and Employment.

Elsammami, Z., Scown, O., & Hackney, R. (2001). A case study of the impact of a diffusion agent on SMEs adoption of a Web presence. In *Proceedings of the 6th UKAIS conference* (pp. 562-576). University of Wales Institute, Cardiff 26-28 April 2000.

Feuer, E., Messnarz, R., & Sanchez, N. (2002). Best Practices in E-Commerce: Strategies, Skills, and Processes. In *Proceedings of the eBusiness and eWork 2002 Conference proceedings, Challenges and Achievements in E-Business and E-Work* (Part 1, pp. 109-116). Amsterdam: IOS Press.

Fielden, S. L., Davidson, M. J., Dawe, A. J., & Makin, P. J. (2003). Factors inhibiting the economic growth of female-owned small businesses in North West England. *Journal of Small Business and Enterprise Development, 10*(2), 152–166. doi:10.1108/14626000310473184

Fischer, E., Reuber, R., & Dyke, L. (1993). A theoretical overview and extension of research on sex, gender and entrepreneurship. *Journal of Business Venturing, 8*(4), 151–168. doi:10.1016/0883-9026(93)90017-Y

Fitzsimons, P., & O'Gorman, C. (2007). *Entrepreneurship in Ireland. The Global Entrepreneurship Monitor 2007: The Irish Report*. Dublin, Ireland: Enterprise Ireland.

Fitzsimons, P., & O'Gorman, C. (2008). *Entrepreneurship in Ireland. The Global Entrepreneurship Monitor 2008: The Irish Report*. Dublin, Ireland: Enterprise Ireland.

Forfás. (2005). *Broadband Benchmarking* [online], Retrieved May 22, 2008 from http://www.forfas.ie/publications/forfas_annrpt05/platforms/broadband.html

Forfas (2007). *Towards developing a policy for Entrepreneurship in Ireland*. Dublin.

Gender Equality Unit. (2000). *Women and Men in Ireland as Entrepreneurs and Business Managers*. Dublin, Ireland: Department of Justice, Equality and Law Reform.

Glancey, K. (1998). Determinants so Growth and Profitability in Small Entrepreneurial Firms. *International Journal of Entrepreneurial Behaviour and Research, 4*(1), 18–27. doi:10.1108/13552559810203948

(2002). *Goodbody Economic Consultants*. Dublin, Ireland: Entrepreneurship in Ireland.

Group, S. I. A. (2001). *Policy and Planning on Developing Women in Enterprise*. Dublin: Report to Enterprise Ireland.

Hampton, A., Cooper, S., & McGowan, P. (2007). *Female Entrepreneurial Networks and Networking in Technology-Based Sectors*. Institute for Small Business & Entrepreneurship, 7-9 November 2007, Glasgow, Scotland.

Henry, C., & Johnston, K. (2003). *State of the Art of Women's Entrepreneurship in Ireland: Access to Financing and Financing Strategies*. Ireland: Centre for Entrepreneurship Research, Dundalk Institute of Technology.

Henry, C., & Kennedy, S. (2002). *Search of a New Celtic Tiger-Female Entrepreneurship in Ireland*. Dundalk Institute of Technology.

Holden, L. (2007, August 1). Gaining the confidence to go it alone. *The Irish Times*.

Irish Chamber of Commerce, (2002). *SME eBusiness Survey*.

Jacobs, G., & Dowsland, W. (2000). The Dot-Com Economy in Wales: A Long Road Ahead. In *Proceedings of the 5th UKAIS Conference, University of Wales Institute* (pp. 590-596). Cardiff 26-28 April.

Jones, P., Beynon-Davies, P., & Muir, E. (2003). eBusiness Barriers to Growth within the SME Sector. *Journal of Systems & Information Technology, 7*(1).

Kjeldsen, J., & Nielsen, K. (2000). *Women Entrepreneurs Now and in the Future*. Danish Agency for Trade and Industry.

Lewis, R., & Cockril, A. (2002). Going global - remaining local: the impact of ecommerce on small retail firms in Wales. *International Journal of Information Management, 22*, 195–209. doi:10.1016/S0268-4012(02)00005-1

Marlow, S. (2002). Self-employed women: Apart of, or apart from, feminist theory? *Entrepreneurship and Innovation, 2*(2), 83–91. doi:10.5367/000000002101299088

Marlow, S., & Carter, S. (2004). Accounting for change: professional status, gender disadvantage and self-employment. *Women in Management Review, 19*(1), 5–17. doi:10.1108/09649420410518395

Marlow, S., Carter, S., & Shaw, E. (2008). Constructing female entrepreneurship policy in the UK: is the US a relevant benchmark? *Environment and Planning. C, Government & Policy, 26*, 335–351. doi:10.1068/c0732r

Marlow, S., & Patton, D. (2005). The financing of small businesses- female experiences and Strategies. In Davies, M., & Fielden, S. (Eds.), *International Handbook on women and small business entrepreneurship*. Cheltenham, UK: Edward Elgar.

McClelland, E. (2003). Following the pathway of female entrepreneurs. *International Journal of Entrepreneurial Behaviour & Research, 11*(2), 84–107. doi:10.1108/13552550510590527

McGowan, P., & Henry, C. (2004). Special Issue: Female Entrepreneurship-A Research Agenda. *International Journal of Entrepreneurial Behaviour and Research*.

Minniti, M., & Bygrave, W. D. (2003). *National Entrepreneurship Assessment: United States of America Executive Report*. Kansas City, MO: Kauffman Foundation.

Moussi, C., & Davey, B. (2000). Internet-based electronic commerce: perceived benefits and inhibitors. In *Proceedings of the Australian Conference on Information Systems, Information Systems Management Research Centre*. Queensland, University of Technology, Brisbane, December 6-8.

Nath, R., Akmnakigil, M., Hjelm, K., Sakaguchi, T., & Schultz, M. (1998). Electronic Commerce and the Internet: Issues, Problems and Perspectives. *International Journal of Information Management, 18*(2), 91–101. doi:10.1016/S0268-4012(97)00051-0

OECD. (2004). *Promoting entrepreneurship and Innovative SME's in a global economy: Towards a more responsible and inclusive globalisation*. Paper presented at 2nd OECD Conference of Ministers responsible for Small and Medium-sized enterprises (SME's), Istanbul, Turkey, 3-5 June 2004.

Oppedisano, J. (2004). Giving back: women's entrepreneurial philanthropy. *Women in Management Review, 19*(3), 174–177. doi:10.1108/09649420410529889

Orhan, M., & Scott, D. (2001). Why women enter into entrepreneurship: an explorative model. *Women in Management Review, 16*(5), 232–247. doi:10.1108/09649420110395719

Philipsen, R.L. & Kemp, R.G.M. (2003). *Capabilities for Growth An Exploratory Study on Medium-Sized Firms in Dutch ICT Services and Life Sciences*. Scales Paper N200313 EIM Business and Policy Research.

Report of the Small Business Forum. (2006). *Small Business is Big Business. The Small Business Forum*. Forfás Dublin.

Richardson, I., & Hynes, B. (2007). Women in engineering and technological entrepreneurship: exploring initiatives to overcome the obstacles. In Carter, N.M., Henry, C., Cinnedie, B. Ó., & Johnston, K. (Eds.), *Female entrepreneurship implications for education, training and policy*. Routledge.

Sakai, K. (2002). *Global Industrial Restructuring: Implications for Small Firms*. STI Working Papers 2002/4, OECD, Paris (www.oecd.org/sti/working-papers.)

Small Business Service. (2003). *A Strategic Framework for Women's Enterprise*. London: DTI Small Business Service.

Smith, J., & Webster, L. (2000). The Knowledge economy and SMEs: a survey of skills requirements. *Business Information Review*, *17*(3), 138–146. doi:10.1177/0266382004237656

Stevenson, L. A. (2003). Against all odds: The entrepreneurship of women. *Journal of Small Business Management*, *24*(4), 30–36.

Still, L. V., & Timms, W. (2000). Women's business: the flexible alternative workstyle for women. *Women in Management Review*, *15*(5/6), 272–283. doi:10.1108/09649420010372931

Storey, D. J. (1994). *Understanding the Small Business Sector*. London: Routledge.

Straeder, T., & Shaw, M. (2000). Electronic Markets: Impact & Implication. In Shaw, M., Blanning, R., Straeder, T., & Whinston, A. (Eds.), *Handbook on Electronic Commerce* (pp. 77–98). London: Springer-Verlag.

Valiulis, M., Drew, E., Humbert, A., & Daverth, G. (2004). *Springboard for Women in Business Initiative Wicklow Chamber of Commerce*. Centre for Gender and Women's Studies, Trinity College Dublin.

Van Akkeren, J., & Cavaye, A. (1999). Factors affecting Entry Level Internet Adoption by SMEs: An Empirical study. In *Proceedings from the Australasian Conference in Information Systems* (pp. 1716-1728).

Villanueva, J., & Pavone, C. (2007). *The Effect of Entrepreneurial Motives on Growth: A Study of Women Entrepreneurs*. Babson College.

ENDNOTE

[1] A previous version of this chapter was published in the International Journal of E-Politics, Volume 2, Issue 1, edited by Celia Romm, pp. 17-29, copyright 2011 by IGI Publishing (an imprint of IGI Global).

Chapter 3
Women in IT Careers:
Investigating Support for Women in the Information Technology Workforce[1]

Elaine K. Yakura
Michigan State University, USA

Louise Soe
California State Polytechnic University, USA

Ruth Guthrie
California State Polytechnic University, USA

ABSTRACT

Research on the declining numbers of women in the Information Technology (IT) workforce focuses on 'filling the pipeline' by attracting women into IT disciplines at colleges and universities. This research looks at the other end of the pipeline, examining both barriers and support structures that have helped women persist in their IT careers. The chapter draws from extended interviews with 38 women who have been successful in planning their careers and navigating a male-dominated industry. It focuses on what women cited as barriers and as areas of support, in response to open-ended questions about their careers and career paths. The interviewees, drawn from nine industry sectors, represented a wide breadth and depth of experience. Half were at the professional level, and half at the managerial or executive level in organizations that varied from single-person consulting firms to large institutions.

INTRODUCTION

This chapter discusses the results of open-ended interviews with women in various stages of careers in Information Technology (IT). The research focuses specifically on support structures, formal and informal, that women identified and the perceived benefit of these support structures. The objective of this chapter is to examine the perspective and experience of women in industry. Specifically what they can teach us about what helped them in their careers, mentoring program effectiveness and work accommodations for women. Practical experiences can inform women seeking careers in IT about biases they face prior to entering the workforce. These experiences can also help industry and academia programs prepare women before they enter the IT workforce.

BACKGROUND

Women are underrepresented in IT college programs (Computer Information Systems, Computer Science, Management Information Systems), and the numbers of women in IT careers is declining. Many attribute this to the declining interest of women seeking degrees in technology related disciplines, a phenomenon dubbed the 'pipeline' problem (Blickenstaff, 2005; Camp, 1997; Soe & Yakura, 2008). Mitigation strategies have resulted in changes to IT academic curricula, the establishment of mentoring and role model programs, as well as the development of women's networking programs to reduce feelings of isolation among women. The assumption is that if IT curricula are more accessible to people without programming backgrounds, more women may be attracted to the discipline (Sloan & Troy, 2008). Coursework that emphasizes the context and implications of the use of technology, may better attract and retain women students than ones that emphasize technology for technologies' sake, (Stiller & LeBlanc, 2003). Mentoring programs and anti-isolation programs help to support women in IT academic and professional careers (Simard, Henderson, Gilmartin, Schiebinger, & Whitney, 2008). However, despite these efforts to attract and retain women, the number of women in IT careers is at an all time low, and dropping quickly.

There is ample evidence of this decline. In 1996, the Information Technology Association of America reported that women made up 41% of the IT workforce, but by 2002, the percentage dropped to 34.9% (Hollis, 2003). An NCWIT report (Ashcraft & Blithe, 2009) gives slightly different figures (in 1991, 36% of the IT jobs were held by women, down to 24% in 2008), but the downward direction is the same. The NCWIT report also noted that women seem to face a mid-career 'fight or flight' moment, in which many of them opt out of the IT profession, in spite of indications that 74% of them were highly satisfied with their careers.

Many researchers in the overlapping areas known as STEM or SET (Science, Engineering, Technology) education have focused on the reasons for the dearth of women in these disciplines, either as students or faculty (Bystydzienski & Bird, 2006). Sappleton and Takruri-Rizk (2008) described the approaches of different disciplines to study the problem, and explained that the reasons for the under-representation appeared complex. Guzman, Stam, and Stanton (2008) pointed out distinct differences in IT occupational culture, which suggests that while IT careers are similar to SET or STEM careers, one cannot assume they are identical.

In research on women and IT careers, the issue of "opting out" by women who have successfully navigated finding an IT job is an interesting one. Some researchers concluded that the decision of women to start families explained this opt out (Armstrong, Riemenschneider, Allen, & Reid, 2007). Other studies demonstrated that men and women differed little in what attracted them to IT careers or what factors they valued if they persisted in them (McKinney, Wilson, Brooks, O'Leary-Kelly, & Hardgrave, 2008; Kuhn & Joshi, 2009). In an interview with Computer-World (Melymuka, 2005), Dorie Culp explained women's attrition by saying, "Our research shows that work/life balance is an excuse women give when they leave so they can leave gracefully. But the reality of why they leave is the culture -- the way it marginalizes women."

Addressing similar issues for women in engineering, Faulkner (2007) asserted that it was not a failure to socialize women into the discipline that needed to change. Rather the discipline needed to change so that it no longer devalued women. In her ethnographic study of engineers in a building design engineering consulting company, Faulkner

(2007) described in detail how the identities and practices of engineering professionals conform to (and not conform to) understandings of gender: "Thus, many men engineers cleave to a technicist engineering identity because it feels consistent with versions of masculinity with which they are comfortable" (p. 350). She argued that engineering culture can and should change, and that "as a profession [it] must find ways to foreground and celebrate heterogeneous understandings of engineering and heterogeneous engineering identities" (2007 p. 351).

Several empirical models examine the motivations of IT professionals. Schein (1990) introduced the concept of career anchors, and identified ten anchors that could describe people's motivations and decisions in their career choices. IT scholars adopted this model to investigate the issue of motivation for technical work, finding several different and conflicting results. The number of motivators and their importance to IT professionals varied. Crepeau (1992) found that career anchors were independent of each other and that lifestyle and creativity were not relevant to IT professionals. Sumner, Yager, and Frankie (2005) found four relevant career anchors: creativity, autonomy, identity and variety. Attempts to understand the differences through clustering of the anchors created several more taxonomies of career motivators, particularly focusing on what women want (Ferratt, Enss, & Prasad, 2006; Ituma, 2006). Quesenberry and Trauth (2007) examined women in IT careers, and found that women had very different career anchors depending upon their individual context and that over time, women's career anchors changed. Later studies by Trauth and Quensenberry (2008) examined populations of women in different geographic and cultural regions, finding that the context and socio-cultural factors played a significant role in determining choices women make in their careers. Trauth, Queensberry, and Morgan (2004) proposed that the theory of individual differences was a better framework for gender and IT. A framework examining personal data, influencing factors (e.g. education and life experiences) and environmental context (e.g. policy and economics, culture) allowed for deeper insight into a complex issue that had traditionally lumped all women together in the world of IT.

Research into mentoring programs and other institutional policy programs in IT mirrored other industries. Much of the research on mentoring focused on the benefits that accrued to protégés, as well as their mentors, at the individual level (Wanberg, Welsh, & Hezlett, 2003). In one recent study, Dougherty, Dreher, Arunachalam, and Willbanks (2009) examined career outcomes for male and female protégés with mentors, as well as men and women without mentors. They found that female protégés with high-ranking mentors received greater compensation than either males or females who had non-high-ranking mentors. Even those with non-high-ranking mentors received greater compensation than those without mentors. Clearly, mentor relationships could be of great benefit. Some negative aspects of mentoring relationships are also evident (Wanberg, Welsh, & Hezlett, 2003), such as a lack of mentoring expertise and mismatches within a mentor-protégé dyad (Eby, McManus, Simon, & Russell, 2000). Eby and McManus (2004) identified malevolent deception as another possible problem in mentoring relationships, including intentional deceit by the protégé. Thus, while mentoring relationships could be beneficial to both mentor and protégé, as well as benefit the organizations of which each individual is a part, mentoring relationships (like any relationship) could also be a source of negative experiences.

Other studies that look at women IT professionals showed how different social and policy support structures and accommodations opened the doors for female career success. However, the support for female success at work is a complex topic (Halpern, 2005). If women have flextime, they can work while their children are at school or asleep, and stay at home with sick children.

However, often women with flexible schedules work many extra hours, are perceived as not shouldering the same workload as their male counterparts and are passed over for promotion. Family leave, a great benefit to men and women, can lead to perceptions of weakness and liability. Women are reluctant to take advantage of these programs and for good reason. While legally there may be no bias, perceptually there is a stigma associated with leaving your job for your children (Kimmel & Ameundo-Dorantes, 2004).

This chapter seeks to explore the barriers to success that women with successful IT careers said they faced, as well as the areas of practical support these women IT professionals identified when they talked about mentoring programs and work accommodations given to women.

METHODOLOGY

The data in this qualitative study came from interviews with 38 women with IT careers in Southern California. A qualitative approach seemed more suitable for issues related to IT careers because it allowed these women to describe their experiences to us, without prompting and in their own words. Collecting oral histories as primary documents of women's experience at work uncovered a variety of different phenomena and gave insight into the complex issues of the IT workplace that impeded and supported the success of these women.

The initial list of women contacts were graduates of the Computer Information Systems and graduate business programs at Cal Poly Pomona University. A snowballing technique (Berg, 2001), located additional interviewees from local companies, professional organizations, and LinkedIn. The sample included women from "extreme" cases (in career longevity, age, organizational level) as well as "typical cases" to achieve "maximal variation in the sample" (Flick, 1998). The resulting sample represents different ethnic cultures, different age levels, a wide range of organizational levels, and different industries (see Table 1).

After completing a form advising them of confidentiality precautions taken in this research, the women spent from 60 to 90 minutes reflecting

Table 1. Women in IT Sample Group Demographics

Industry	%	Self-identified Ethnicity	%
Engineering	20%	Caucasian	55%
Financial or Insurance Services	18%	Asian/Pacific Islander/East Indian	32%
Big-4 or Management Consulting	13%	Hispanic	13%
Education (not teachers)	10%		
Entertainment	8%	**Highest Level of Education**	**%**
Health Industry	8%	Baccalaureate degree (BA or BS)	55%
IT Products or Services	3%	Masters level (MBA, MS, MSMBA)	37%
Other	20%	Doctor of Philosophy	5%
		Some college but no degree	3%
Current Organizational Level		**Demographics**	
Professional	42%	Average Age	38 years
Managerial	47%	U.S. citizen	75%
Other (unemployed)	10%	Women with Children	54%
		Women with dependents of any kind (children, parents, disabled siblings)	61%

on several different aspects of their careers. In this chapter, the themes of barriers and enhancers to their careers are reported along with the types of support they received from firms and mentors. Interviews were recorded and then transcribed and coded, using the qualitative data analysis software Atlas-ti 5.0. Coding used top-down and bottom-up (in-vivo) coding (Lewins & Silver, 2007). Top-down codes reflected research assumptions covering issues such as cultural references at the societal, occupational, organizational, and workgroup levels, and other recommendations for improving the status of women in IT, such as mentoring and networking. A grounded approach (Strauss & Corbin, 1990) using in-vivo coding or bottom-up coding looked for recurring themes that evolved during the study that emerged from the data. For example, a surprising number of capable women ascribed their success to "luck" and "good fortune." Further coding identified and separated positive and negative quotations about the same topic, and identified other issues.

FINDINGS

Several themes emerged during the interviews. Women shared personal and work experiences,

Table 2. Most frequently named barriers to women's success

Career Barriers	Total N=38	%
Occupational Cultures / Subcultures	24	63%
Stressful, Demanding Work, Long Hours / Travel	21	55%
Hostile Work Environment	21	55%
Relationships: Boss, co-workers, subordinates	19	50%
Worklife balance issues	19	50%
Poor Infrastructure, processes, technology change	16	42%
No room at the top: Old Boys' Club	15	39%
New in organization, new boss, newly promoted	13	34%

and talked about barriers they faced (and still face) as well as characteristics in themselves and the profession that aided success. This chapter examines cultural barriers in organizations and the IT occupation, as well as elements that mitigated those barriers for these successful women, such as mentoring, workload and other firm-provided special accommodations for women.

CAREER BARRIERS AND ENHANCERS

In coding the interview dataset, themes were identified using note taking and tagging features of the software. Tables 2 and 3 list the most common categories of career barriers and enhancers and the numbers of women who spoke about the listed category in the table. Women, were asked to tell their career stories, and then asked broad questions about their success in the IT profession and were not prompted to speak about these particular categories. The quotations that follow are taken from transcripts of the interviews; interviewees are identified by number only.

Table 3. Most frequently named sources of support for women's success

Types of Support	Total N=18	%
Boss or Mentor	26	68%
Networks	24	63%
Workplace Accommodations / ability to set boundaries around work times / telecommuting	22	58%
Team / Co-worker support	21	55%
Luck or good fortune	19	50%
Parental support	16	42%
Job choice for worklife balance	16	42%
Spousal support	14	37%
Teacher support	14	37%
Workaholic / loves work	10	26%
Being Measured by performance not hours	5	13%

Women at all levels of their organizations recognized the importance of mentoring and networking, understanding the politics of the organization, and having a sponsor or boss who would promote and make opportunities known to her. Earlier in their careers, women identified a need for education and for seeing the 'bigger picture' in the projects on which they worked. They also were aware that women should explicitly ask for what they want, to make their career expectations clear to their managers. A few women mentioned 'giving a little extra.' By this, they meant that it was advantageous to foster good relationships with people, and deliver extra high-quality technical services to the client. It could also make the job more interesting, as a cost analyst at an engineering firm said,

... it can be something that is not officially assigned, but you can deliver more than what they want. If you can show them that you can do something extra, above and beyond what is needed.... it's meaningful because it's more interesting than what the job calls for. (R20)

One woman in risk management worked for a woman manager who continually alienated employees in other departments. The interviewee said she had to build bridges behind her manager's back, in order to get information she needed to do an effective job. She did not believe in "burning her bridges."

People in network security knew our perspective but they didn't really care to help us out at times because they didn't want to deal with my boss. I would build relationships with them and they would tell me something was going to happen. It would be a little diamond they were handing me. Then I would have to develop it and then say that I heard that something was happening and that would really start the wheels rolling. But, if I didn't have that relationship with them, they never would have given us the heads-up. (R15)

All of the women felt that some form of teamwork, being able to solve the client's problem, and understanding the business, were important career boosters. Several of them felt that doing different jobs within a firm or changing jobs gave them the advantage of having a broader perspective. A devastating job loss could lead to an even better opportunity. These women knew that women needed to display confidence to ensure that the people above them were aware of their capabilities and accomplishments.

Most women gave the advice "Don't take it personally."

So, if someone puts your idea down, and I have been in situations where someone says "I think that's ridiculous." To be able to say that you don't agree, but it's your thought and you are putting it out for consideration. Being able to be gracious, even when you are on the line of fire, which happens a lot. (R39)

When an organization was too top heavy and had no women executives, or at most a token woman or two, women saw fewer opportunities for promotion. They also identified specific career paths with 'dead ends' or having no clear path to the top. One woman who became the manager of an IT Audit group, recognized that managers of the audit group never advanced to the VP level in the firm. However, these jobs could be appropriate for women when they had young families, since the demands for longer hours were lower.

Women felt the strains of worklife balance. Women who worked in consulting firms often needed to travel for work, and only saw their families on weekends.

The bottom line is that if you travel this much, you cannot do it. You have to face reality...It was a choice that I had to make. What is my top priority? Is it the family and the baby? Or is my top priority my consulting work? I chose the family

and the child because ... one of them has to be the first, whichever one is the most critical and important at the time. (R17)

One woman manager, with several children including a disabled son, took less pay in a more flexible organization, so she could deal with family emergencies.

I can use my project management skills and get everyone on a routine, but that only goes so far, until somebody misses a bus or somebody has a medical emergency or daycare flakes out or whatever happens. (R14)

Women reported not 'looking right' for the job and having to prove themselves on the job in ways men did not. Women who looked young especially had difficulties.

I was with the company for a little over three years there. What I noticed is when I was applying for the job, a lot of people were very skeptical of two things. They're not used to having an IT audit manager who is a woman and I look awfully young for the job. ... One of the directors told me, ... [if a] man tries so hard and fails, they get praise for it, because they tried. But, when we try so hard and fail, it's because we're a women and we don't know what we're doing. I, my first reaction to that comment was that I thought she probably exaggerated, but, dealing with these people long enough, I can see that happening and I can understand why most of these women have the mentality that they have, that they have to prove themselves so hard. Their role is constantly challenged. Their ability to get the job is constantly challenged. And these men are not shy about challenging all these things. (R33)

The woman who made the following observation was not especially young, but she was the lone woman in a "macho" networking group. She worked with the telephony system, which the men viewed as lower status:

... even though I had the exact same degree as some of the guys in the networking department, and I did telephony and they did data switches, and I kept up with them in grades, and everything was equal, I was not equal. I presented that to my boss, but he didn't do anything. It might have just been him, but I am quite certain that people above him felt the same way. (R19)

Another barrier was a lack of meaningful work that yielded visible accomplishments and enhanced the woman's reputation. One high-level manager tried to figure out how to build a career as a 'direct generator of revenue,' rather than playing a support role to the business. Work directly related to the business units is higher in status than support or "overhead" work. Several interviewees mentioned that it was better to work in a group that was not part of the IT department. By moving their careers to a higher status position, or moving into an organization where IT is the business focus, women can gain more power and exposure in their careers. One woman advanced after she remained and 'saved the day' when her company was having difficulties, and the people above her left for other jobs.

FORMAL AND INFORMAL MENTORING

Few of the women in this sample worked in organizations with formal mentoring programs. Women from engineering, the Big 4,[2] and the insurance industry said the mentoring programs were ineffective because of lack of interest of the mentor and lack of fit to the mentee's career. At one Big 4 consulting firm, employees got two mentors: a 'buddy' (usually not within her department) and a 'performance manager'. If the woman got a performance manager who advocated for her,

Table 4. Informal Mentoring Types Identified by Women During Interviews

Informal Mentoring Type	Description	Number of Women
Guide	A trustworthy person who can give advice. Ad Hoc Guide: sought out for particular situations. Women may have multiple ad hoc mentors, depending on the problems and situations Long Term Guide: provides advice for many years, even when both move to different locations	7
Role Model	Short term person who women admire. A "strong woman" either within the organization or at a client's organization.	4
Boss Advocate	Manager or boss who actively promotes and advocates for the women. Can be a key to launching one's career into the fast-track.	3
Defender	Manager or peer who stands up for women when ideas are discounted.	2
Teacher	Professors from universities and colleges that still give guidance to graduates.	2

then she could gain greater access to higher-profile billable project hours. Mentoring of some kind seems to be an essential part of Big 4 culture all the way to the partner level, although mentors are responsible for multiple mentees. Most women at Big 4 firms praised their mentors (often their first managers who continued to help their careers even after they switched managers). One young woman used the opportunity of reviewing her mentor (part of his performance review) to resolve successfully an issue that bothered her. Another young woman described an unfortunate situation in which her mentor ignored her, and then asked her to lie about his performance. She soon left the firm, but she did not "burn any bridges," a phrase commonly heard in these interviews.

The definition of mentor varied. Women found informal mentoring much more successful. Women earlier in their careers identified women as informal mentors. They self-selected women who were leaders in the organization as mentors, referring to the "strong woman" who was a role model. Sometimes their contact with this woman was temporary and short-lived, but the woman inspired them. (Some women also talked about mothers as "strong women" who provided important support to them.) Women who had worked in the IT industry longer identified mostly men who were their bosses. When a woman had a boss who was an advocate for her, she felt more supported and had better access to opportunities.

From the interviews five types of informal mentor relationships emerged: guide, role model, boss advocate, defender and teacher (Table 4).

Many women (24 of 38) recognized the importance of networking, either socially or professionally, to gain access to opportunities that could help their careers or where they could help others.

One woman who was put on the "fast track" by her first boss (who was also fast-tracked), described what he did for her:

I didn't know what kind of management job I would be walking into but when one opened up, he was right there pushing me into it. He promoted me. That's where things started to move for me. I was put into more high visibility jobs. More exposure, and I got creditability as a result...

My ex-manager had about 25 years in the company, so he would help me understand what the agendas were, what the politics were, what the major issues were. It could be easy to take a side and then get caught up in it. Certainly if I wanted to change positions, he was there to advise me. Managing employees, because he had managed people for 20 of the 25 years he was there, and he would help me. That was really valuable to me. (R18)

Typically, women maintained these early manager-mentoring relationships throughout their careers, going back to the same informal mentor, even when they no longer worked together.

One woman described why she preferred a female mentor in these terms.

But the ones I get the most out of are women in leadership positions, because our motivations for leadership are different. Men can work all the hours in the world and be aggressive and women cannot necessarily do that, no matter what their challenges are. The barriers they have are a little higher, to get into those positions. They have to do things differently than the men. (R14)

WORKLOAD AND WORK ACCOMMODATIONS FOR WOMEN

Differences in workload were related to industry and professional area in IT. The workplace accommodations for women also varied greatly. Twenty-one women mentioned the stress that a heavy workload and required travel put on their lives.

Internal IT auditors tended to work normal workweeks. Managers and women in financial and Big 4 consulting companies worked 60-70 hour workweeks, often traveling. Women with overlong workweeks worried about the effects on their families.

It was very tough during the first year of Sarbanes Oxley, very tough, because lots of the middle staff jumped ship. They were seniors but they could get director positions because everybody wanted someone with Sarbanes Oxley experience. So we lost the middle resource, the assistants. When we were out at a job on an audit, it was just me and an intern. Or me and a staff, who had no experience. So not only did you have to stay there and train them and explain things to them, you still had to do your own job, which meant I had to do everything, the senior job, the manager job, because they were not there. It was a very difficult year. That's part of what drove me to leave. I felt like I was punishing my family. (R38)

Several women described their hours as cyclic, requiring more hours during certain times of year. Several women said they were 'workaholics' and liked to take work home. One woman, a telecommunications manager, indicated,

I sit with my laptop on my lap every night and work. Thank God, he (the husband) doesn't care. We watch a movie together, and I will be working while we watch the movie. I like to multitask... I also have an iPhone, so during dinner I was reading resumes because I am on a recruitment committee, which ones I want to call for an interview. I worked like a fool last week. Normally, I like to at least have Sunday off and maybe part of Saturday. If I had to think about it, 12 to 15 hours a day. I get so much email it is insane. To my employees I always write back thanks, so they know I read their email. (R19)

One woman in the healthcare industry said:

My bosses, the old one, and the current one, they don't believe in watching the hours. They expect you to be professional and do what is necessary to get the work done. So, to the extent I get my work done, I work a 40-hour week. In almost 4 years ... I can count with my fingers how many times I've had to work more than a 40-hour week. (R33)

Another manager, who had a 25-year career, found that her non-profit healthcare company focused on getting the work done, not on the location or time of day. She managed 200 employees who telecommute. The organization saved office space with this arrangement.

One of my direct employees has a child with sickle cell disease. Every three weeks the child goes to

the hospital for blood transfusions. I understand how that is. But life goes on. When she told me about that, we sat down and figured out what to do. She is very capable, she is very intelligent, she has a very strong background, she produces very high-quality work. I told her to manage her schedule, that when she needs to go, she should go. It has worked out very well. You know what happens? Every time she takes her daughter to the hospital for a blood transfusion, she sits next to her daughter with her laptop, her air card, her palm pilot, cell phone, still on the conference, still on email, still reachable, and she gets the work done. I told her, to me, I am not going to keep track of where you are. As long as you figure out how you are going to get it done and balance it, we can make it happen. (R17)

In most other cases, organizations did not have official policies on telecommuting. Accommodations for women, especially those with children, varied widely. Managers decided on telecommuting, through informal agreements with employees, usually those they trusted. Women indicated they were more productive working off-site because they did not have so many interruptions. However, some felt isolated and missed the social interaction. While interviewees were asked about accommodations for promoting women or creating equal opportunity, very few mentioned any programs other than ones required by law, such as family leave.

Women spoke about the benefits of taking disability instead of family leave. A professional woman who worked in the health industry, said

When I was on maternity leave, which was twice since I have been here, my boss had to call me once, and he was really upset because he had to call me. I was on disability and I was not supposed to be working. The thing with family leave is that they offer it, but they do not have to guarantee your job. If you are in fear of losing your job, you wouldn't want to take that. On disability, they have to give you your job back when you come back. But not with family leave. I wasn't really nervous about leaving, so I was able to take all of that time...I have never heard of a man taking family leave (here). When my husband took it, it was when it first became available. Six weeks of it was going to be paid, just like disability.... He had to educate his Human Resources Department about it. (R16)

A few women mentioned that telecommuting was possible, depending upon your boss. Non-managerial interviewees talked about work as non-accommodating. Generally, if women work from home or start a family, they run the risk of not being promoted or not being seen as serious about their work. Women worried about what impact family might have on their careers and how work might impact their families. One young woman, who had just begun her career at a 'laid-back' company stated:

Another weakness, well not weakness, but something that is going to hold me back a little is having kids and being intimidated about getting married, and having to have time off. But I think, well, males have to do that do. But I think that being a female, I think companies look down on it. I have been struggling with that, and telling myself that everyone gets married, and everybody has time off when they have a child, so I won't be discriminated against. I guess I feel guilty about wanting to take that much time off to have a kid so that part is a little in my mind, but I am dealing with it. It is new to me. To be gone from work for 3 months, to take care of a child and then coming back to a workplace, that's a little scary for me. But that could be in any field. (R37)

This sentiment reverberates with that of another, experienced woman:

But when I had my daughter, I was a high level manager, about to become a senior manager. I

didn't get it that year and I knew why. Because I had just had my baby and I was on maternity. And I felt like I was punished for that. Nobody had told me that I wouldn't be getting it. They were prepping me to get it. I made sure that as soon as my medical leave was over, that I went back to work right away. (R38)

Women who were more established, identified other options that worked for them and seemed to have more flexibility. One company let employees accrue paid time off, although, from the description, employees did not take it unless the company didn't have enough hours and wanted employees to stay home.

The woman in the healthcare industry who manages 200 employees, works two days a week at the company and telecommutes the rest. Another high-level manager commutes between Los Angeles and Phoenix, working on-site for three days and from home the rest.

DISCUSSION

Several interesting themes emerged during the interviews with the women in IT. The women who participated in this study articulated several barriers and enhancers to their careers. The answers agree with Quesenberry and Trauth's (2007) assertion that individual experiences, temporal situations and environmental factors influence how women make career decisions. It is also apparent that the obstacles identified by most women were related to organizational culture and work environments that were chilly towards them. They had to find ways to 'fit' into these cultures, even though they saw them as potentially career limiting. Women found ways to become accepted (being non-threatening, not taking things personally, working harder than others, etc.) or they sought work situations that were more friendly. These are really social accommodations made by women to 'fit' with the IT culture. Women are good at this as they may have had experiences in accommodating culture throughout their technical experiences in life. Positive support structures that women identified mostly related to people, bosses, mentors and the ability to work with teams.

The topic of mentoring yielded surprising results. Mentoring programs were rare at many firms. At some of the large consulting companies, where mentoring was institutionalized, the results were mixed. Criticisms of mentoring programs often cite the lack of bond between the mentor and mentee as a primary reason for lack of successful mentoring (Wanburg, Welsh, & Hezlett, 2003). However, advocacy was taking place for many of the women in this study. Many women in the study had bosses or more senior colleagues who gave them career guidance, which sounds like informal mentoring. However, the relationship is clearly more than that. Several of the women maintained these relationships, long after they no longer worked for these bosses or these organizations. Most of the informal and formal mentoring relationships focused on the earlier stages of the women's careers. Given that there is a large drop-off of women at managerial levels, mid-career mentoring might help women move into management positions.

The most common work accommodation made for women was family leave, even though this is an accommodation made for men too. When asked to describe other formal programs for women, very few were mentioned. Some women from large firms gave examples of formalized groups (e.g, groups for women or people of color) that their firms organized. Though these groups may diminish feelings of isolation, they do not give women access to male groups (golf buddies). Moreover, similar to the situation with formal mentors, women may not form a meaningful bond with others in the group. This means that for some women, these groups might not be useful at all.

Flex time is the second most common accommodation reported by women. For some women this was especially helpful when children were small or ill. However, many agreed that this had the disadvantage of working longer hours and still having to face the stigma of not being viewed as dedicated to their work. Formal work accommodations or structural arrangements (Sappleton & Takruri-Rizk, 2008) are band-aids for the real problems that women face. Firms claim that they make accommodations for women, even though the same structural arrangements exist for men, but the underlying culture remains unchanged. Consequently, women must find informal ways to accomplish the same result. Women must establish their own informal networks for career advice and for finding new jobs. Several women, once established in their careers, found informal ways to gain flexibility without stigma. For example, several interviewees described days that they telecommute 'off the books.' These working arrangements were informal, between them and their immediate bosses. No formal policy on telecommuting was recognized. Again, these are accommodations made based upon performance, rather than on the basis of gender or the need for childcare.

FUTURE RESEARCH DIRECTIONS

As stated at the beginning of the chapter, this research is a preliminary exploration of some of the issues related to the retention of women in IT. To allow us to conduct qualitative, open-ended interviews, the sample size was deliberately small. Future research must confirm whether or not some of the issues that these women discussed are also applicable to women in larger samples.

This research focuses on women in IT careers. Interviewing men about their experiences may provide more insight. Much research has been done that confirms the IT work environment is not very supportive of women. Interviewing men about their experiences, particularly at entry level jobs, may show that they do not find the work environment particularly friendly either, even if they find it easier than women to fit in.

The sample of women in IT for this research was from Southern California. A broader sample might inform discussions about regional differences.

The sample of industries and types of IT work represented by the women who participated in the study is very broad. In future studies, the examination of issues of gender and technology might be more easily described if industry or type of IT work were narrowly defined.

CONCLUSION

Support structures help women aspire to and maintain a technology career. Further, the right support can assist women in finding a career that is rewarding and offers opportunities for advancement.

The women interviewed in this study were highly satisfied with their careers and remarkably resilient when talking about difficulties they faced on the job. The more experienced felt that earlier obstacles and failures prepared them for later success. They were confident that if they were fired or decided to leave, they would find work somewhere else. Overall, they found the IT profession difficult to transition into, because they did not 'know the dance steps.' Once they figured them out, they had satisfying careers.

This study uses a grounded theory approach to develop themes for women's success in the IT field. Differences between industries and how women approached their careers started to emerge. For example, Big 4 consulting firms, not known for having many women in long-term careers, can be a great training ground for women early their careers. Interestingly, the Big 4 firms now realize that when they lose women, they lose expertise and dedication. They therefore have begun offering women part-time opportunities when they begin their families. However, one woman noted that part-time at her Big 4 firm was 40 hours per week.

For women seeking more supportive work environments, IT audit as a profession and, healthcare as an industry stood out as being better than other industries. More work is needed to develop greater insight into issues of work, culture and equity.

This research has several practical implications for women entering IT careers. The job they find might not 'fit' their expectations and needs. However, women do find ways to adjust to male-dominated environments and enjoy lucrative and rewarding careers. Entry-level women may also benefit from informal mentors and professional advocates who may assist in creating opportunities for them. Contrarily, if no advocate or informal mentorship is possible, women should consider transferring into an area that is more supportive.

Organizational implications of this study affirm findings of other research such as that by Faulkner (2007): IT organizations and IT occupational cultures are not women-friendly. Creating a program for women, while helpful, does not address the underlying problem, but rather serves simply as a surface or band-aid solution. Women may be reluctant to take advantage of programs that give co-workers the perception that they need special favors to be successful in the world of IT. In this regard, IT organizations have a long way to go in making the cultural changes necessary to be open to everyone.

REFERENCES

Armstrong, D. J., Riemenschneider, C. K., Allen, M. W., & Reid, M. F. (2007). Advancement, voluntary turnover and women in IT: A cognitive study of work-family conflict. *Information & Management*, *44*, 142–153. doi:10.1016/j.im.2006.11.005

Ashcraft, C., & Blithe, S. (2009). *Women in IT: The facts*. Boulder, CO: National Center for Women & Information Technology. Retrieved from http://www.ncwit.org/pdf/NCWIT_WomenInITFacts_FINAL.pdf

Berg, B. (2001). *Qualitative Research Methods for the Social Sciences*. Boston: Allyn & Bacon.

Blickenstaff, J. (2005). Women and science careers: Leaky pipeline or gender filter? *Gender and Education*, *17*, 369–386. doi:10.1080/09540250500145072

Bystydzienski, J. M., & Bird, S. R. (2006). *Removing barriers: Women in academic science, technology, engineering, and mathematics*. Bloomington: Indiana University Press.

Camp, T. (1997). The incredible shrinking pipeline. *Communications of the ACM*, *40*(10), 103–110. doi:10.1145/262793.262813

Crepeau, R. G., Crook, C. W., Goslar, M. D., & McMurtey, M. E. (1992). Career anchors of systems personnel. *Journal of Management Information Systems*, *9*, 145–160.

Dougherty, T., Dreher, G., Arunachalam, V., & Willbanks, J. (2009). The powerful mentor effect: Differential career returns for males and females. In *Academy of Management Best Paper Proceedings*. Chicago, IL: Academy of Management.

Eby, L., & McManus, S. (2004). The protégé's role in negative mentoring experiences. *Journal of Vocational Behavior Psychology*, *65*, 255–275. doi:10.1016/j.jvb.2003.07.001

Eby, L., McManus, S., Simon, S., & Russell, J. (2000). The protégé's perspective regarding negative mentoring. *Personnel Psychology*, *57*, 441–447.

Faulkner, W. (2007). 'Nuts and bolts and people': Gender-troubled engineering identities. *Social Studies of Science*, *37*, 331–356. doi:10.1177/0306312706072175

Ferratt, T., Enns, H., & Prasad, J. (2006). Employment arrangements, need profiles, and gender. In Trauth, E. M. (Ed.), *Encyclopedia of Gender and Information Technology* (pp. 242–248). Hershey, PA: Idea Group Publishing. doi:10.4018/978-1-59140-815-4.ch038

Flick, U. (1998). *An Introduction to Qualitative Research*. Thousand Oaks, CA: Sage.

Guzman, I., Stam, K., & Stanton, J. (2008). The occupational culture of IS/IT personnel within organizations. *Database, 39*(1), 33–50.

Halpern, D. F. (2005). How time-flexible work policies can reduce stress, improve health and save money. *Stress and Health, 21*, 157–168. doi:10.1002/smi.1049

Hollis, E. (2003). ITAA: Fewer women and minorities entering IT workforce, June 2, 2003, Certification Magazine. Retrieved March 25, 2010, from http://www.certmag.com/read.php?in=265

Ituma, A. (2006). The internal career: An explorative study of the career anchors of information technology workers in Nigeria. In *Proceedings of the ACM SIGMIS Conference on Computer Personnel Research* (pp. 205-212). Claremont, CA: ACM Press.

Kimmel, J., & Amuendo-Dorantes, C. (2004). The effects of family leave on wages, employment and the family wage gap: Distributional implications. *Journal of Law and Policy, 15*, 115–142.

Kuhn, K., & Joshi, K. (2009). The reported and revealed importance of job attributes to aspiring information technology professionals: A policy-capturing study of gender differences. *Database, 40*(3), 40–60.

Lewins, A., & Silver, C. (2007). *Using Software in Qualitative Analysis: A Step-by-step Guide*. Los Angeles, CA: Sage.

McKinney, V., Wilson, D., Brooks, N., O'Leary-Kelly, A., & Hardgrave, B. (2008). Women and men in the IT profession. *Communications of the ACM, 51*(2), 81–84. doi:10.1145/1314215.1340919

Melymuka, K. (2005, April 18). What IT women want, a virtual roundtable of high achievers talks about what today's women bring to IT and what they expect in return. *ComputerWorld*. Retrieved from http://www.computerworld.com/s/article/101088/What_IT_Women_Want

Quesenberry, J., & Trauth, E. (2007). *What do women want? An investigation of career anchors among women in the IT workforce*. In the SIGMIS Proceedings on Computer Personnel Research (pp. 122–127). St. Louis, MO: Association for Computing Machinery.

Sappleton, N., & Takruri-Rizk, H. (2008). The gender subtext of science, engineering, and technology (SET) organizations: A review and critique. *Women's Studies, 37*, 284–316. doi:10.1080/00497870801917242

Schein, E. H. (1990). *Career Anchors: Discovering Your Real Values*. San Francisco, CA: Jossey-Bass.

Simard, C., Henderson, A., Gilmartin, S., Schiebinger, L., & Whitney, T. (2008). *Climbing the technical ladder: Obstacles and solutions for mid-level women in technology*. Retrieved December 3, 2009, from http://anitaborg.org/files/Climbing_the_Technical_Ladder.pdf

Sloan, R., & Troy, P. (2008). CIS 0.5: A better approach to introductory computer science for majors. In *Proceedings of the 39th SIGCSE Technical Symposium on Computer Science Education*. Portland, OR: Association of Computing Machinery.

Soe, L., & Yakura, E. (2008). What's wrong with the pipeline? Assumptions about gender and culture in IT work. *Women's Studies, 37*, 176–201. doi:10.1080/00497870801917028

Stiller, E., & LeBlanc, C. (2003). Creating new computer science curricula for the new millennium. *Journal of Computing Sciences in Colleges, 18*(5), 198–209.

Strauss, A., & Corbin, J. (1990). *Basics of Qualitative Research: Grounded Theory Procedures and Techniques*. Newbury Park, CA: Sage.

Sumner, M. Yager, S., & Frankie, D. (2005). Career orientation and organizational commitment of IT personnel. In *Proceedings of the ACM SIGMIS Conference on Computer Personnel Research* (pp. 75-80). Atlanta, GA: Association of Computing Machinery.

Trauth, E., Quesenberry, J., & Morgan, A. J. (2004). Understanding the under-representation of women in IT: Toward a theory of individual differences. In *Proceedings of the ACM SIGMIS Conference on Computer Personnel Research* (pp. 114-119). Phoenix, AZ: Association of Computing Machinery.

Trauth, E., Quesenberry, J., & Yeo, B. (2008). Environmental influences on gender in the IT workforce. *Database, 39*(1), 8–32.

Wanberg, C., Welsh, E., & Hezlette, S. (2003). Mentoring research: A review and dynamic process model. *Research in Personnel & Human Resources Management, 22*, 39–124. doi:10.1016/S0742-7301(03)22002-8

ENDNOTES

[1] A previous version of this chapter was published in the International Journal of E-Politics, Volume 2, Issue 1, edited by Celia Romm, pp. 30-44, copyright 2011 by IGI Publishing (an imprint of IGI Global).

[2] Big 4 refers to large accounting and IT consulting firms. Specifically KPMG, Deloitte-Touche, Ernst & Young and Pricewaterhous Coopers. An IT employee at one of these firms would typically be assigned to different projects implementing large scale IT solutions for other large businesses.

Chapter 4
Gender Segregation and ICT:
An Indo-British Comparison

Sunrita Dhar-Bhattacharjee
University of Salford, UK

Haifa Takruri-Rizk
University of Salford, UK

ABSTRACT

Gender segregation in science, engineering, construction, technology (SECT) is a common persistent feature, both in India and U.K. Even though culturally the two countries differ in various ways, under-representation of women in SECT is widespread and a cause for general apprehension and in recent years this has attracted centre stage in the study of gender, work and family. In this chapter we discuss our research findings of a comparative study undertaken between India and Britain in the ICT sector. With twenty seven interviews with ICT professionals in the two countries, we discuss their views on ICT education, recruitment and employment practices, work-life balance, changing gender relations, opportunities for progression and retention in the two countries taking into consideration women's role in power and politics in the both countries; how 'public' and 'private' patriarchy shapes women's position in the labour market, with an essential backdrop of 'patrifocality' in the Indian context.

INTRODUCTION

In the UK, historically, science, engineering and technology did not rank very highly as an occupation and there are several explanations. In the mid 1960's, scientists and engineers were ranked below dentists, university lecturers, company directors, solicitors and only just above primary- school teachers, unlike in India where an engineering degree undoubtedly enhances social status and increases chances of employability. More recently, India's IT industry has been growing at a very fast pace with the adoption of economic liberalization policies and emerging as the 'most watched test of global capitalism'. Information communication technology (ICT) driven growth and development

DOI: 10.4018/978-1-60960-759-3.ch004

in India has been observed to have skipped the middle stages of traditional economic development models and 'leapfrogged' to technology driven stages of economic development. Yet, a gendered occupational structure in science, engineering, construction and technology (SECT) is very much predominant, like the UK. A gendered occupational structure is often assumed to exist due to social inequality and gender equality is measured by the Gender Empowerment Measure (GEM). It is generally assumed or implied that greater empowerment of women would reduce gender segregation. But, there are exceptions; in countries where the degree of women's empowerment is greater, the level of gender segregation is also greater. Interestingly, Sweden with a higher GEM also has higher gender segregation than Japan with a lower GEM.

In this chapter, we focus on the 'universalistic theorisations' and 'particularistic explanations' to study women's position of employment in SECT, with a focus on the ICT sector. We look at the role of social and labour market policy in the two countries, how this shapes gender relations and the modification needed to develop an equitable gender division of labour in ICT. We explore the working practices and working cultures of IT companies as a factor in causing the under representation of women engineers in terms of recruitment, progression and retention.

Why India and UK?

The Gender Empowerment Measure (GEM) takes into account the female share of parliamentary representation; proportions of legislators, senior officials, managers, professional and technical employees who are women; and the ratio of female to male earnings. It is generally assumed or implied that greater empowerment of women would reduce gender segregation. The Table 1 (GEM measure for UK and India) below shows the GEM for India and UK, highlights the breakdown of the components that make the GEM, i.e., the percentage of seats held by the parliament, percentage of female legislators, senior officials, and managers, the percentage of female professional and technical workers, the ratio of estimated female to male earned income. This data is derived from the UNDP report, 2009, however, we must mention that the GEM for India was not available on the UNDP report. This was only available from the report developed by the Ministry of women and child development, Government of India, 2009. The GEM scores measured by UNDP HDR 1998, were very low, and this is the reason why the Govt of India calculated the GEM using the indicators as given in the Table 2 (GEM scores for India, 1996, and 2006.)

UK is already higher in the ranks with regards to the GEM score and India is not, but India's annual GDP growth (ranging from 5-7%) has been very promising. This clearly means British women are more empowered than Indian women. As it is evident in the case of Sweden and Japan, a higher empowerment does not necessarily relate to less segregation. The aim of our research is to find out whether Britain's ICT sector is more gendered than India's and whether this leads to more segregation in ICT related jobs. The variables that are used in the research are education, recruitment practices, salary, work-life balance, employment practices, changing gender relations, opportunities for progression, retention rates.

The Table 2 below shows the GEM scores for India, 2006 and 1996. It is interesting to note that the scores are highest for PI at 0.573 and lowest for PoERI at 0.231 in 1996. The increase has been smallest for PI from 0.573 in 1996 to .625 in 2006 and the largest for EI from 0.443 in 1996 to 0.546 in 2006.

Women's Contribution to Science, Engineering and Technology: Historical Analysis

Women's theoretical engagement with western science has been philosophically varied ranging from

Table 1. GEM measure for India and UK

HDI Index	UK	India
GEM: Rank and value	15 and 0.79	Not Available and 0.497 (2006)
Seats in parliament held by women % of total	20	9
Female legislators, senior officials and managers (% of total	34	
Female professional and technical workers (% of total)	47	
Ratio of estimated female to male earned income	0.67	0.32
Year women received the right to vote and stand for election	1918, 1928	1935 and 1950
Year a woman became a Presiding officer of Parliament or one of its houses for the first time	1918, 1928	1935, 1950
Women in ministerial positions	33	10

gnosticism and alchemy through to Cartesianism and Newtonian theory.(Rowbotham,1995) If we look at the literature of women who contributed to the development of science and technology, it is not surprising that women have been obliterated from history; Sheila Rowbotham showed a new direction in the analysis of female approaches to technology.

"Rather than viewing history in terms of an undifferentiated structure of patriarchy, it is possible to see women emerging intellectually in some periods and forced into retreat in others" (Rowbotham 2006, pp 36)

Women's contribution to science existed since 3000 BC. Women doctors were found in Egypt as early as 3000 BC and ancient babylonian women perfumers developed chemical techniques which were used amongst alchemists in Alexandria as early as the first century AD (Margaret Alic, 1986). Traditionally women played a key role in food gathering, collecting fuel, finding nutrition for crops and animals, organic recycling, collection fodder, producing, processing, marketing and preparing food along with performing other household duties. They had the ability to spot healthy plants and preserve their seeds for sowing for the next year. Alic mentioned the story of the Arab slave girl, called Tawaddud, whose intelligence and knowledge 'outwitted the readers of the Koran, doctors of law, and medicine, scientists, and philosophers...' (ibid) It has also been seen that in the Byzantine Empire, a succession of women rulers were scientists.

"In China women engineers and Taoists adepts pushed science and technology forward at a steady rate... Women studied at the medical school in Baghdad and female alchemists followed the teachings of Maria the Jewess. If Moslem women scholars are not recorded in the historical text, their existence is at least testified to by stories from the Arabian Nights." (M. Alic 1996, pp 47)

Table 2. GEM scores for India, 1996, and 2006.

Year	PI	EI	PoERI	GEM
2006	0.625	0.546	0.319	0.497
1996	0.573	0.443	0.231	0.416

Note: PI= Index of Political participation and decision making power; EI= Index of economic participation and decision making power; POERI= Index of power over Economic resources and GEM = Gender Empowerment Measure

As Rowbotham points out historically, class has been a very important factor which enabled women to enter the world of science. The Byzantine Emperor Alexius's daughter, Anna Comnena wrote a book called 'The Alexiad' which contained detailed description of military technology and weapons. Also elsewhere in Japan, Empress Shotuku-Tenno produced the earliest printed documents in any country (George Sharton, 1927). In the 11-12 century, Trotula of Salerno worked as a female physician and had several writings on women's health, e.g. diseases of women, treatments for women (many of these were of Muslim origin), and women's cosmetics. These texts were a major source of information on women's health in medieval Europe. Secondly the crafts skills which the women possessed were also considered to be an important factor for women's entry into this field. Textile crafts where women in many cultures spun silk, wool and linen were originally female trades and were closely related to domestic and household duties. Huang Tao P'o, well known as the 'inventor of loom' from China was the textile technologist who brought the knowledge of cotton growing, spinning and weaving from Hainan to the Yangtze (ibid). Thirdly, family connections were equally important in women's contribution and entry into science. Hypatia of Alexandria (AD 370) was the daughter of the well known astronomer and, mathematician, Theon. She designed a plane astrolabe for measuring the positions of the planets and the sun to calculate the time and zodiac sign.

There is also evidence that the famous mathematician, Pythagoras of Samos, c.582-500 B.C formed a community of teachers and students. It was one of the teachers, Theano, who married Pythagoras (when he was old) was a renowned healer who believed that the 'human body in microcosm reflected the macro universe.' (ibid)

Nevertheless, women's entry and contribution to science has a long history. However, as Swasti Mitter pointed out,

"The technological innovations become commercially successful if and when the creator of the innovation could make use of political, economic and legal networks. Thus the dominant group in a society determines the shape and direction of a society's techno-economic order - and the image of an inventor has almost always been male." (Mitter, 1995, pp 4)

Understanding Gender Segregation

There are various approaches to understanding gender segregation. To understand the theoretical conceptualisation of what gender is as a social phenomenon, Ridgeway and Correll, 2004, put forward that gender is not, "primarily an identity or role that is taught is childhood and enacted in family relations." Instead, "Gender is an institutionalized system of social practices for constituting people as two significantly different categories, men and women, and organizing social relations of inequality on the basis of that difference" (ibid, pp 510) They contend that the whole gender system and the social structure of inequality and difference which branches from it rests on the widely established cultural beliefs about the distinguishing characteristics of men and women and how they are expected to behave.

Acker, back in 1990 pointed out that organizational structure is not gender neutral. Their gendered nature is partly masked through nature of work and the universal worker is actually a man. Acker contends organisations are not only gendered and patriarchal where women only occupy positions on the lower ranks. As Joan Acker argued, it is part of the larger strategy of control in industrial capitalist societies and is built on a deeply embedded substructure of gender difference. Between 1995 and 1996, Lee undertook a research to find out why science, mathematics and engineering disciplines loose potential women and framed these trends as identity- acquisition issues. Lee examined the links between gender, self-concepts and focused on how these contributed to the student's interest in science, mathematics and engineering interests and its effect on the student's 'educational trajectories.' According to Lee, on average, girls' self-concepts are more like their perceptions of same-sex others than those of boys, and more unlike their perceptions of other science students. The discrepancies between self-concepts and perceptions of those in science-related disciplines were associated with lower interest in those disciplines. The discrepancies explained some differences, by sex, in interests. Lee used the identity theory to explain how men and women interrelate in any social relational context.

Similarly, Hacker used observatory and exploratory studies in the form of in-depth interviews with a group of young men and women engineers and discussed issues around family, friendship, childhood intimacy. Hacker found that young men studying humanities were able to communicate better emotionally than their male counterparts in engineering. The young men studying engineering learned to value control over natural – emotions, feelings, intimacy. As Cockburn, points out as an effect of this process, later in life, women are naturally perceived as better carers, while men are perceived as naturally occupying the economic role.

A number of theoretical explanations to understand how gender structures the social and economic life from a theoretical and empirical standpoint have been put forward. According to Gregory and Windebank (2000) there are two sets of theorisations/explanations that explain women's position in employment (paid and un-paid work.) - 'universalistic theorisations' and 'particularistic explanations'. Universalistic theorisations '... seek to provide universal theoretical models for understanding the gender division of labour across societies based on over-arching analyses of social structures or economic behaviour.' For instance, in Europe, after the two major World Wars and due to the low number of men, women's role outside the 'home' became prominent. A large number of women started working in paid employments and gradually gained/earned recognition. Gradually, over the years, national and state policies were designed in a way that was inclusive of women's welfares, e.g. provision of childcare facilities and others. In this sense, economic deprivation acted as an impetus for women to seek employment. Particularistic explanations, '...does not involve an over-arching analysis of social structures, is based more often on empirical research and focuses on one or more explanatory variable.'(Gregory and Windebank, 2000, pp 102)

Till the 1960's women's work was assumed to be unimportant and was therefore excluded from mainstream work. The 'unisex' worker was given the main attention. The economic importance of domestic labour was not given any importance and women's responsibility in terms of domestic labour was deemed as biologically natural in sociology. Women's role as a homemaker was regarded as natural and 'indispensable' for the stability of the society (Parsons, 1955). The Parsonian view of the ideal family conceptualised gender relations in terms of sex roles: men performed the instrumental role and mainly catered for work outside home while women performed the expressive role of looking after the internal needs of the family. These sex roles should be kept intact so that there is peace and harmony between the occupational structure and the kinship system. However, soon after the Parsonian view received several criticism from feminist circles and the question of women's oppression and exclusion from the labour market were highlighted and questions as to why women were primarily responsible for household care, what role does household care/domestic labour have on the society and how this relates to women's position in paid work were brought forward. A number of theoretical explanations (both universalistic and pluralistic) to explain women's employment were developed – Micro-economic theorizations (Human Capital, Rational Choice, Hakim's Preference theory), Marxist- Feminist, Liberal-Feminist, and Labour Market Segmentation.

The neo-classical economic theories of human capital: Becker (1964), Mincer (1966), Polacheck (1981) primarily focussed on the worker's position in the workforce. It is assumed that the choices that the workers make both from the demand and supply sides of the labour market, determines their position in the labour force. Workers are viewed as a stock of human capital, which is a combination of experience and qualification. Worker's productivity is directly proportional to their qualification and experience, which means the more qualified and experienced the workers are their human capital too increases. The more productive workers earn more money and gain more senior positions within the labour force. According to this theory, women make different decisions from men. Compared to men, they spend less time in education and training and devote majority of their time to household and domestic care. This results in lower productivity and lesser human capital than men, which is why many employers are hesitant to invest in women's human capital. Women rationally choose to focus on child bearing and domestic work, and sacrifice better qualification or valued experiences in exchange of domestic work, thereby reducing their human capital. While men rationally choose more demanding work experience or more valued qualifications to maximise the earning of family as a unit, as a result human capital of the man increases, which implies the segregation between the man and woman increases in terms of their human capital.

Becker (1964) further developed the Rational choice theory by developing the conceptualisation of the household as a production unit within the neo-classical model to explain the growing number of women remaining in the labour force. He justified gender divisions in terms of the choices couples make to maximise the family's well-being. This was explained in terms of two types of production: market and domestic. According to him, couples choose rationally to divide their time between domestic and outside work depending on their economic circumstances, abilities and preferences to maximise their well-being. This has similarities with the Parsonian view, we discussed earlier. Further explanations were suggested by Lemmenicier (1980) and Sofer (1986). According to them married women with caring responsibilities rationally choose domestic work than paid outside work to compensate the cost of expensive childcare.

Hakim's (1996, 1998, 2000) Preference theory is the refinement of the rational choice and human capital theories. She adopted a Beckerian approach to explain the nature of women's employment in the labour force. She identified three basic types of women in relation to their family and work commitments, (work centered, home centered and adaptive women.) According to the Marxist Feminists, women have a different place in the production/reproduction process to men because of their role in domestic labour. They emphasise that women's oppression within capitalism have a material basis, which is domestic labour. Further theories developed from the notion that women's employment patterns were to some extent determined by capitalist relations. Marx's 'cyclical reserve army' theory implies that women are pulled into the labour force by capitalism when there is an economic boom and returned to the family in times of recession. Braverman's theory on the other hand explained that women spent a much longer time in employment not as a result of cyclical variations but due to the fact that Taylorist-inspired managerial strategy led women to take up newly deskilled jobs. Secondly, there was a progressive shift of household tasks to the factory due to the emergence of machines replacing some of the basic household duties.

The Marxist segmentation theory explains the divisions between by gender, ethnicity and social class within the labour market in terms of the outcome of the struggle between capital and labour. It divides the society into two types of labour markets, the primary market which provides the top jobs associated with stability, good pay and secondary market, which is characterised by 'precarious and unstable employment. (Collin and Young 2000)

Thus Marxist Feminists explain why capitalism acts as a primary source for subjugating and oppressing women and their secondary status in the capitalist society due to their continuing responsibility for domestic labour, which is often deemed as 'unproductive' in the Marxist sense. Firstly because domestic work cannot be exchanged in the market against a wage and secondly' it does not work directly with the capital's means of production to produce commodities which have a calculable exchange value from which surplus value can be directly extracted (Gregory and Windebank, 2000). It is interesting to note that Marxists Feminists analysis does not highlight the fact that women's participation in the labour market can be examined as part of a patriarchal system and whether gender relations are part of the capitalist mode of production, patriarchal mode of production or both (Hartman 1981, Walby 1990, 1997).

The Feminist Marxists propagated that women's oppression in terms of their employment and domestic work was linked to capitalism and patriarchy and that women's domestic labour needed to be looked at from the angle of the patriarchal system. Women's primary responsibility for domestic and household duties puts them in an unfavourable and disadvantaged position. According to Hartmann, patriarchy is "a set of social relations between men, which have a material base, and which, though hierarchical, establish or create interdependence or solidarity among men that enable them to dominate women."(Hartmann, 1981: pp) The 'material base upon which patriarchy rests' is fundamentally men's control of women which is manifested in terms of men excluding women from participating in the labour force, and ensuring women are the primary carers for domestic and household duties and childcare. Women are denied access to economically productive resources as a result of domestic and household responsibilities, thus lacks the opportunities to acquiring training and upskilling and on the other hand men form an alliance with capital (Walby, 1990). The evidence for this is cited in the development of capitalism, as well as the working class response to its problems in the form of demands for protective legislation and family wage. The main logic behind was that both were fought for by male workers in order to benefit them by putting women into household duties where they could both service men and be controlled sexually by them. (German, 2006)

In Britain, the development of capitalism destroyed domestic production and forced men, women and children to forcibly enter the factory system. This changed the dynamics of the traditional family setting as women were working longer hours and depending on older children to look after the younger ones. And this had a devastating repercussion on the reproduction of the working class. As a result of this, infant mortality reached horrific levels, (as Marx stated in Capital) because women were working long hours away from home. Marx and Engels described the horrific early factory system in, 'The condition of the Working Class in England', how it pulled the old pre-capitalist family apart as more and more members from each family became 'wage labourers'. It was also during this time that the demand for protective legislation and family wages came up. Walby further advanced the theories of patriarchy and developed the concept of 'private' and 'public' patriarchy and put forward gender relations as an autonomous relation from capital. She describes private patriarchy as being based on the household where women are denied all access to paid employment and are dependent on the father or husband as the patriarch. Whereas in the public patriarchy, women are subordinated within the structures other than the household, and are segregated from men, given lower pay and status. Walby points out that there is a shift from private to public patriarchy, mostly in industrialised nations, which arose as result of the capitalist interest of employing more women who were able to readily available for cheap labour and also partly due to a feminist movement. The conflict within capitalism lies within the patriarchal structure and private patriarchy.

Blackburn et al, 2002 put forward the social reproduction and changing gender relations theory in contract to the theories discussed earlier, which they state as '*essentially static, and unable to explain a real situation which is continually changing and developing.*' They take into account three important processes which have crucial influence on gender segregation and inequality which would possibly capture its dynamic nature. They are firstly, the substantial expansion of education in last 50 years, secondly considerable change in occupational structures and thirdly an increase in adult female participation in the labour force which has become relatively easier, especially for women with childcare needs. In order to understand and theorise why women entered the labour force, we need to focus on the women's familial and social roles and their relation to the labour market. Traditionally women were in charge of the sole responsibility of rearing the children, taking care of domestic responsibilities and providing an ideal atmosphere where the man of the family could progress and succeed in their profession. Women's multiple role as an obedient daughter, skilful, efficient and capable wife, and a caring mother as predisposed by the patriarchal notion of family puts them at a disadvantage to be a skilled worker.

Key Shapers for Equality and Diversity Laws for UK and India

Immediately after independence from the British rule in 1947, India adopted its comprehension Constitution in 1950. Through the Directive principles of the State policy, it provided the guidance for India's future development. This document laid down a number of welfare activities which were mandatory for state and private organisations either by legislation or trade practices. A number of legislations were passed that have a special implication on women's participation in the labour market and balancing their domestic responsibilities. In the UK, in the late 1980's the business case for equal opportunities became prominent and accepted in the business world, as the financial benefits of good equal opportunities became increasingly visible. Table 3 summarises some of the notable legislations related to women's work and family in UK and India.

Methodology

For the purpose of this international comparative research (ICR), we have used a mix of qualitative and quantitative methods in the form of questionnaires and semi-structured interviews. We have utilised Glaser's (1992) Grounded Theory (GT) methodology. GT is inductive in nature, whereby theories, issues or themes and sub-themes emerge from an ongoing process of an analysis of the data. In the context of the Grounded theory methodology, a sample of ten is considered as good as the main intention of the research is to develop a model and not to test it. Qualitative research methods were initially developed in the social sciences to enable researchers to study social and cultural phenomena. The motivation for doing qualitative research, as opposed to quantitative research, comes from the observation that, if there is one thing which distinguishes humans from the natural world, it is our ability to talk! Qualitative research methods are designed to help researchers understand people and the social and cultural contexts within which they live. (Anne, Ryan, 2006). Qualitative data sources include observation and participant observation (fieldwork), interviews and questionnaires, documents and texts, and the researcher's impressions and reactions. Some researchers have suggested combining one or more research methods in the one study called triangulation, Gable (1994), Kaplan and Duchon (1988), Lee (1991), Mingers (2001) and Ragin (1987).

Table 3. Notable legislations relating to women's work and family in UK and India

British Acts and Legislations	Description	Indian Acts and Legislations	Description
Sex discrimination Act 1975 (amended in 1986, 2005)	A person is not treated unfavourably because of their gender or marital status.	Maternity Benefit Act of 1961	Entitles women to six weeks of leave with full pay, before and after delivery. Women are allowed to take 2 nursing breaks in addition to normal breaks till the child is 15 months old.
Race Relations Act 1976	A person is not unfairly treated because of his or her race.	Factories Act of 1948	Chapter V – Welfare An employer is required to provide crèche facilities where more than 30 full –time permanent employees are employed with children below the age of six years. Prohibition of women workers at night shift – women should not be allowed to work in a factory except between the hours of 6 a.m and 7 p.m.
Disability Disctimination Act1995(amended in 20060	To end end discrimination against disabled people, and giving rights in the areas of employment, access to goods, facilities and services.	Right to protection from sexual harassment at the workplace 1869 (revised again 1997)	Sexual harassment in the workplace was extended to include unwelcome sexually determined behaviour – verbal non-verbal conduct of a sexual nature, like physical contact, advances, sexual favours, bantering,
Equal pay Act 1970 (amended 1983)	Every employment is deemed to include an 'equality clause' which guarantees both sexes the same money for doing the same or broadly the similar work.	Employees State Insurance Act of 1948	Provided relief in case of medical emergencies/problem for workers.
Age Discrimination 2006	Employees have the right to be informed of ther expected retirement date and of their right to request to work longer.	Plantation Labour Act of 1951	
Equality Act 2006	Makes unlawful discrimination against the grounds of religion, or belief, creates a duty on public authorities to promote equality of opportunity and prohibit sex discrimination	Mines Act of 1952	
		Equal Remuneration Act of 1976	Providing men an women doing the same or broadly similar jobs, the right to get equal wages.

Our research is informed by the critical and interpretivist epistemologies. The ICR has an interpretive stance and a critical edge to justify and explain the causes of gender segregation in the two countries. As Kvasny and Richardson point out, "critical research should be praxis oriented – combining theory and action (praxis) to create a scholarship which may lead to more equitable social change." From an interpretive perspective, we assume that people create and associate their

Figure 1. Research model

A higher GEM empowers women but is not indicative of decreasing segregation (hypothesis)
There are three elements in the hypothesis, - GEM,empowerment and segregation.
UK has a higher GEM, implies women are more empowered than women in India
India's GEM is lower than that of UK's. This implies that women are less empowered than women in UK.
Our research is to focus on the third element of the hypothesis, segregation, and whether this is true in the ICT profession as there is an evidence of gender segregation in ICT in both the countries.
If this is true, then how will segregation and empowerment be compared? The variables that are used in the research are education, recruitment practices, salary, work-life balance, employment practices, changing gender relations, opportunities for progression, retention rates.

own experience and understanding as they interact with the world around them. From a critical perspective, we aim to remain as social critiques.

The research model below in Figure 1 shows the major variables that underlies the research. The variables that are used in the research to compare empowerment and segregation are education, recruitment practices, salary, work-life balance, employment practices, changing gender relations, opportunities for progression, retention rates.

Equivalence in Terms and Concepts and Questionnaire Design

In the UK, a bachelor's degree in technology or engineering is not essential for a technical job in the ICT sector, whereas in India it is. Therefore, we approached candidates who had a bachelor's degree either in technology or engineering in the both countries for equivalence. In terms of recruitment to ICT professions, both in India and UK, graduates primarily depend on campus recruitments. A range of companies are available depending on the name and prestige of the educational institution as well as geographical location. Generally speaking, due to globalization, there isn't any major noticeable difference of the nature of companies visiting the campus for recruiting undergraduates. The only noticeable measure when undertaking comparative studies would be to select a base line for starting salaries for the two countries. The base line for the starting salary in each country was selected after an initial literature review and liaising with university recruitment officers for ICT and fresh graduates in the two countries. The local currency was used in the questionnaire designed for employees. In terms of linguistic ambiguity and organisational policies relating to women's progression in the two countries, there wasn't much discrepancy. The terms and concepts used in the UK and India relating to work-life balance, flexible working, flexi-time, equal opportunities policies, are similar and therefore this was also reflected in the questionnaires.

Access and Sample Size

The participants both men and women were drawn from a random and diverse age group between 21-65 years that represented different layers in the ICT workforces in order to gather extensive information about working practices and organisational cultures and the way these practices and cultures influence ICT professionals; analyse the findings of the studies conducted and compare the results to existing research and national data where available. The participants were given the option to complete the questionnaire on their own time. Time taken approximately was between 15-20 minutes. All participants were made aware of the nature of the research beforehand and were informed of the confidentiality and anonymity of the research data as well as the voluntary nature of investigation.

Analysis

Our analysis was based on the fieldwork that focuses both on the personal accounts of female and male ICT professionals, and on the statistical analysis of aggregate questionnaire data. We have subdivided the analysis in three categories, the first section focuses on the demographics of the respondents, their gender, age, working patterns, nature of employment and personal information with regards to their living patterns, take up of state and private childcare facilities, and caring responsibilities. The second section focuses on the image of ICT in the two countries, the nature of technical skills required for entry in this sector, national statistics on the number of women in ICT – comparative analysis. The third section focuses on the experience of women in the workplace, organisational policies available, take up and awareness of policies – an evaluation of how policies are translated into practice, women's coping strategies and incentives, benefits and motivations of working in this sector . The fourth section focuses on the structure of power and how women's empowerment could benefit progression and retention rates of female employees.

Demographics

We interviewed seven female engineers from India and five from the UK. There were six male engineers from India and nine from the UK. The data was collected between a eight to ten month period, this was also followed by a visit to India for field work for 4 weeks in between. The majority of the Indian women, we interviewed were in the age group 18-25 years, and 31-45 years. The majority of the men in India were in the age group 18-25 years and 26-30 years in the UK.

In terms of work experience, the majority of the Indian women respondents had about 5-10 years of experience in the industry. And the women in the UK had more or less the same. We received a good response from the men in the UK who had a work experience of 10-20 years. It is interesting to note that most of the respondents from India were either working full-time or in contract work, whereas in the UK, we found there were 2 respondents who were working part-time. Majority of the men in the both countries were in full-time employment. This shows a very interesting pattern of work which is prevalent in the two countries. Only three male respondents from the UK were temporary workers and two female respondents from Indian and UK were temporary workers and not in permanent employment. One common ground for both the women in the two countries was their constant juggling between balancing office work and family life. Indian women relied upon their parents and in-laws to look after the children, while women in British women were mostly reliant on private childcare facilities. Even though opportunities for women in the IT sector were much better compared to other sectors in India, women IT professionals were mainly responsible for managing the work at home, even if their partners were working in the same organisation. This could be primarily due to the patrifocal nature of the Indian society. (Gupta and Sharma, 2003) There is a lack of affordable childcare provision in the UK, as a result of which women have an impact on their ability to progress in this sector. It is estimated that a women spend an average of 11.5 years for caring responsibilities while men spend an average of

1.3 years. One woman working in the IT sector in India mentioned,

"Oh yes, I have two kids, one two years and the other five years. They are both looked after by my parents and in-laws. They (parents and in-laws) come over and stay with us from time to time and they have taken the responsibility to look after them till they reach the school going age." Indian, woman, ICT professional.

In India, the 'nuclear' family unit, one that comprises of parents and kids is now on the rise in urban areas. Most of the respondents of this study were from urban areas and mostly came from metropolitan cities of Mumbai, Bangalore, Kolkata, Delhi, Chennai and Hyderabad and therefore majority of them were in their nuclear family units with partners and kids. Very few of the respondents in India were living with their parents and none in the UK were living with their parents. When asked if crèche facilities were provided by the organisation, she mentioned,

"Not yet..but I know there has been talks about it. But I would prefer to leave my children with grandparents – you know the love they receive from them will not be equivalent to anywhere else." Woman ICT professional, India.

This trend is an accepted norm, in India, especially when both parents are working full-time. However, the main responsibility mainly lies with the grandmother. In the UK, each week almost a quarter of the families with children less than fifteen years use a grandparent to provide childcare (around 1,740,000 families). The average care provided by grandparents is about 15.9 hours per week. In early Feb, 2010, Baroness Deech, emphasised that children should be forced to care for their elderly parents and grandparents as a payback for the 'free' childcare support they provide.

Image of ICT

Ever since the Government of India introduced the five year plans after independence in 1947, there has been an emphasis on developing the science and technological base in India. As a result of this, the best science and engineering institutions were set up all over India, e.g the Indian Institute of Technology and other regional engineering colleges. An attempt was made to draw the best brain drain towards science, engineering and technology. The best jobs were associated with SECT so was the pay. The ICT sector is also the sector where more women are concentrated compared to other engineering sectors, as the ICT sector is considered more gender neutral. Our participants confirmed the same.

"The IT sector is definitely a preferred sector to work for women, compared to other engineering sectors – you know good work environment, good pay, flexible working patterns..." Woman ICT professional, India.

"I think about 30% of all the technical employees in our organisation are women - ...yes they are in technical roles and if you include the admin staff, of course it will be more." Woman ICT professional, India.

In the UK, the ICT industry is a dynamic industry. However, women remain severely under-represented in this sector. Women make up 46% of the overall UK workforce but only 18% of people in IT are female (Technology Counts IT & Telecoms Insights 2008, e-skills UK, January 2008). Despite the obstacles faced by the IT industry in attracting, progressing and retaining female graduates, the opportunities for women to enter employment in the IT sector have never been greater in recent times. And more so with the forthcoming Equality Bill, which is predicted to be passed in October 2010. Women are in demand and add to diversity to teams. (Inside Careers,

Information Technology 2007/8, published in partnership with the British Computer Society (BCS), 2007). However, although it is illegal in both the countries to positively discriminate anyone in recruitment, many employers sometimes refrain from recruiting women. This is primarily due to their duty towards household and child rearing responsibilities. According to Collinson et al (1990), many employers preferred to employ men instead of women irrespective of educational credentials for technical positions primarily because of the reason that high level jobs originally belonged to men, and women's primary commitment towards caring responsibilities makes them less committed.

In the west, success in the ICT field is strongly associated with the histories and behavior of the "boywonder icon." The male model of computing assumes that students must have a fascination with the machine quite early on in life. We did not find technology as closely tied to masculinity in India as it is in Britain. Indian women IT professionals are regarded as technologically competent in their work as men. Both men and women ICT professionals were of the opinion that a strong base or background in mathematics and science was needed for a successful career in the IT sector.

The participants were asked to specify who influenced them to study ICT and from the responses we received there was a clear distinction between the responses of women and men in the UK. Most of the men in the UK stated 'no one' influenced them to study ICT, whereas the women mentioned about parental influence and other influence from elder sisters and brothers. In India, mostly all the women stated parental and other family members' influence was the major decision. The responses from men were more geared towards influence from school teachers as well as parents. However, through an analysis of the qualitative data through interviews with the participants, we were able to gain a better understanding. One male engineer from the UK revealed his science teacher was particularly an inspiration for him.

"...at school I was good at science, ...my science teacher was very good." Male ICT professional, UK.

A women engineer from India explained,

"My parents always wanted me to become a doctor, my dad is a doctor, by the way. My elder sister went to IIT, and I wanted to be an engineer as well." Woman ICT professional, India.

This shows that ICT jobs in India are considered gender neutral and girls are equally encouraged to pursue it. In the UK, ICT is seen more of a gendered occupation mainly dominated by men. Earlier research in the UK shows that often career advisors discourage young girls to take up science and engineering careers (Takruri-Rizk et al, 2006) We were interested to find out what motivated the participants to study ICT and majority of the respondents both from India and UK were of the opinion that an interest in SECT (science, engineering, construction and technology) was the main factor to pursue a career. Interestingly, women engineers from India mentioned about job opportunities, prestige and good pay.

Women's Experience in the Workplace: Retention and Progression

The ICT culture in the west is increasingly competitive, individualised, sexualised and gendered (Kvande, 1999; Griffiths & Moore & Richardson, 2007)The literature shows that workplaces are not always ideal for women's progression and well being in the sense that women often have to 'adjust' and fit in the male dominated work culture and sometimes even conceal their femininity. (ibid) The equal opportunities approach often do not encourage changing the existing structures to accommodate individuals and their diverse needs, but tends to individualise this problem whilst homogenising women's needs. Research shows that women's career progression reaches its peak when women are in the age group of 35 to 45 years, after that there is a steady decline (Takruri-Rizk al e, 2008). One interviewee who recently joined the manufacturing industry in India in an ICT role explained,

" I am not sure if I did the right thing. I don't think I will be able to sustain this for long." When asked why, she answered, *"I am the only female trainee here and after my training period I'll have to work really long hours – sometimes on site, sometimes on call...Not sure if I would like to do this for long."*

With further conversation, it was evident that she was concerned about security and safety issues. Being the only female trainee, she was intimidated with the dominant male work culture, late working hours and personal safety and security issues -working on site and on call meant working away from the office a lot of times. She mentioned about the flat which she would have got after her training was over and the availability of a company car, but somewhere things didn't just turn up the way she expected. Curious to know if she has been harassed, or faced discrimination in any way, she was asked about her experience on the first day at work. She candidly replied she was mistaken to be an administrative staff, had a lovely day otherwise. She was asked what her plans were for the next five years. She said without any thought,

"I think I'll move into management – that'll be much better for me." Woman ICT professional, in a manufacturing firm, India.

We would like to make a distinction here, we have seen women working in a core ICT firm are happier, more relaxed than women working in IT but in either manufacturing or other heavy engineering industries. According to Liff and Ward (2001), research, male managers were more concerned about the time away from their family, not being able to spend quality time, while women were more concerned about the exhaustion from work. Our interviews with women reveal the same, especially when this is a 24/7 industry and sometimes very demanding. One interviewee from UK, mentioned,

"I am working part-time – three days, but I'm expected to be checking emails five days a week. Where does that lead to?" Woman ICT professional, UK

Interestingly, none of the Indian women ICT professionals were working part-time. We have enquired whether this was a forced decision, one woman revealed,

"No, I wanted to come back full-time." We enquired whether part-time policies were available in the organisation, she mentioned,

"Yes, probably there are. I'm not sure. There are lots of policies. I do not know anyone who has come back part-time after maternity leave. They either left because their husbands were relocated or didn't come back." Woman ICT professional, India.

There was a distinct lack of awareness of policies amongst the ICT professionals in India. As we have discussed earlier, in Table 1, the Factories Act of 1948 makes it mandatory for organisations with more than thirty women employees to provide crèche facilities. However, we did not find a single organisation where there was a provision for crèche facilities. As Rajadhyaksha and Sinha (2008) put forward, that employers often by-passed the legislation requiring them to provide crèche facilities if 30 or more women were employed in the workplace. They either employ fewer than 30 women as permanent employees with the rest as part-time or contract labour. We felt that the Indian

Table 4. Comparative analysis – Responses from Indian and British IT professionals

Strongly agreed (Likert scores between 4-5)	Strongly disagreed (Likert scores 1-2)
ICT profession is dominated by men	
Women and men are equally suited to ICT profession	Men promote themselves better than women
Women face more barriers than men in recruitment procedures	
Part-time employees are not valued as full-time employees	
Long hours culture persists in this sector	
Female managers are effective as male managers	
Less family and caring responsibilities would contribute to women's progression in the sector	

IT professionals were hesitant to claim their rights and generally seemed too satisfied with whatever was offered to them. We also found that the number of temporary and contract workers were more easily found in India. Participants in the UK were more aware of the organisational policies and rights, compared to the Indian ICT professionals. When asked about Equal opportunities policies within the organisation, the participants from both the countries unanimously agreed that these policies were there. But we doubt whether these were being implemented. There wasn't any noticeable difference of gender pay gap in the ICT sector in India. In fact none of the women respondents mentioned about any gender pay gap. One interviewee from India mentioned,

"We have a very straightforward salary structure. There is no provision of a man getting paid more than a woman doing the same job. They are paid equal. But it's different in other sectors. There is some gender pay gap." Woman ICT professional, India.

Table 4 shows the questionnaire responses from the participants of both of the countries. The left column represents the statements which were used in the questionnaire for rating, where a score of 1 = strongly disagreed and 5 = strongly agreed. Interesting to note that the statement, 'men promote themselves better than women' was strongly disagreed by the respondents from both the countries. Valuing part-time employees in SECT was another statement which was agreed by respondents from both the countries. Surprisingly, none of the participants from India were on a part-time contract and it was evident from the interviews that although organisations did offer part-time provisions for their employees, the take up was nil. Through further discussions on this in the interviews, it was apparent that part-time work was clearly seen as a threat for the employee's stability and therefore it remained as a area 'not ventured'. The issue of women facing more barriers than men in recruitment is another theme which is common in both the countries.

It is interesting to note that majority of the Indian respondents disagreed that men are often paid more than women in the same role, more women quit this sector due to inequality and discrimination, more women quit this sector due to gender pay gap, and more women quit this sector due to lack of progression. This is contradictory to what the respondents from the UK mentioned. Another interesting thing is that most of the respondents strongly believed that education and entry in SECT profession is very competitive, also unlike to that of the British respondents.

The British respondents agreed that majority of the women occupy administrative positions and that women are bullied, harassed and discriminated in the workplace. This has not been

Table 5. Indian context

Strongly agreed (Likert scores between 4-5)	Strongly disagreed (Likert scores 1-2)
Women and men are paid the same starting salary in this sector	Majority women occupy administrative positions in ICT profession
	Women work harder than men to receive the same degree of recognition
Equal pay reviews	Women are bullied, harassed and discriminated in the workplace
Education and entry in ICT profession is very competitive	Men are often paid more than women in the same role
	More women quit this sector due to inequality and discrimination
	More women quit this sector due to gender pay gap
	More women quit this sector due to lack of progression
	Women in technical roles often more feel isolated than men

observed from the Indian respondents. To delve into this more, I contacted the British respondents who strongly agreed with this statement and found that that most of them had witnessed some sort of discrimination towards their women colleagues. But very few actually experienced it first hand.

Women in Power and Politics: How this Affects Women's Empowerment

Evidence from various studies earlier suggests there is a strong link between the positive effects of having a diverse Board and the importance of 'critical mass' of women in any organisation. A number of studies examining the performance of British as well as European top companies by the McKinsey management consulting firm in 2007& 2008 revealed that,

"...organisational performance increases sharply once a threshold of at least 3 women on management committees, with an average membership of 10 people, is reached. Below this threshold, no significant difference in organisational performance was observed." (Zalevski and Kirkup, 2007/2008)[1]

There has also been a recent debate whether organisations are facing acute problems during recession if they are primarily headed by men. Questions as to if more women were in top decision making positions the situation could have been better is also on the rise. For instance, in Iceland, where the only solvent Bank headed by two women, Kristin Petursdottir and Halla Tomasdottir raised everyone's eyebrow as this has significantly increased the support for Johanna

Table 6. British context

Strongly agreed (Likert scores between 4-5)	Strongly disagreed (Likert scores 1-2)
Majority women occupy administrative positions in SECT profession	Education and entry in SECT profession is very competitive
Women and men are offered equal access in terms of career advancement	The starting salaries in the SECT sector is excellent
Part-time employees are not valued as full-time employees	Physical strength is very important in SECT profession
Women are bullied, harassed and discriminated in the workplace	Appointments and promotions are not influenced by one's gender
Equal pay reviews	
Regular job appraisals	
Effective networking and socialising skills would contribute to career progression	

Sigurdardottir in an election for a new Prime Minister of Iceland[2]. Men's social behaviour at work as it has been pointed out has an effect on this situation, while women's behaviours and actions have a much soothing and positive effect. Another study by Singh and Vinnicombe (2005) reported that organisations with more women directors had more transparency, was better managed overall in terms of organisational excellence in accountability, internal policies and most likely to score higher than other organisations with less women on top.

However, despite evidence showing the need for more women in Board level, in reality women are heavily under-represented in decision making positions, which in turn have an effect on the glass ceiling, faced by women in middle and lower positions of the employment structure. According to Sealy, Vinnicombe and Singh's report (The Female FTSE Report 2008) [3] women held only 11.7% of all directorships and 4.8% of executive directorships in the FTSE 100 companies, while men in senior positions still occupied 22.0% of the companies.

Even during the times of Voltaire (i.e. 18th century), women luminaries have graced the fields ranging from literature to physics, e.g. Voltaire paid a glowing but perhaps insensitive tribute to his friend, collaborator and lover, the Marquise Emilie du Châtelet, a leading advocate of Newtonian physics, 'Never was a woman so learned as she. She was a great man whose only fault was in being the woman' (cited in Noble, 1992: p. 199). Du Châtelet wrote to Frederick of Prussia, 'Do not look upon me as a mere appendage. I am in my own right a whole person, responsible to myself alone for all that I am, all that I say, all that I do' (Rowbotham, 2005)

Women holding high political offices represent a milestone. Western democracies have seen women in top (Angela Merkel in Germany now, Margaret Thatcher in UK in the past). South Asian politics has been consistently dominated by strong women leaders (Indira Gandhi in the past, Sonia Gandhi now in India; Sheikh Hasina in Bangladesh; Benazir Bhutto in Pakistan; Sirimavo Bandaranaike and Chandrika Kumartunga in Sri Lanka). Assessment of how these women political leaders have shaped the prevailing gender debate in their respective countries' calls is critical to understand the impact of women political leaderships on gender segregation.

The process of reaching equal representation of women in political institutions and parliament has taken a longer time in the Nordic countries while in other countries the transition to democracy has contributed to a more rapid development. The overall change in society has opened and increased opportunity to promote women's participation in political life. But in both contexts, the struggle to reach a higher representation of women in parliament and to let the women elected make a greater impact has met stiff resistance. A strong group of women negotiating for their rights as equal members of the society is required even in countries led by strong woman. There has been some progress though. The average number of women in parliament worldwide has increased from 11.8 percent in 1998 to nearly 16 percent in 2005[4]. If we look at the data of the total number of MP's in the UK, we will see that in total there is 19% female representation; Labour 27% female (lead by a man); Conservative 9% female (led by a man) and the cabinet comprises of 7 out of 31 women (22.5%) representation.

In India, if we look at the Trinomool Congress, led by the current Railway minister, Mamata Bannerjee and the Bahujan Samaj Party led by Mayawati, the number of women in parliament is twice the national average. It is most likely because it is headed by women leaders who built these political parties and is still in charge of the overall political affairs. On the other hand, the Conservatives in The UK have considerably less number of women MP's compared to Labour, but there is a conscious effort to tackle this problem through women only shortlists for parliamentary seats in the 2010 elections.

Table 7. UK MP's by gender

Party	Male	Female	Total
Chairman of Ways and Means	1		1
Conservative	175	18	193
Democratic Unionist	8		8
First Deputy Chairman of Ways and Means		1	1
Independent	5		5
Independent Conservative	1		1
Independent Labour		1	1
Labour	255	94	349
Liberal Democrat	54	9	63
Plaid Cymru	3		3
Respect	1		1
Scottish National	7		7
Second Deputy Chairman of Ways and Means	1		1
Sinn Fein	4	1	5
Social Democratic & Labour Party	3		3
Speaker	1		1
Ulster Unionist		1	1
Total	519	125	644

Table 8. Indian MP's by gender

India
In India: 59 women MPs out of 552 i.e. 11%
Congress: 23 women MPs out of 206 i.e. c. 11% (Led by a woman)
BJP: 13 women MPs out of 116 i.e. c. 11% (Led by a man)
Trinamool Congress: 4 women MPs out of 19 i.e. 21% (Led by a woman)
Bahujan Samaj Party: 4 women MPs out of 21 i.e. 19% (Led by a woman)
Cabinet (including attendees): 2 out of 33 (i.e. 6% female)

The cause of gender empowerment is often blocked by poor implementation. For example, the first Commission on the Status of Women in India, established in 1972, recommended the constitution of statutory all-women panchayats (village councils) at the village level to promote the welfare of women, although this recommendation was not implemented by most provincial governments. The National Perspective Plan for Women (1988–2000) recommended the reservation of at least 30 percent of the total seats for women in the local government institutions. The provision of reserved seats in local government for women under the 73rd and 74th amendments to the Indian Constitution was a key initiative during this phase. The Women's National Commission was established in 1995 and has overseen the expansion of the quota system in India. More recently the Indian Govt has elected its first ever female speaker, Meira Kumar, the daughter of a former deputy prime minister and a member of the Dalit caste, once known as an Untouchables. Her election is certainly indicative of greater acceptance of women's leadership in India.

Media could play a very important role to limit and reduce the sexism deterring women to progress in the higher echelons of political power. For example, the UK House of Commons had a discourse characterized by a formal set of titles, modes of address and rules of debate, as well as a barracking, sexist and scatological 'humour', from many years of male domination, that women MPs found offensive, especially when it was used on them. Also, familiarity with speaking and debating techniques can help to curtail such sexist heckling. British women politicians have some success in using the media to draw attention to the sexism in the House of Commons by revealing these practices to women in the media. The result has been a series of press and broadcasting items about the childish and sexist behaviour of male MPs. The public, previously unaware of this, disapproved of their MPs' behaviour.

Between 1992 and 1997 four Labour women MPs in the UK were shadow ministers for women. All four were appointed to important government positions (two at cabinet level) when Labour won the election in 1997. They proved reliable advocates of gender equality and were supporters of women's concerns in their departments. Their numbers expanded in 2001. This indicates (a) that such positions need not be a ghetto for women, but may instead be a means of advancement, and (b) that they can facilitate the mainstreaming of gender sensitivity across government.

After all, it is more difficult politically to come out against equality for women than it is to prevent equality issues from getting onto the agenda in the first place. An example in the UK is the Sex Discrimination Candidates Act of 2002 which permits political parties to use affirmative action to increase the number of women MPs and candidates. The bill was passed with all-party support and almost no dissent because its opponents were silent, unwilling to oppose women's representation in public. Also, research from the UK shows that the interventions of women MPs are more likely than those of men to refer to examples of how policies and decisions affect individuals and families, while male MPs invoke abstract concepts such as citizens or constituents*. Another important representation strategy is to expand definitions of representation to include all public decision-making bodies, and to campaign for women's inclusion in the senior civil service and the judiciary. Such campaigns have been undertaken in Austria, Finland, the Netherlands, the UK and elsewhere. A difficult but necessary further step will be the extension of such demands to the private sector, as has happened in Finland.

CONCLUSION

It has been predicted that the UK economy will require an average of 141,000 new IT professionals every year for the next four years (Technology Counts IT & Telecoms Insights 2008, e-skills UK, and January 2008). Women's contribution and participation is therefore vital in this sector. There is an economic argument for creating a more diverse workforce and the forthcoming Equality Bill makes an obligation for all public bodies to comply with the necessary Equality Impact assessments. We have seen, historically women's participation and contribution to science were reliant on class, craft skills, family connections as well as community support. We would emphasise that these factors still play a major role in attracting women in both countries. The ICT sector is not an exception. Respect for diversity is does not make sense until and unless the disadvantaged groups have access to political and economic networks. (Swasti Mitter, 1995).

What would really make a difference would be to have more women in power, for example in the cabinet and in the civil service, who are in a position to influence the policy making directly in areas such as ICT and 'Digital Britain'. The benefits of gender diversity on Company Boards have proved to be very effective. Research indicates that gender diverse Boards have a positive impact on the performance of the company. According to a research by McKinsey, organisational performance increased sharply once a threshold of at least 3 women on management committees, with an average membership of 10 people, was reached (McKinsey, 2007, 2008). It has been seen that companies with three or more women on their Boards had stronger than average profits (Catalyst, 2007). A diversity in the Board is linked to good governance credentials, including more attention to audit and risk oversight and control, and greater consideration for the needs of a variety of stakeholders (Conference Board of Canada, 2002). When there are 3 or more women in the Board, it is more likely to ensure effective communication between the company and its stakeholders and proves to be more accountable (Conference Board of Canada, 2002). And a gender-diverse Board provides stricter monitoring of their performance (Adams and Ferreira, 2008).

Britain has a much high GEM measure than India, but having a higher GEM, only tells a part of the story in terms of equating progress in gender relations. We would like to point out that specifically in the IT sector in India, there seems to be evidence of a better working environment than the other sectors. Interesting, none of the respondents working in the IT sector gave any evidence of a gender pay gap, whereas this was different in the other sectors. In the UK, this was not the case. In terms of training and personal development, the respondents from UK mostly agreed that training schemes for both men and women in the sector was great. Employment practices in both the countries have similar trends – long working hours, de-valued part-time work, juggling with paid and un-paid work. Discrimination, harassment and bullying has been evident in previous researches (Adam et al 2006, 2007, Griffiths 2006, 2007, Takruri-Rizk et al 2006, etc), and some participants mentioned it from both the countries. Interestingly, equality diversity policies as well equal opportunities policies, has often been erroneously assumed and its operational approach has been linked to a mere 'tick box' system and often not even that. For instance, in India as per the Factories Act of 1948, any organisation employing more than thirty full-time employees are required by law to provide an in-house crèche facility for its employees. Not surprisingly, none of the organisations we studied provided one. In spite of this, in light of the recent growth of intellectual capital in India and its ICT- led growth and economic development of the urban areas, India with a lower GEM score than the UK, shows a better picture in terms of gender segregation in ICT – in terms of numbers, well-being and satisfaction.

ACKNOWLEDGMENT

A previous version of this chapter was published in the International Journal of E-Politics, Volume 2, Issue 1, edited by Celia Romm, pp. 45-67, copyright 2011 by IGI Publishing (an imprint of IGI Global).

REFERENCES

Acker, J. (1990). Hierarchies, Jobs, Bodies: A Theory of Gendered Organizations. *Gender & Society*, *4*(2), 139–158. doi:10.1177/089124390004002002

Adam, A., Richardson, H., Tattersall, A., & Keogh, C. (2004). *WINWIT: Women in North West Information Technology, ESF Report*. Salford University of Salford, Informatics Research Institute.

Adams, R. B., & Ferreira, D. (2008). *Women in the Boardroom and Their Impact on Governance and Performance*. Retrieved January 2010 from http://ssrn.com/abstract=1107721

Alic, M. (1986). *Hypatia's Heritage: A history of women in science from Antiquity through the Nineteenth Century*. Boston: Beacon Press.

Anastasopoulos, V., Brown, D. A. H., & Brown, D. L. (2002). *Women on Boards: Not Just the Right Thing... But the 'Bright' Thing*. Conference Board of Canada.

Antony, P., & Gayathri, V. (2008). Ricocheting gender equations: Women workers in the call centre industry. In Saith, A., Vijayabaskar, M., & Gayathri, V. (Eds.), *ICTs and Indian social change: diffusion, poverty, governance* (pp. 291–381). Los Angeles: Sage Publications.

Ballington, J., & Karam, A. (Eds.). (2005). *Women in Parliament: Beyond Numbers*. IDEA.

Cockburn, C. (1983). *Brothers: Male Dominance and Technological Change*. London: Pluto.

Cockburn, C. (1985). *Machinery of Dominance. Women, Men and Technical know-how*. London: Pluto Press.

Cockburn, C. (1999). Caught in the Wheels: The High Cost of Being a Female Cog in the Male Machinery of Engineering. In MacKenzie, D., & Wajcman, J. (Eds.), *The Social Shaping of Technology* (pp. 55–66). Buckingham: Open University Press.

Collinson, D., Knights, D., & Collinson, M. (1990). *Managing to discriminate*. London: Routledge.

Desvaux, G., Devillard-Hoellinger, S., & Baumgarten, P. (2007). *Women matter: Gender diversity, a corporate performance driver*. McKinsey & Company.

German, L. (2006). Theories of Patriarchy. *International socialism*.

Gregory, A., & Windebank, J. (2000). *Women's Work in Britain and France: Practice, Theory and Policy*. Macmillan. doi:10.1057/9780230598515

Gupta, Namrata and Sharma, A.K. (2003). Gender Inequality in the Work Environment at Institutes of Higher Learning in Science and Technology in India. *Work, Employment and Society, 17*(4), 597-616.

Hacker, S. L. (1981). The Culture of Engineering: Woman, Workplace and Machine. *Women's Studies International Quarterly, 4*(3), 341–353. doi:10.1016/S0148-0685(81)96559-3

Hartmann, H., & Bridges, A. (1981). The unhappy marriage of Marxism and feminism: towards a more progressive union . In Sargent, L. (Ed.), *Women and revolution: A Discussion of the Unhappy Marriage of Marxism and Feminism*. Boston: South End Press.

Joy, L., Carter, N. M., & Wagner, H. (2007). The bottom line: Corporate performance and women's representation on boards. *Catalyst*. Retrieved from http://www.catalyst.org/publication/200/the-bottom-line-corporate-performance-and-womensrepresentation-on-boards

Keogh, C., Moore, K., Tattersall, A., Griffiths, M., & Richardson, H. (2006). Managing Diversity or Valuing Diversity in Gender and the IT Labour Market . In Neiderman, F., & Ferratt, T. (Eds.), *IT Workers: Human Capital Issues in a Knowledge-Based Environment*. Hershey, PA, USA: Information Science Publishing.

Kvasny, L., Payton, F., Mbarika, V., Amadi, A., & Meso, P. (2008). Gendered Perspectives on IT Education and Workforce Participation in Kenya. *IEEE Transactions on Education, 51*(2), 256–261. doi:10.1109/TE.2007.909360

Kvasny, L., & Richardson, H. (2006). Critical Research in Information Systems: Looking Forward, Looking Back. *Information Technology & People, 19*(3), 196–202. doi:10.1108/09593840610689813

Lee, J. D. (2002). More than Ability: Gender and Personal Relationships Influence Science and Technology Involvement. *Sociology of Education, 75*(4), 349–373. doi:10.2307/3090283

Liff, S., & Ward, K. (2001). Distorted views through the glass ceiling: the construction of women's understandings of promotion and senior management positions. *Gender, Work and Organization, 8*(1), 19–36. doi:10.1111/1468-0432.00120

Lovenduski, J. (2005). *Feminizing Politics*. Oxford: Polity Press.

Meaney, M., Devillard-Hoellinger, S., & Denari, A. (2008). *Room at the top: Women and success in UK business*. McKinsey&Company.

Michie, S., & Nelson, D. L. (2006). Barriers women face in information technology careers: Self-efficacy, passion and gender biases. *Women in Management Review, 21*(1). doi:10.1108/09649420610643385

Mitter, S. (1994). On organising women in casualised work: a global overview . In Rowbotham, S., & Mitter, S. (Eds.), *Dignity and Daily Bread: New Forms of Economic Organising among Poor Women in the Third World and the First*. London: Routledge. doi:10.4324/9780203422946_chapter_1

Mitter, S., & Rowbotham, S. (Eds.). (1995). *Women encounter technology: Changing Patterns of Employment in the Third World*. London, New York: Routledge. doi:10.4324/9780203208618

Radhakrishnan, S. (2008). Examining the "Global" Indian Middle Class: Gender and Culture in the Silicon Valley/Bangalore Circuit. *Journal of Intercultural Studies (Melbourne, Vic.)*, *29*(1), 7–20. doi:10.1080/07256860701759915

Rajadhyaksha, U., & Smita, S. (2004). Tracing a timeline for work and family research in India. *Economic and Political Weekly*, *39*(17), 1674–1680.

Ridgeway, C. L., & Correll, S. J. (2004). Unpacking the Gender System: A Theoretical Perspective on Gender Beliefs and Social Relations. *Gender & Society*, *18*(4), 510–531. doi:10.1177/0891243204265269

Rowbotham, S. (2006). Feminist approaches to technology . In Grewal, I., & Kaplan, C. (Eds.), *An Introduction to Women's studies gender in a transnational world*. McGraw-Hill.

Ryan, A. B. (2006). Post-Positivist Approaches to Research. In M. Antonesa, H. Fallon, A.B. Ryan, A. Ryan, T. Walsh, & L. Borys (Eds.), *Researching and Writing your Thesis: a guide for postgraduate students* (pp. 12-26). MACE: Maynooth Adult and Community Education.

Sarton, G. (1927-48). *Introduction to the History of Science* (3 v. in 5). Carnegie Institution of Washington Publication no. 376. Baltimore: Williams and Wilkins, Co.

Sealy, R., Vinnicombe, S., & Singh, V. (2008). *The Female FTSE Report 2008*. Cranfield School of Management.

Singh, V. (2008). *Transforming Boardroom Cultures in Science, Engineering and Technology Organizations*. Research Report Series for UKRC No. 8.

Takruri-Rizk, H. (2006). *Women in North West Engineering. ESF Report: University of Salford*. Informatics Research Institute.

Thatchenkery, T., & Stough, R. R. (2005). *Information Communication Technology and Economic Development – Learning from the Indian Experience*. Edward Elgar Publishing Ltd.

Walby, S. (1990). *Theorising Patriarchy*. Oxford: Blackwell.

Walby, S. (1997). *Gender Transformations*. London: Routledge.

ENDNOTES

[1] http://www.ukrc4setwomen.org/downloads/research/Research_Briefing_12_2009_07_08.pdf

[2] http://www.spiegel.de/international/europe/0,1518,620544,00.html accessed on 4.11.09

[3] http://www.som.cranfield.ac.uk/som/dynamic-content/media/2008%20Female%20FTSE%20Report.pdf, accessed on 4.12.09

[4] Women in Parliament beyond numbers 2002 report

Chapter 5
Gender and E-Marketing:
The Role of Gender Differences in Online Purchasing Behaviors

Erkan Özdemir
Uludag University, Turkey

ABSTRACT

The increasingly widespread use of the Internet and the increasing participation of females in business have been significant changes leading to society today. As females have begun to have a more important role in business life, in addition to their crucial role within the family in the decision-making process of making purchases, they have had more financial independence. Additionally, the gap in Internet use in favor of males has also begun to narrow. This makes it necessary for e-marketers that carry out some or all of their transactions online to be fully aware of the effects of gender differences in online purchasing behaviors. This chapter consists of a literature review on the subject of Internet usage and online purchasing behaviors with a focus on gender-based differences. Accordingly, the aim of this chapter is to explore gender-based differences in Internet usage and online purchasing behavior and to suggest some e-marketing strategies for e-marketers. Additionally, this chapter provides a foundation on which to build future studies.

INTRODUCTION

The Internet brings about significant changes in all functions of businesses. In particular, marketing activities of businesses are deeply affected by these changes. In an attempt to support traditional marketing activities, businesses not only use the Internet but also exclusively carry out their business operations online. Therefore, e-marketing constitutes a significant part of marketing activi-

DOI: 10.4018/978-1-60960-759-3.ch005

Copyright © 2012, IGI Global. Copying or distributing in print or electronic forms without written permission of IGI Global is prohibited.

ties of businesses (Krishnamurthy, 2006). Many factors help the Internet market to prosper as well. While some of these are related to technological advances, some others have to do with managers' perceptions of the Internet and the lifestyles of consumers. As Internet shopping becomes more of a popular activity among the Internet users, electronic commerce is growing day by day (Donthu & Garcia, 1999). Therefore, as Feinberg and Kadam (2002) stated, business is moving online, not as a matter of choice, but as a matter of necessity. This shift occurs because the Internet has very seriously altered the consumer-seller relationship and steered the balance of power in favor of consumers through interactive features such as personalization, customized content and virtual communities. The Internet creates an extreme competitive marketplace in which consumers have shopping choices that they have never had before (Kim & Kim, 2004). This e-market environment offers retailers special challenges and leads them to revise their marketing strategies to secure more targeted customers.

As businesses have conducted marketing activities over the Internet, the concept of e-marketing has emerged. E-marketing can be defined as the use of the Internet and information technologies in order to conduct marketing activities (Krishnamurthy, 2006). Jayawardhena et al. (2003), on the other hand, defined e-marketing as the use of computerized processes in an attempt to integrate consumer information related to customer features, their choices and operating records from the marketing database. Perner and Fiss (2002: 38) defined e-marketing as "the concentration of all efforts in the sense of adapting and developing marketing strategies into the web environment." It is evident that e-marketing is not just a design of a web page set up by the business to advertise, as it can easily be misunderstood by many people. Strauss and El-Ansary (2004), on the other hand, describes e-marketing in a broader sense. According to Strauss and El-Ansary (2004), e-marketing is the application of a broad range of information technologies, such as transforming marketing strategies to create more customer value using the following strategies: more effective segmentation, targeting, differentiation, and positioning strategies; more efficiently planning and executing the conception, distribution, promotion and pricing of products, services, and ideas; and creating exchanges that satisfy individual consumer and organizational customers' objectives. E-marketing efforts have an impact on traditional marketing as well. While e-marketing efforts, on the one hand, enhance the efficiency of the traditional marketing functions, the e-marketing technologies have, on the other hand, transformed many of the marketing strategies.

Innovations in today's business life do not only take place in the areas of technology and the Internet. In comparison to the past, business life has in recent decades grown away from being a male-dominated arena, and females have begun to take more a bigger part in it. While some professions, apart from those such as teaching, working as a secretary and nursing, were dominated by males in the past, today females have become involved in all kinds of jobs and professions. The fact that females are more educated today enhances their chance of being financially independent, and this independence offers more freedom to them in the area of spending as well (Ahlström et al., 2001). This freedom is one of the most important reasons for choosing females as the target market. Another reason why females are targeted is the dominant role of females within the family in decision making for purchases. Therefore, knowing the purchasing differences between the two genders is crucial in terms of whether traditional businesses and businesses operating some or all of their transactions online are successful. Essentially, the variable of gender is regularly used by firms' marketing departments in segmenting consumers. However, marketing executives do not have comprehensive information on gender regarding the content of this segmentation. Therefore, knowing the role of the gender differences of the consumers in their purchasing behavior and developing appropriate marketing/e-marketing strategies for these differences and applying them are crucial in terms of the success of the businesses.

This chapter, which is based on a literature review, aims to explicate the role of gender-based differences in online purchasing behaviors and identify the e-marketing strategies and the points to consider for businesses conducting some or all of their trade online. In order to realize this aim, first the topic of gender-based differences in Internet usage is addressed. Then, online purchasing behaviors are reviewed in terms of gender differences, and e-marketing strategies based on gender-based differences for e-marketers are offered. In the conclusion section, some key points are discussed, and some suggestions are offered for the future research. The main value of this chapter is *an integrated review* in terms of gender-based differences in Internet usage and online purchasing behaviors, presenting the e-marketing strategies from the gender perspective.

BACKGROUND: INTERNET-RELATED GENDER DIFFERENCES AND GENDER DIFFERENCES IN ONLINE PURCHASING BEHAVIORS

Internet-Related Gender Differences

There are some gender differences in terms of access to Internet, Internet usage, attitudes towards the Internet, the frequency of Internet usage, self-assessment of Internet competency and confidence towards the Internet (Durndell & Haag, 2002; Hargittai & Shafer, 2006; Joiner et al., 2005; Li & Kirkup, 2007; Peng et al, 2006; Tsai et al., 2001; Wu & Tsai, 2006). However, some other studies show that there is a change in the younger generation. These studies suggested that the gender gaps are narrowing or even disappearing (Lin & Yu, 2008; Rainer et al, 2003; Tsai & Lin, 2004; Tsai & Tsai, 2010; Volman et al, 2005).

According to Park (2009), there are differences among generations regarding the gender gap. The gender difference was greater in the adult group than in the adolescent group in the case of the concentration of Internet usage.

According to Bimber (2000), some gender effects exist in Internet use. Due to some gender-based reasons and combination of those reasons, females are less extensive Internet users than males. In his study, Bimber (2000) stated that there were some gaps between males and females in terms of access to the Internet, but these gaps stemmed not only from gender-based factors, but also from socio-economic and other factors. He argued that these gaps would shrink over the long run as the differences of education and income between males and females gradually declined. According to Teo and Lim (1997), males found the Internet more exciting than females and thought about using it much more as well. In their subsequent study, Teo and Lim (2000), in addition to the gender differences, found differences between age groups in their Internet use. According to Teo and Lim's (2000) study in Singapore, females under 21 often spent more time on their messaging activities compared to males. However, no significant differences were found among people over 21.

In the study in which Ono and Zavodny (2003) examined the dimension of gender differences in various Internet use measurements in the 1997-2001 periods, they found that females used the Internet much less compared to males during the late 1990s but that this gap began to disappear beginning in 2000. The findings of this study also reveal that females used the Internet much less infrequently and extensively. Jackson et al. (2001), on the other hand, stated that no gender differences emerged in their study regarding the overall use of the Internet. Shaw and Gant (2002) found no gender differences in Internet use in their study and explained that the reason for the lack of gender differences was the fast narrowing of the gender gap in the use of the Internet among the younger generation. As is clearly seen in the studies done and illustrated here, as of the year 2000, the gaps in the use of the Internet among the younger generation have gradually been diminishing and disappearing. However, Heimrath and Goulding (2001) stated in their study that the traditional gender role of the females restricted their Internet participation. The male and female respondents in their study indicated that females had less spare free time of their own and that males, compared to females, had greater exposure to the Internet due to their jobs and professions. According to the findings of this study, social roles play a key role in females' participation on the Internet. Therefore, social conditions have both a direct effect on the environmental experiences and an indirect one on the perception of females of use of the Internet.

There are some gender-based differences related to the purpose of Internet use. Males regard the use of the Internet more highly compared to females because they think it enhances their job performance and labor productivity and provides information for them to make better decisions and enhance their effectiveness (Teo & Lim, 1997). In their study among university students and library workers, Heimrath and Goulding (2001) found that while males used the Internet for games, business, competition, computer software updates and professional awareness, females used it for shopping, banking, and new and various areas of information searching. In their study in which Orviska and Hudson (2009) examined Internet usage in EU member countries using Eurobarometer data for 2004 and 2005, they found that males are significantly more likely to use the Internet for banking, e-commerce, information and leisure. According to Garbarino and Strahilevitz (2004), in comparison to males (females 38%, males 27%), females spent most of their online time on using e-mail to stay in touch with others. Similarly, the other studies also reported that females, in comparisons to males, used e-mail with a stronger motivation for interpersonal communication and that males, compared to females, used the web more in order to access information (Akman & Mishra, 2010; Fang & Yen, 2006; Jackson et al., 2001; Tsai & Tsai, 2010; Weiser, 2000).

Gender Differences in Online Purchasing Behaviors

Convenience is one of the main reasons why consumers make online purchases (Jayawardhena et al., 2003). Males particularly find online shopping more convenient than females (Hui & Wan, 2007). However, while the aspect of convenience helps to enhance online purchases of consumers, the aspect of security prevents Internet purchases from rising. This is basically because females perceive online purchasing to be more risky than males (Garbarino & Strahilevitz, 2004). In other words, males have more confidence in online shopping than females and find it more practical and convenient (Rodgers & Harris, 2003). According to Shim et al. (2001), influences on consumers' intention to shop online include such factors as payment security, privacy, security and product guarantees. Dennis et al. (2002) found that shoppers were still concerned about online buying. Should e-retailers operating in the online environment wish to take full advantage of e-marketing activities, they have to find ways of altering the public's perception of their trustworthiness (Kolsaker & Payne, 2002). As the perceived risk of online buying drops, consumers' willingness to buy rises. Therefore, while e-marketers target female consumers in particular, they should also make an effort in order to minimize females' risk perception of online shopping (Garbarino & Strahilevitz, 2004).

In the study in which Donthu and Garcia (1999) compared the Internet shopper and nonshopper, they concluded that no gender differences existed between Internet shoppers and nonshoppers. Teo and Lim (1997) found, on the other hand, some significant gender differences in online purchasing activities in their study. According to Teo and Lim (2000), the probable reason for these differences was due to males being more experienced in computers and computing compared to females, leading to the possibility that they are better able to carry out such kinds of activities. In the online research carried out by Chang and Samuel (2004), they found some statistically significant associations between the demographic characteristics of Internet users (gender, age, income and location) and the frequency of online purchases and expenditures of the consumers. Accordingly, the percentage of males with five or more online purchases is higher than females. Moreover, the percentage of males spending 100 Aus$ and above per online transaction is higher compared to females (Chang & Samuel, 2004). According to Van den Poel and Buckinx (2005), gender is a significant demographic variable related to online purchasing behavior. According to this study, males, in comparison to females, have a greater tendency to make an online purchase in their next web site visits. In addition, Cyr and Bonanni (2005) show that males spend more time and money on online purchases than the amount of time spent by females.

Online consumers and traditional consumers follow a similar pattern of purchasing behaviors. However, they perform this faster and more thoroughly in online shopping compared to traditional purchasing (Jayawardhena et al., 2003). However, there are also some gender-based differences in online shopping behavior. For instance, if females are the fundamental purchasing agents in their families, they take more pleasure in shopping compared to males. In this sense, the positive image of in-store purchases is greater for females compared to males. On the other hand, if females are shopping merely out of necessity, their perception of online shopping is more positive compared to in-store purchases (Alreck & Setle, 2002). Females regard in-store purchases more positively in terms of hedonic benefit (Dholakia & Uusitalo, 2002). Some other research results also support this finding. For instance, according to the study of Rodgers and Harris (2003), males found the experience of online shopping more satisfying than females. In other words, the females taking part in that particular study were emotionally less satisfied with online-shopping compared to males.

One of the fundamental reasons why online purchases are not made is the preference of the consumers to browse the products (Teo, 2002). Because females particularly fail to interact physically like touching when shopping online, they dislike online shopping. In other words, they are emotionally less satisfied with online shopping (Hui & Wan, 2007). However, when it is compared with traditional shopping, there is a consensus that online shopping has attractive attributes. The attractive attributes of Internet shopping includes time and money savings; convenience and easy accessibility; access to an alternative range of products and being able to choose from a wide variety; the availability of information for making a purchase and ordering decisions (Kim & Kim, 2004).

E-MARKETING STRATEGIES WITH REGARD TO GENDER-BASED DIFFERENCES

It is commonly observed that females and males are different in their shopping orientation. Particularly in married households, shopping is still considered a gendered activity. Females are faced with shopping together with other household tasks as a challenge; as a result, they have some negative feelings/attitudes towards shopping. In particular, time-pressured working females constitute a market segment for e-marketers to monitor and target (Dholakia & Uusitalo, 2002). As is clearly seen, gender as a variable of segmentation can be taken up together with other segmentation variables, and homogenous segments as target markets can be selected. In this particular part of the study, taking into account the gender-based differences reviewed before and the online purchasing behaviors based on these differences, e-marketing strategies are presented with regards to marketing mix (4P: product, place, price and promotion) elements for business and some significant points in practice for the e-marketers to consider are emphasized.

Product Strategies

Because the product element of the marketing mix cannot be physically illustrated, the most important instrument at this point is the web sites on which the product is displayed. The significant attributes of web sites, such as convenience and variety together with detailed information about the products, should be presented to prospective customers. This presentation is important because the online shoppers do not evaluate the quality of a product solely on its price and brand (Donthu & Garcia, 1999). Today the Internet has a great potential for improving the image of a product, establishing brand recognition and structuring brand value. The interactive web sites with detailed graphic images attract the attention and interest of potential consumers (Aldridge et al., 1997).

When businesses offering services on the Internet and their e-marketers single out females as target consumers, they must find ways of meeting their touching needs. Females prefer and like the physical evaluation of products, such as seeing and feeling the product, before they buy it (Dittmar et al., 2004). The inability to feel the product can be compensated by establishing a more interactive web site, providing more information or using new Internet technologies allowing for virtual senses of tasting and touching (Hui & Wan, 2007). The contents of the web sites may influence the attitudes of the consumers depending on their genders. For instance, in their study on the effects of gender on attitudes towards web site design, Simon and Peppas, (2005) found that males had more positive attitudes compared to their female counterparts for all web sites examined in their study. Moreover, they also found that males had more positive attitudes regarding both rich and plain web sites when compared to females.

Place Strategies

While the place element of the marketing mix can be accomplished by the distributor and retailers in traditional marketing, this element in the context of e-marketing can occur through the businesses' web sites. This presents businesses opportunities for the "manufacturer direct-to-customer" approach (Kalyanam & McIntyre, 2002). Just as the attributes of the distributor, its speed and image, and the way the products are presented to the customers, are important in the distribution channel in traditional marketing, the web site of a business that partially fulfills the duties of the distribution channel online is equally important. For instance, Chen and Lee (2005) found in their study that the image of the web site had an impact on purchasing behaviors and intention of the customers. Teo and Lim (2000), on the other hand, found that compared to males, females attached more significance to animation and multimedia features such as sound effects and background music.

Dabholkar and Sheng (2009) carried out a study on the control perceived in consumer reactions and the role of gender and their effects on the download delays among university students in the United States. According to the findings of this study, there were gender differences in the attitudes towards the download delays. Accordingly, males were less tolerant of online delay and focused more on download speeds. Therefore, when a download delay occurs, males easily switch from the existing site to another one. Based on this information, it is possible to say that the positive effects of the perceived speed of the web site on consumers' attitudes and intentions are more powerful among males in comparison to females. While males react directly to the download delays, females do not pay much attention to a slow process as much as males. However, if this slow process makes females feel a loss of control, they react negatively. The findings of this study demonstrate that the males' reaction was based on the perception of speed and the females' reaction was based on the feeling of control and assurance throughout the process. Therefore, e-marketers have to focus on minimizing the download delays of the web sites whose users are predominantly males. In order to accomplish this, complicated and rich web site content may have to be sacrificed. It is only in this way that positive reactions to the web sites can be elicited from male online users. For the web sites whose users are predominantly females, e-marketers may produce web sites with rich content. Soothing colors, relaxing music and information on delays for female online users, even though they may cause delays, may be used. Even after one delay, female online users will not have the feeling of the loss of control.

There are also gender-based differences regarding the duration of web site visits. For instance, in their study, in which they examined the factors affecting the web site visit duration, Daneher et al. (2006) found that females had a tendency to visit web sites for longer periods of time compared to males. The reason for this difference is rooted in the fact that female consumers pay greater attention to details and wish to be better informed. The fact that female consumers spend more time on browsing a web page in order to search more before they make up their minds for purchasing something reveals that female consumers go through a more intensive process of information gathering (Brunel & Nelson, 2003). Brain lateralization and attributes reveal gender-based differences in terms of information gathering and assessment. Therefore, females will presumably assess a web site in greater detail, while males will presumably think about an e-commerce site in terms of global attitude. The reason why males are perfectly satisfied with their online shopping behavior is that the information processing and emotionality in the right hemisphere of the brain has a great role to play in holistic assessment. On the other hand, the fact that females process information in their left brain hemisphere and therefore more detailed assessments affecting females' feelings about web sites are made compared to males cause females to have lower emotional gratification from online shopping (Rodgers & Harris, 2003).

Price Strategies

The price element of the marketing mix has a greater role to play for males compared with females in their purchasing decision (Myers, 1994). Although convenience and the services provided are crucial for male consumers, price and product quality are much more important for male consumers (Torres et al., 2001). Although male consumers pay more on average compared to females, they have higher price flexibility. Therefore, when male consumers are targeted, discounts can be very effective in increasing demand (Mazumdar & Papatla, 1995). Furthermore, because male consumers like to bargain for price, businesses should offer them opportunities for bargaining to make them feel that they are the winning party (Otnes & McGrath, 2001). When it comes to deciding the price, one family member, either the male or the female party, depending on the type of the product or role structure in the family, may be the dominant party. If the product is to be purchased based on an economical decision, male consumers will make this decision, being more dominant in such decisions. The purchasing decisions, in which females are traditionally more dominant, are made by the females themselves (Mellot, 1983).

The e-marketing environment makes it possible for e-marketers to utilize such pricing methods as forward auctions, reverse auction, dynamic pricing and "name your own price" (Kalyanam & McIntyre, 2002) It is quite helpful for consumers to make price comparisons in the Internet before going shopping, and both genders are content with this particular feature of the Internet. For instance, Hui and Wan (2007) concluded that both male and female respondents felt that they got a better deal in online shopping. Businesses may choose to utilize the appropriate pricing method using their databases. For instance, businesses may offer different prices to different customer segments through dynamic pricing (Strauss & El-Ansary, 2004). More independent, self-confident, competitive and externally motivated males are more willing to take financial risks in comparison to females. Therefore, if they are not content with a product or a service, they complain less than females (Mitchell & Walsh, 2004).

Promotion Strategies

The promotion element of the marketing mix can be considered from the point of view of sub-elements of advertising and sales. The gender-based differences give rise to differences in the perception of the advertising messages and their different assessment. For instance, the information processing strategies of males and females are different. When processing advertising messages, females are more interested in the messages' content and claims and are more sensitive to the particulars compared with males. Males' processing of the message is driven more by the overall message themes and schema. Consequently, females use a detailed processing strategy, and males use a schema-based strategy (Meyers-Levy & Maheswaran, 1991). This gender-based difference in information processing also means that the information to be provided in the web sites by the e-marketers has to be different depending on the gender targeted. Because the visual depictions are suitable for processing in the right hemisphere of the brain, when males are targeted as prospective consumers, the technical information may be presented in graphic form. When females are targeted as prospective consumers, because they process the information in their left brain hemisphere, they may be presented the product information in the form of verbal prose (Rodgers & Harris, 2003). For instance, according to a study done by Putrevu (2004), the reactions of males and females to similar printed advertising were found to be very different. Accordingly, males are more influenced by comparative, simple and product-oriented advertising.

The significant differences between males and females in terms of human relationships also has an impact on the way in which the product is presented or on how and where the emphasis is to be placed in advertising. Females are more involved in human relations compared to males, which means females are more open to empathy (Holden & Holden, 1998). Because females have a tendency to make communication contact and are more human-oriented, in order to attract the attention of female consumers, a strategy of focusing on the user rather than the product and informing the consumer about the benefits of the product/service rather than its features has to be adopted (Wilson, 2004). Furthermore, because females prefer to interact with other people, community-based strategies will also have an impact on female consumers (Walker, 2001). Therefore, advertising targeting female consumers should portray females in sincere and close relationships with males and females or in a noncompetitive relationship with others (Prakash, 1992). Males and females also have distinct way of regarding and assessing advertising messages and gender descriptions of the opposite gender. For instance, while females welcome the advertising of brands for the male gender, men do not welcome the advertising of brands for the female gender (FitzGerald & Arnott, 1996). There are also some differences in consumers' reactions to gender descriptions in advertising. While females react positively to gender descriptions, males, on the contrary, react less positively (Orth & Holancova, 2004).

E-marketers use the Internet for online sales promotions as well. For instance, businesses send out digital product samples such as music or software and electronic coupons to consumers, and conduct sweepstakes on various subjects (Strauss & El-Ansary, 2004). Promotional e-mails are extensively used by e-marketers as sales promotion and advertising. There are no gender-based differences in preferences for receiving the promotional e-mails. Both genders illustrate similar types of attachment and awareness for e-mail advertising. However, females, who greatly feel the need to establish and maintain social relationships, forward the promotional e-mails to their friends more than males (Phillip & Suri, 2004).

E-marketers may also use other marketing tactics over the Internet. In order to attract special-interest users, they may create a community-building web site. In this method, the business invites potential customers to chat and send e-mails in their web site in order to direct their attention to the web site. Businesses, therefore, obtain the e-mail addresses of the users of those groups for their e-mail marketing campaigns. Through community building, e-marketers may generate social networks that will enhance the relationship with the customers (Strauss & El-Ansary, 2004). Females are especially targeted in this strategy because females not only focus on maintaining relationships, but also contact and stay in touch with one another more often. Most online communities in recent years have been for female consumers (Garbarino & Strahilevitz, 2004), and females send and receive messages more. For instance, when females under 21 were compared to males in a study in Singapore, Teo and Lim (2000) found that females spent more time on messaging activities. E-marketers, in order to take advantage of this particular preference of females, may effectively utilize online social networks allowing e-mails and the exchange of information and ideas.

In their study on Korean consumers, Kwon et al. (2007) found that the community aspect was better appreciated by the married respondents. The effects of females' tendency to build communities and establish relationships set them apart from males in terms of giving and receiving advice and recommendations. For instance, Garbarino and Strahilevitz (2004) concluded that there was a substantial drop in the risk perceived by females, in comparison to males, after recommendations from friends. Furthermore, while recommendations from friends had a great impact on the online purchases of females, these recommendations did not have such an impact on males. What these findings mean for e-marketers is that positive word of mouth has to be encouraged in order for these findings to have a positive impact on female consumers. In addition, e-marketers can also generate rewarding programs that will enable existing web site users to recommend it to other potential users. Additionally, the link of "tells a friend about this web site" can be placed on the web site and will enable users to relate their shopping experience to other people. The positive opinions and useful information of family and friends are likely to enhance the level of confidence in consumers, which will, in turn, have a positive impact on the purchasing intention of the consumers (Howard, 1989). Consequently, when e-marketers target online female consumers, e-marketers can utilize online forums and chat rooms and provide incentives for online female consumers to share their experiences with other online consumers to enhance social and interpersonal experiences in online shopping (Zhou et al., 2007; Hasan, in press).

E-marketers should make an effort to establish a relationship through both web sites and customer representatives with existing or prospective customers based appropriately on gender differences. This strategy is necessary because the relationship capital is a strategic value in terms of e-marketing. As Strauss and El-Ansary (2004) stated, e-marketers should attach greater importance to establishing and maintaining a relationship with the customers than to the land, property and financial entities of the business because this relationship capital will constitute the foundation of future business. Because female consumers, in comparison to male ones, have greater numbers of and stronger interpersonal and brand relationships, they are more loyal to the brands of their choice (Ndubisi,

2006). Therefore, it is all the more important to establish long-term relationships with female consumers compared to male ones.

CONCLUSION

In addition to the role of females in the decision-making process of making a purchase for the family and as result of females gaining more financial independence as a result of their ever-increasing participation in business life, marketers are forced to consider females an important target market. As the gender-based gap in the use of Internet narrows and as general Internet use becomes less of a male-dominated activity, it is all the more important for e-marketers that carry out some or all of their transactions online to be fully aware of the effects of gender-based differences in females' online shopping behavior. Gender as a segmentation variable in traditional marketing should be used as an important segmentation variable in e-marketing as well because significant gender-based differences exist in online purchasing behavior.

The attitudes and behaviors towards Internet usage also differ on the basis of gender. In fact, gender-based differences in Internet usage have begun to diminish and disappear in recent years. Therefore, despite the existence of differences in Internet use observed in previous studies (Bimber, 2000; Teo & Lim, 1997; Teo & Lim, 2000), later studies (Jackson et al., 2001; Ono & Zavodny, 2003; Shaw & Gantt, 2002) have found no such differences. This particular finding comprises one of the fundamental reasons why e-marketers in this day and age choose females as well as males as one of the target markets. Accordingly, being aware of the gender-based differences in the use of the Internet discussed in this study and considering the gender differences in online purchasing behaviors will have a direct impact on the success of e-marketing strategies. Both gender-based differences and sub-segments such as working females, mothers and full-time housewives have also to be taken into consideration in e-marketing strategies. For instance, as the time pressure among working females increases, working females turn to online shopping. Therefore, in addition to gender differences, time savings and the convenience aspect have to be taken into account as well.

While e-marketers form their e-marketing strategies, they should provide more detailed information about the products offered as they target females as consumers, and this information should be verbal. The web sites that provide information about the products should have rich content and advanced Internet technologies allowing features such as virtual tasting and touching should be utilized. Moreover, some multimedia features such as sound effects, background music and animation should be used; however, they should not let females feel as if they have lost control as they are browsing the site or as they are conducting the purchasing transaction. When females are targeted as online consumers, the price of the products should not be the sole criterion; rather, price has to be considered together with other purchasing factors. Advertising messages, on the other hand, have to be detailed, verbal and inclusive of social relationships. Therefore, when females are targeted as online consumers, alternative applications such as e-mail advertising, social groups and word-of-mouth advertising could be used.

When e-marketers target males as online consumers, they should offer results-oriented, technical information illustrating the functions and benefits of the products quantitatively and/or offer graphics in a more plain and simple fashion. When males are targeted as online consumers, the prices of products should be used as an important factor in purchasing decisions. Because products may be priced with regard to a competitive method of pricing, such opportunities as price cuts and bargaining over the price should be offered. Advertising geared toward males should be compared with other products and/or with the products of rivals and should be competition-oriented. Moreover, advertising messages should be plain and presented in a quality-oriented fashion.

In this study, the role of gender differences in online purchasing behaviors has been examined, and some important points to consider in terms of gender-based strategies in e-marketing among businesses that conduct part or all of their business transactions online have been discussed. Future studies can empirically examine the sensitivity to online prices in reaction to online pricing methods and whether gender-based differences exist in understanding the messages of online advertising and during the process of online purchasing.

REFERENCES

Ahlström, S., Bloomfield, K., & Knibbe, R. (2001). Gender differences in drinking patterns in nine European Countries: Descriptive findings. *Substance Abuse*, *22*(1), 69–85. doi:10.1080/08897070109511446

Akman, I., & Mishra, A. (2010). Gender, age and income differences in internet usage among employees in organizations. *Computers in Human Behavior*, *26*(3), 482–490. doi:10.1016/j.chb.2009.12.007

Aldridge, A., Forcht, K., & Pierson, J. (1997). Get linked or get lost: Marketing strategy for the internet. *Internet Research: Electronic Networking Applications and Policy*, *7*(3), 161–169. doi:10.1108/10662249710171805

Alreck, P., & Setle, R. B. (2002). Gender effects on internet, catalogue and store shopping. *Journal of Database Marketing*, *9*(2), 150–162. doi:10.1057/palgrave.jdm.3240071

Bimber, B. (2000). Measuring the gender gap on the internet. *Social Science Quarterly*, *81*(3), 868–876.

Brunel, F. F., & Nelson, M. R. (2003). Message order effects and gender differences in advertising persuasion. *Journal of Advertising Research*, *43*(3), 330–341. doi:10.1017/S0021849903030320

Chang, J., & Samuel, N. (2004). Internet shopper demographics and buying behavior in Australia. *Journal of American Academy of Business*, *5*(1/2), 171–176.

Chen, W., & Lee, C. (2005). The impact of web site image and consumer personality on consumer behavior. *International Journal of Management*, *22*(3), 484–496.

Cyr, D., & Bonanni, C. (2005). Gender and website design in e-business. *International Journal of Electronic Business*, *3*(6), 565–582. doi:10.1504/IJEB.2005.008536

Dabholkar, P. A., & Sheng, X. (2009). The role of perceived control and gender in consumer reactions to download delays. *Journal of Business Research*, *62*(7), 756–760. doi:10.1016/j.jbusres.2008.06.001

Danaher, P. J., Mullarkey, G. W., & Essegaier, S. (2006). Factors affecting web site visit duration: a cross-domain analysis. *JMR, Journal of Marketing Research*, *43*(2), 182–194. doi:10.1509/jmkr.43.2.182

Dennis, C., Harris, L., & Sandhu, B. (2002). From bricks to clicks: Understanding the e-consumer. *Qualitative Market Research: An International Journal*, *5*(4), 281–290. doi:10.1108/13522750210443236

Dholakia, R. R., & Uusitalo, O. (2002). Switching to electronic stores: Consumer characteristics and the perception of shopping benefits. *International Journal of Retail & Distribution Management*, *30*(10), 459–469. doi:10.1108/09590550210445335

Dittmar, H., Long, K., & Meek, R. (2004). Buying on the internet: Gender differences in on-line and conventional buying motivation. *Sex Roles*, *50*(5-6), 423–444. doi:10.1023/B:SERS.0000018896.35251.c7

Donthu, N., & Garcia, A. (1999). The internet shopper. *Journal of Advertising Research*, *39*(3), 52–58.

Durndell, A., & Haag, Z. (2002). Computer self efficacy, computer anxiety, attitudes towards the internet and reported experience with the internet, by gender, in an East European sample. *Computers in Human Behavior*, *18*(5), 521–535. doi:10.1016/S0747-5632(02)00006-7

Fang, X., & Yen, D. C. (2006). Demographics and behavior of internet users in China. *Technology in Society*, *28*(3), 363–387. doi:10.1016/j.techsoc.2006.06.005

Feinberg, R., & Kadam, R. (2002). E-CRM web service attributes as determinants of customer satisfaction with retail web sites. *International Journal of Service Industry Management*, *13*(5), 432–451. doi:10.1108/09564230210447922

FitzGerald, M., & Arnott, D. (1996). Understanding demographic effects on marketing communications in services. *International Journal of Service Industry Management*, *7*(3), 31–45. doi:10.1108/09564239610122947

Garbarino, E., & Strahilevitz, M. (2004). Gender differences in the perceived risk of buying online and the effects of receiving a site recommendation. *Journal of Business Research*, *57*(7), 768–775. doi:10.1016/S0148-2963(02)00363-6

Hargittai, E., & Shafer, S. (2006). Differences in actual and perceived online skills: The role of gender. *Social Science Quarterly*, *87*(2), 432–448. doi:10.1111/j.1540-6237.2006.00389.x

Hasan, B. (in press). Exploring gender differences in online shopping attitude. *Computers in Human Behavior*.

Heimrath, R., & Goulding, A. (2001). Internet perception and use: A gender perspective. *Program: electronic library and information systems*, *35*(2), 119-134.

Holden, L., & Holden, A. C. (1998). Woman to women: social marketing and idea to the new world. *Psychology and Marketing*, *15*(2), 175–193. doi:10.1002/(SICI)1520-6793(199803)15:2<175::AID-MAR5>3.0.CO;2-9

Howard, J. A. (1989). *Consumer behavior in marketing strategy*. Englewood Cliffs, NJ: Prentice Hall.

Hui, T. K., & Wan, D. (2007). Factors affecting internet shopping behaviour in Singapore: Gender and educational issues. *International Journal of Consumer Studies*, *31*(3), 310–316. doi:10.1111/j.1470-6431.2006.00554.x

Jackson, L. A., Ervin, K. S., Gardner, P. D., & Schmitt, N. (2001). Gender and the internet: Women communicating and men searching. *Sex Roles*, *44*(5/6), 363–379. doi:10.1023/A:1010937901821

Jayawardhena, C., Wright, L. T., & Masterson, R. (2003). An investigation of online consumer purchasing. *Qualitative Market Research: An International Journal*, *6*(1), 58–65. doi:10.1108/13522750310457384

Joiner, R., Gavin, J., Duffield, J., Brosnan, M., & Crook, C. (2005). Gender, internet identification, and internet anxiety: Correlates of Internet use. *Cyberpsychology & Behavior*, *8*(4), 371–378. doi:10.1089/cpb.2005.8.371

Kalyanam, K., & McIntyre, S. (2002). The e-marketing mix: A contribution of the e-tailing wars. *Journal of the Academy of Marketing Science*, *30*(4), 483–495. doi:10.1177/009207002236924

Kim, E. Y., & Kim, Y. K. (2004). Predicting online purchase intentions for clothing products. *European Journal of Marketing, 38*(7), 883–897. doi:10.1108/03090560410539302

Kolsaker, A., & Payne, C. (2002). Engendering trust in e-commerce: A study of gender-based concerns. *Marketing Intelligence & Planning, 20*(4/5), 206–214. doi:10.1108/02634500210431595

Krishnamurthy, S. (2006). Introducing e-markplan: A practical methodology to plan e-marketing activities. *Business Horizons, 49*(1), 51–60. doi:10.1016/j.bushor.2005.05.008

Kwon, H. J., Joshi, P., & Jackson, V. P. (2007). The effect of consumer demographic characteristics on the perception of fashion web site attributes in Korea. *Journal of Fashion Marketing and Management, 11*(4), 529–538. doi:10.1108/13612020710824580

Li, N., & Kirkup, G. (2007). Gender and cultural differences in internet use: A study of China and the UK. *Computers & Education, 48*(2), 301–317. doi:10.1016/j.compedu.2005.01.007

Lin, C. H., & Yu, S. F. (2008). Adolescent internet usage in Taiwan: Exploring gender differences. *Adolescence, 43*(170), 317–331.

Mazumdar, T., & Papatla, P. (1995). Gender difference in price and promotion response. *Pricing Strategy & Practice., 3*(1), 21–33.

Mellot, D. W. (1983). *Fundamentals of consumer bahaviour*. Oklahoma: Pen Well Publishing Company.

Meyers-Levy, J., & Maheswaran, D. (1991). Exploring differences in males' and females' processing strategies. *The Journal of Consumer Research, 18*(1), 63–70. doi:10.1086/209241

Mitchell, V. W., & Walsh, G. (2004). Gender differences in German consumer decision-making styles. *Journal of Consumer Behaviour, 3*(4), 331–346. doi:10.1002/cb.146

Myers, G. (1994). *Targeting the new professional woman: How to market and sell to today's 57 million working women*. Chicago: Probus Publishing Company.

Ndubisi, N. O. (2006). Effect of gender on customer loyalty: A relationship marketing approach. *Marketing Intelligence & Planning, 24*(1), 48–61. doi:10.1108/02634500610641552

Ono, H., & Zavodny, M. (2003). Gender and the internet. *Social Science Quarterly, 84*(1), 111–121. doi:10.1111/1540-6237.t01-1-8401007

Orth, U. R., & Holancova, D. (2004). Men's and women's responses to sex role portrayals in advertisements. *International Journal of Research in Marketing, 21*(1), 77–88. doi:10.1016/j.ijresmar.2003.05.003

Orviska, M., & Hudson, J. (2009). Dividing or uniting Europe? Internet usage in the EU. *Information Economics and Policy, 21*(4), 279–290. doi:10.1016/j.infoecopol.2009.06.002

Otnes, C., & McGrath, M. A. (2001). Perception and realities of male shopping behavior. *Journal of Retailing, 77*(1), 111–137. doi:10.1016/S0022-4359(00)00047-6

Park, S. (2009). Concentration of internet usage and its relation to exposure to negative content: Does the gender gap differ among adults and adolescents? *Women's Studies International Forum, 32*(2), 98–107. doi:10.1016/j.wsif.2009.03.009

Peng, H. Y., Tsai, C. C., & Wu, Y. T. (2006). University students' self-efficacy and their attitudes toward the internet: the role of students' perceptions of the Internet. *Educational Studies, 32*(1), 73–86. doi:10.1080/03055690500416025

Perner, P., & Fiss, G. (2002). Intelligent e-marketing with web mining, personalization and user-adpated interfaces. In Perner, P. (Ed.), *Data Mining in E-Commerce, Medicine, and Knowledge Management* (pp. 37–52). Springer Verlag.

Phillip, M. V., & Suri, R. (2004). Impact of gender differences on the evaluation of promotional emails. *Journal of Advertising Research, 44*(4), 360–368.

Prakash, V. (1992). Sex roles and advertising preferences. *Journal of Advertising Research, 32*(3), 43–52.

Putrevu, S. (2004). Communicating with the sex: Male and female responses to print advertising. *Journal of Advertising, 33*(3), 51–62.

Rainer, R. K., Laosethakul, K., & Astone, M. K. (2003). Are gender perceptions of computing changing over time? *Journal of Computer Information Systems, 43*(4), 108–114.

Rodgers, S., & Harris, M. A. (2003). Gender and e-commerce: An exploratory study. *Journal of Advertising Research, 43*(3), 322–329. doi:10.1017/S0021849903030307

Shaw, L. H., & Gant, L. M. (2002). Users divided? Exploring the gender gap in internet use. *Cyberpsychology & Behavior, 5*(6), 517–527. doi:10.1089/109493102321018150

Shim, S., Eastlick, M. A., Lotz, S. L., & Warrington, P. (2001). An online prepurchase intentions model: the role of intention to search. *Journal of Retailing, 77*(3), 397–416. doi:10.1016/S0022-4359(01)00051-3

Simon, S. J., & Peppas, S. C. (2005). Attitudes towards product website design: A study of the effects of gender. *Journal of Marketing Communications, 11*(2), 129–144. doi:10.1080/1352726042000286507

Strauss, J., & El-Ansary, A. I. (2004). Integrating the "e" in e-marketing. *Journal of Business & Economics Research, 2*(8), 69–80.

Teo, T. S. H. (2002). Attitudes toward online shopping and the internet. *Behaviour & Information Technology, 21*(4), 259–271. doi:10.1080/0144929021000018342

Teo, T. S. H., & Lim, V. K. G. (1997). Usage patterns and perceptions of the internet: The gender gap. *Equal Opportunities International, 16*(6/7), 1–8. doi:10.1108/eb010696

Teo, T. S. H., & Lim, V. K. G. (2000). Gender differences in internet usage and task preferences. *Behaviour & Information Technology, 19*(4), 283–295. doi:10.1080/01449290050086390

Torres, I. M., Summers, T. A., & Belleau, B. D. (2001). Men's shopping satisfaction and store preferences. *Journal of Retailing and Consumer Services, 8*(4), 205–212. doi:10.1016/S0969-6989(00)00024-2

Tsai, C. C., & Lin, C. C. (2004). Taiwanese adolescents' perceptions and attitudes regarding the internet: Exploring gender differences. *Adolescence, 397*(156), 725–734.

Tsai, C. C., Lin, S. S. J., & Tsai, M. J. (2001). Developing an internet attitude scale for high school students. *Computers & Education, 37*(1), 41–51. doi:10.1016/S0360-1315(01)00033-1

Tsai, M. J., & Tsai, C. C. (2010). Junior high school students' internet usage and self-efficacy: A re-examination of the gender gap. *Computers & Education, 54*(4), 1182–1192. doi:10.1016/j.compedu.2009.11.004

Van den Poel, D., & Buckinx, W. (2005). Predicting online-purchasing behavior. *European Journal of Operational Research, 166*(2), 557–575. doi:10.1016/j.ejor.2004.04.022

Volman, M., van Eck, E., Heemskerk, I., & Kuiper, E. (2005). New technologies, new differences: Gender and ethnic differences in pupils' use of ICT in primary and secondary education. *Computers & Education, 44*(1), 35–55. doi:10.1016/S0360-1315(04)00072-7

Walker, T. (2001). Wooing female consumers reaps rewards for health plans. *Managed Healthcare Excutive, 11*(3), 42–45.

Weiser, E. B. (2000). Gender differences in internet use patterns and internet: Application preferences: A two-sample comparison. *Cyberpsychology & Behavior*, *3*(2), 167–178. doi:10.1089/109493100316012

Wilson, M. (2004). A world of differences. *Chain Store Age*, *80*(9), 126.

Wu, Y. T., & Tsai, C. C. (2006). Developing an information commitment survey for assessing students' web information searching strategies and evaluative standards for web materials. *Journal of Educational Technology & Society*, *10*(2), 120–132.

Zhou, L., Dai, L., & Zhang, D. (2007). Online shopping acceptance model – A critical survey of consumer factors in online shopping. *Journal of Electronic Commerce Research*, *8*(1), 41–62.

KEY TERMS AND DEFINITIONS

E-Marketing: E-Marketing is the marketing of products and services over the internet by businesses. E-marketing is also referred to as online marketing, internet marketing, i-marketing and web marketing.

Marketing Mix: Marketing mix is composed of the combination of marketing variables controlled by the marketing executives in order to meet the wants and needs of the targeted consumers. The traditional marketing mix is composed of the elements of product, pricing, place, and promotion and is also known as 4P.

Marketing Strategy: Marketing strategy is a process adopted by a marketing department in order to fulfill the aims of increasing the sales of the business and gaining a sustainable competitive advantage. Marketing strategy comprises particular strategies such as target markets, marketing mix and the level of marketing expenditure.

Online Consumer: Online consumer is the consumer that searches and makes the purchase of the product and services on the internet; in other words, s/he is the consumer who carries out the process of making the purchase in the electronic media.

Online Purchasing: Online purchasing refers to the purchasing of products and services over the internet by consumers. Due to reasons such as the development of the internet technologies and paying more attention to online security, more and more consumers today are making online purchases.

Online Shopping: Online shopping is the process consumers go through to purchase products or services over the Internet and used for business-to-business (B2B) and business-to-consumer (B2C) transactions.

Web Site: A group of similar web pages linked by hyperlinks and managed by a business, organization, or person. A web site may include text, graphics, images, audio and video files, and hyperlinks to other web pages. In the case of e-business web sites, the products or services may be purchased at the web site itself, by entering credit card number or other payment information into a payment form on the web site. E-business web sites offer products and services for online sale and enabling online transactions for such sales.

Chapter 6
Women in Information Communication Technologies

Olca Surgevil
Dokuz Eylul University, Turkey

Mustafa F. Özbilgin
Brunel University, UK & Université Paris-Dauphine, France

INTRODUCTION

Individual participation in the information economy is affected by a large number of variables such as education, skill, income, race / ethnicity, language, gender and disability (OECD, 1999). Some of these variables such as education and skill are considered legitimate in most countries, while other variables based on arbitrary criteria such as gender, race and disability are viewed as illegitimate in a growing number of countries. Employment in the Information Communication Technology (ICT) sectors around the world has also been affected by gender discrimination and stereotyping practices.

The concept of technology gained popularity in 1950s and it has traditionally been associated with men. Technology was usually assumed as to be men's work and most definitions reinforce such a masculine ideology (Eriksson-Zetterquist, 2007). Most feminist historians and sociologists of technology have emphasized strong alignments between technology and masculinity especially in the area of engineering (Oudshoorn et al., 2004; Kusku et al., 2007).

The number of female workers has been gradually increased in the field of ICT. However, the number of women users of ICT is not growing at a sufficient pace to allow women to influence the

DOI: 10.4018/978-1-60960-759-3.ch006

development of ICT. Thus, recruiting and retaining more women in ICT profession or making the ICT work environment more welcoming for women is still a significant challenge in industrialized societies (Faulkner&Lie, 2007).

In this study women's attendance to ICT field and the qualifications (fundamental capabilities and training) which they need to get into the sector are discussed in the light of the pertinent literature. Drawing on interdisciplinary insights, this chapter questions the implications of numerical feminization in the context of gendered cultures and processes of work in the information technology (IT) sector, and proposes some directions for future research.

WOMEN IN ICT SECTOR

After the information technology boom in the 1990s, an extremely high demand for IT workers occurred in the United States (Little, 1999; Forson&Ozbilgin, 2003). There was an increase in the number of women using Information Communication Technologies (ICTs). But the same cannot be said for the *working* women in ICT professions. It is possible to have an optimistic view when we talk about women *and* ICT (women as users), but the pessimistic view becomes more dominant when we talk about women *in* ICT (women within the ICT professions) (Faulkner&Lie, 2007).

Women account for about 25% of technology workers in the European workforce, and about 20% in the United States workforce. While the majority of women are employed in routine and speciality requiring work fields, men are engaged in analytical and managerial fields. Also only 5% of upper management posts in IT industry are held by women. Unfortunately, the pace is changing rather slowly in women's access to jobs in ICT sectors (US Department of Labor, 1975–1990).

There are several pioneering women professionals in the history of IT. For example, 150 years ago Ada Lovelace developed the conceptual framework for programming; 60 years ago six women programmed the first electronic computing machine, the ENIAC; 50 years ago Grace Hopper developed the first compiler. But today, only a small proportion of highly paid and interesting jobs are held by women in this sector (Gutek, 2006).

Although there is a tendency to consider gender as a minor issue in adoption of technology, access and use of computers; today there is recognition that engagement with ICT is still gendered in several ways (Selwyn, 2007). In other words, despite earlier suggestions that IT would be less sex-typed than other sectors, today we can see that IT is as sex-typed as other traditional sectors (Gutek, 2006).

REASONS FOR UNDER-REPRESENTATION AND GENDER DISCRIMINATION IN IT SECTOR

IT is a male-dominated industry and most of the high technology oriented people are male (Horrigan, 2007). In other words, women are under-represented in IT field (Allen et al., 2006). This sector also suffers from a class divide between highly sought after knowledge workers and highly dispensable temporary workers, and women are placed at the lower proletariat end of the spectrum in the sector, so this class divide is also gendered (Gutek, 2006).

In most countries of the world, men outnumber women in science, technology, engineering and mathematics (STEM) fields (Lee, 2002). Some scholars have attempted to provide human capital arguments for under-representation of women in the sector. For example, Little (1999) argued that computer science demands mathematical skills which women are less qualified. Some scholars even revert to essentialist arguments about gender differences in terms of knowledge, skills and abilities and argue that, women do not possess requsite mathematical and spatial abilities in order to be successful in the field of IT. However, essentialist approaches to gender and IT are ill equipped to account for the significance of relations of power as well as social constructions of such relations which underpin how IT becomes gendered. Tatli et al. (2008) examined gendered organizational cultures and practices that create hostile climates for women and found that women continue to experience gendered outcomes in professional occupations including STEM fields. Wilson (2003) identifies a mismatch between the *culture of IT work* and the social construction of *suitable work* for women as a barrier of women's full inclusion to the sector.

Today, computer and communication systems are part of a daily life for majority (Kling, 2000). The proportion of women in computing area is also very low, and there are different reasons for it. For example, just seeing *computing* a masculine area can dissuade women from going into computing. And some other factors can be listed as gender stereotyping of subject choices at school, a lack of role models for women, a lack of confidence and experience with computers and the perception of computers as lacking social involvement (Forson&Ozbilgin, 2003).

Because of the culture that they have been raised, women are generally afraid of computers compared to men (Spotts et al., 1997). Perceptions are also important to create psychological discrimination. For example, perceptually some disciplines such as biology and psychology appeal more to girls, and some others such as engineering and physics more to boys (Lee, 2002).

Men and women are differing from each other in terms of accessing to information, to technology, and to scientific and technological careers. According to Hornig (1992) these differences are not arising solely from institutionalized discrimination but also from cultural differences. These cultural differences can be taken as a result of differential socialization and unequal life experiences, and the source of these cultural values can be located in a historically unequal division of labor.

Women also generally face hiring and promotional barriers in IT sector. Discrimination, the absence of mentoring, work-family challenges, and the structure of work environments are some constraints that prevent equal opportunities for hiring, promotion and advancement to senior management for women (Mujtaba, 2007). Also women are not offered full access and participation in the implementation process of IT and they feel inadequately informed about the process (Zauchner et al., 2000). They tend to receive lower salaries (Truman&Baroudi, 1994) and less favorable promotion chances than men (Igbaria&Baroudi, 1995), and face obstacles to advancement, quality mentoring and choice assignments that men may not encounter (Allen et al., 2004). According to a survey on information systems (IS) demographics, also the 'glass ceiling' keeps women in midlevel jobs in the IS departments (Bretts, 1993).

These reasons which impose barriers to women's advancement can also sometimes lead women to leave the IT positions as well as the field (Ahuja, 2002). Women leave the IT sector due to inflexibility of work schedules, imposition of family responsibilities, higher levels of work stress, discrimination, the problematic attitudes of upper management, supervisors and coworkers, and lack of consistency in workplace policies as they apply to women and men (Allen et al., 2004; Allen et al., 2006). In parallel with this, women generally receive more support from friends and family members when they switch out of SME majors compared to men and they feel more freedom to leave the major if it creates dissatisfaction (Lee, 2002).

To explain the role of women in the field of IT, Ahuja (2002) offers a life-cycle stage model. This model incorporates factors that constrain or enable womens' entry and advancement, and illustrates 'the pyramid structure' of women presence in the field of IT, and shows the effects of barriers for women across three separate dependent variables as *career choice, career persistence* and *career advancement*. These three dependent variables represent the three stages of an IT career. Ahuja (2002) suggests that there are several social and structural factors that affect women's careers in IT:

Social factors: These are social and cultural biases that incorporate both the internal view that women have of themselves (self expectations) and the external view of women (stereotyping) that is held by society in general.

Structural factors: Organizations sometimes can limit the opportunities of its own workers. Lack of role models and mentors and the existing proportion of women in the top management are examples of structural factors.

Gender discrimination in IT sector can be classified into two practices as *access* and *treatment*. While the *access* is related to discrimination in practices of recruitment, *treatment* is related to objective and subjective differences between experience of women and men (Truman&Baroudi, 1994).

THE RELATIONSHIP BETWEEN GENDER AND TECHNOLOGY

It is a common view that there is a complex and multifaceted relationship between gender and technology (Jansson et al., 2007). According to the studies which show the moderating effect of time between gender and technology, as the computers become more widely available and accessible in society and people become more familiar with computer technologies, the gendered effects of computer use can diminish (Selwyn, 2007).

Some people choose to stay as non-users in ICT (Faulkner&Lie, 2007). So it is also important to understand why some people prefer to use the Internet or IT technology as a daily life activity, and why others choose not to use it. This effort can also help to understand the use and adoption differences of information-based technologies, internet or ICTs. It could also be interesting to explore gender divisions among non-users. Indeed, a focus group study which is conducted with the Internet nonusers shows that, women are more likely to believe that the Internet is complicated and hard to understand (Dittmar et al., 2004).

Venkatesh and Morris' (2000) study was one of the first studies that focus on gender difference in IT adoption and usage in the workplace. The study of gender differences involved two competing theoretical approaches such as *essentialists* and *social constructionists*. When essentialists view gender-related attitude as something inside a person and consistent across situations and time, social constructionists regard attributes of gender as a product of social conventions and agreements. In their model, Venkatesh and Morris (2000) adopt the essentialist approach in explaining gender differences. The essentialist theories regard gender as an independent variable to explain the differences in the perceptions and behavior between men and women, and constructionists regard gender as the dependent variable and conceptualize that the perception and behavior of men and women often depend more on with whom they are interacting and their particular situation than on anything intrinsic about the gender they are (Cheung et al., 2002). The weakness of the essentialist approach is that it is predicated on some false stereotypes and unfounded assumptions of sex differences. The social constructionist view tends to reject such simplistic frames that essentialist scholarship provides. Instead, they view gender differences as an outcome of social processes and routines of meaning making.

CAPABILITIES AND SKILLS IN IT FIELD

In most economies, there are often links between capabilities, skills and career outcomes of individuals (Faulkner&Lie, 2007). Also IT sector requires special skills and capabilities. But although there is a need for skilled IT workers, many of them are currently underemployed and are using only a limited range of skills repertoirs. Under-employment is a particularly female phenomenon in the IT sector, where women are required to 'prove themselves' and often are frustrated as a result of being given work below their capabilities (Gutek, 2006).

There are also gendered perceptions of skills and capabilities that affect women in the sector. IT and ICT use and especially computing are often viewed as technical and masculine areas of work, which can lead women to feel disempowered and possibly excluded (Dittmar et al., 2004; Shih, 2006). Similarly, women are excluded from software development as this work is considered to be 'too technical' and a man task. Women and men software developers differ in focus on the client's requirements versus product and process aspects. Indeed, according to essentialist researchers, women are better able to understand clients because of their superior communication skills. So managers who also hold such essentialist assumptions can prefer to employ women in socially oriented tasks such as project or quality management (Eriksson-Zetterquist, 2007).

The proportion of women entering computer science and engineering courses is static or in decline in most western countries. There is also a contradictory picture in most Eureopean countries (Faulkner&Lie, 2007). So it is critical to analyse the causes and consequences of the gender gap in the ICT field, rather than reifying gender stereotypes in the sector.

ALL CHANGE: IT, INTERNET, AND IT WORK

New technologies mean new work opportunities. Thus, IT and e-commerce have the potential to create jobs, and change the nature of work. For example, telework and home businesses are two phenomena, which are closely linked to the capabilities of IT and the growth of e-commerce (OECD, 1999).

Women's advancement to positions with more power and authority in organization and their interest in new technologies are relatively new phenomena. Women with higher aspirations of work and career, and better education will have a greater propensity to use new and advanced technologies compared to the past (Eriksson-Zetterquist, 2007).

As a consequence of this, ICTs become more widely used as a daily part of social and economic life, and the image of computers is becoming less 'techy' and 'only for men'. Especially computer manufacturing and design companies pay attention to the untapped potential of women buyers (Faulkner&Lie, 2007). These efforts are important because we can predict that, as the proportion of women ICT users increase, the proportion of working women in ICT professions may increase.

Everyday life gains a "uni-sex" feature especially with the internet (Selwyn, 2007). This gender equality seems as a consequence of the rapid diffusion of information technologies and internet into the society (Eriksson-Zetterquist, 2007).

Internet, the technological innovation of twentieth century, was first conceptualized in 1974 as a "network of networks" (Loebbecke& Wareham, 2003) and became an important tool for developing new businesses (Forson&Ozbilgin, 2003). Internet can connect people to libraries, schools, and shopping centers, and can be a home entertainment, and become a basic interpersonal communication technology (Kraut et al., 1999). Technology intensive sectors are male dominated and do not create much opportunities for women, but Internet, a sub-division of the technology sector, seems to attract more women (Forson&Ozbilgin, 2003).

While some researchers claim that the Internet use contributes to well-being because it provides some opportunities such as connection and access to information, others argue that the Internet use isolates people from the society and block the social relations and also it makes difficult to find the real information (Jackson et al., 2001).

According to Friedman (2005), the internet has a potential to democratize gender equality advocacy by making rapid interaction and information exchange easier. These easy communication opportunities lead to expand feminist ideals of more horizontal relations within the organizations.

Most of the people have access to internet during the whole day. This easy and wide access proliferate some issues such as working at home, communication, entertainment, and other personal uses (Kling, 2000). As the proportion of personal computing and internet access increase, dramatical changes in people's lives occur and females gain more access to the technology than ever before (Knupfer, 1998; Kraut et al., 1999).

Internet as a multi-faceted technology is accepted as a 'feminine' technology, and it is thought that this feminine image can create conducive conditions for women to get more involved. However no single factor accounts for the increase in self-employment, particularly women's self-employment in internet sector. Cultural attitudes, attitudes towards autonomy, flexibility in working hours and a willingness to take economic risks must not be discounted when considering the setting in which women's entrepreneurship flourished in the internet sector (*cited in* Forson&Ozbilgin, 2003).

Indeed, Braten and Stromso (2006) explain that the levels of participation in internet-based communication activities of males are higher than females. On the other hand, the number of women using computers and involving in IT professions is increasing as a result of the rapid development of the internet (Forson&Ozbilgin, 2003).

There are some gender based differences in internet use. According to Jackson et al. (2001), women and men use the internet equally but differently; females use e-mail, and males use web more than each other, and females feel more computer anxiety, less computer self-efficacy, and less favorable and less stereotypic computer attitudes than males.

Females' behavior of using e-mail more than males is thought to be related with their stronger motive for interpersonal communication, and males' behavior of using web more than females is thought to be consistent with their stronger motive for information (Jackson et al., 2001). Females being heavier e-mail user and being lighter users of the web is also supported by some studies (Kraut et al., 1999; Gefen&Straub, 1997).

E-mail links people to each other and reinforces relationships as a personalized and convenient tool. It is faster than postal mail and has lots of utilities such as automatic replies, distrubition lists, and computerized search and so on. As a form of written conversation, people find communication through e-mail to be relatively interactive compared to classical tools (Kraut et al., 1999).

Whereas e-mail sustains ongoing dialogues and relationships, web information has more bounded properties. And the goal is satisfied with one or few visits in web. E-mail is also psychologically interesting because of containing partial information and personally relevant reinforcement that people want to explore further. For women, using e-mail more than men is definitely related to the need for interpersonal communication (Kraut et al., 1999). Also there is also a basic factor that can affect people's internet or technology use such as to have a computer at home (Jackson et al., 2001).

Whereas men's technology usage decisions are strongly influenced by their perceptions of usefulness, women are more strongly influenced by perceptions of ease of use opposite of men. Also subjective norm effects women's technology decisions, although the effect diminished over time. That means as women consider normative influences more than men, subjective norm does not influence men's decision at any point of time. Briefly, men are more driven by instrumental factors (perceived usefulness) and women are more motivated by process (perceived ease of use) and social (subjective norm) factors (Venkatesh&Morris, 2000).

Indeed, some researchers state that gender gaps in internet use were more pronounced until a few years ago, but appear to be closing now.

For instance, UK users were 85% men and 15% women in 1998, but by 2000 women constituted half of the UK on-line population which followed a similar trend in the United States. Although the proportion of female internet users has become higher, female users may not view the experience as positively as their male counterparts. A stereotypically masculine culture has developed around computer use (*cited in*Dittmar et al., 2004).

CONCLUSION AND STRATEGIES TO INCLUDE WOMEN IN ICT PROFESSIONS

In this chapter, we discussed gender and IT, and provided evidence from research studies to show how IT sector is gendered. To understand these gender differences are also important to make managers and co-workers realize that "the same mode of communication may be perceived differently by the sexes" (Gefen&Straub, 1997, p.398). We also identified essentialist and social constructivist approaches to understanding the role of gender in the sector, and contend that social construction of gender presents a more conducive method for understanding gender and IT as dynamically constructed and changing over time and place. Reflecting on the multidimensional nature of the way IT is gendered, we offer a number of strategies to improve the position of women who are marginalized and excluded from positions of power and influence, and demarcated in terms of access to certain fields and aspects of IT sector below.

Faulkner and Lie (2007) state that, some interventions are needed to deal with gender gaps in digital inclusion. Although there are some efforts to attract and keep women and minorities in computing and IT area, their retention and advancement continues to be a significant challenge (Ahuja, 2002). As a result, companies reassess practices that may lead to turnover, including those related to balancing family and work life (ibid.).

In their study Faulkner and Lie (2007) present the main findings of a European Project titled 'Strategies of Inclusion: Gender and the Information Society (SIGIS)'. This project, including case studies of different types of inclusion strategies from Ireland, Scotland, The Netherlands, Italy and Norway, aims to explore initiatives to include women in ICTs. The SIGIS case studies indicate that there is not a unique and one-size-fits-all strategy to improve gender inclusion in the information society. And as there are different technologies, technology users, and different social groups and settings, multiple and variable strategies are required.

Various cultures and countries provide examples of how IT integration can be effectively accomplished (Mujtaba, 2007). There is a serious lack of policies and programmes in most countries with respect to gender and ICT. In the UK, there are government efforts to get more women into ICT, but gender is virtually absent from some discourses about the digital divide and in Norway public policies are frequently scrutinized to assess whether they help to promote gender equality (Faulkner&Lie, 2007).

Also advertisements sometimes can reinforce the gender stereotyping and stereotypical vision of computers as a male domain. To put the male characters into the authority roles especially businessmen who uses technology and computers, and gains more power, and to put the women into the roles of beautiful and friendly who don't suggest achievement and always placed in the background are examples for that. As a suggestion, these advertisements can place both men and women working together and using technology productively, not to foster gender stereotyping (Knupfer, 1998).

Social identity of designers and desing practices in ICT should also be taken into account because of some design practices can prioritize male users. So involving more women in ICT design process or at least equally presenting with men will contribute to the equality in ICT sector (Oudshoorn et al., 2004).

Hermawati and Luhulima (2000, p. 99-100) listed some suggestions to enhance women participation in Science, Engineering and Technology (SET) field in their study which is conducted in the context of Indonesia: "(a) There should be a strong policy and action to enhance women's participation in SET education, R&D and production activities and other SET careers, (b) SET application can be beneficial to disadvantaged women if the technologies to be introduced are related to the specific practical needs of women and the community at large... (c) A gender perspective should be socialized and integrated in the policy- and decision-making process in SET development and SET application for development. (d) More research and studies should be conducted on the sociocultural factors as well as practical and strategic needs of disadvantaged women in rural and urban areas. The results of this study should be used by policy- and decision-makers in SET development and application for sustainable development. (e) Study of the role of women in SET activities should be expanded, by encouraging and developing gender-sensitive statistical data base and indicators, in the public and private sectors, including the industrial sector and educational institutions. (f) There has to be an explicit commitment to ensure that women have equal opportunities to participate in SET programs. Related to SET activities, the government has to support women's micro enterprise activities through facilitating funding and environmentally-sound technology or other relevant SET innovations. Policies concerning women's access to information and communication should support a national gender and SET policy. (g) Continuous monitoring and evaluation of SET projects and programs based on gender impact assessment are necessary."

Also training is the most popular mechanism used to smooth the transition to new technology and represents the key method for successful knowledge transfer to users, and diffusion of new technologies in the workplace (Venkatesh&Morris, 2000). To have an equal training opportunity for both genders is an important requirement for the workplace.

Loyd and Gressard (1984) list computer anxiety, self-confidence, and pleasure and interest in working with computers as aspects of attitudes toward computers. Some scholars (Durndell et al., 1995; Comber et al., 1997) draw attention to the importance of experience because of the research findings showing when experience is taken into account, the differences between girls and boys related to these aspects (except self-confidence) becomes smaller (Volman&Eck, 2001). Hattie and Fitzgerald (1987) note that, computer performance is related to the computer experience, and students who generally work with computers perform better than students with less experience, and boys generally spend more time in computing activities than girls (Volman&Eck, 2001). Experience is in relation with the access to technology. People who access to technology will spend more time with it, and show more positive attitudes towards it. So it can be said that significant gender differences can exist in regard to equality of access and performance outcomes (Spotts et al., 1997).

According to Faulkner and Lie (2007), access to digital economy is constrained by economic class of individuals. Lower income groups are digitally excluded. The authors argue that simply widening access to ICT does not result in improved access. Development of ICT skills and knowledge is also important for enabling access. Similarly, it is very important to bring the technology to 'where people are' and as some authors argue, "equal access to the Internet is the way to a better world" (Faulkner&Lie, 2007).

As a last sentence, it is believed that there should be more women in IT sector, because women can contribute to the field with their different approaches to work. It is also should be thought as a gender equality issue and an ethical problem. Being a sector which ensures just gender equality is a crucial factor. Giving equal opportunities to every member of society is a requirement of a good ethical effort naturally. So each nation should proactively define educational policies that embrace female population in IT (Mujtaba, 2007).

REFERENCES

Ahuja, M. (2002). Women in the information technology profession: A literature review, synthesis, and research agenda. *European Journal of Information Systems, 11*, 20–34. doi:10.1057/palgrave/ejis/3000417

Allen, M. W., Armstrong, D. J., Riemenschneider, C. K., & Reid, M. F. (2006). Making sense of the barriers women face in the information technology work force: Standpoint theory, self-disclosure, and casual maps. *Sex Roles, 54*, 831–844. doi:10.1007/s11199-006-9049-4

Allen, M. W., Reid, M., & Riemenschneider, C. (2004). The role of laughter when discussing workplace barriers: Women in Information Technology jobs. *Sex Roles, 50*(3/4), 177–189. doi:10.1023/B:SERS.0000015550.92555.7e

Braten, I., & Stromso, H. I. (2006). Epistemological beliefs, interest, and gender as predictors of internet-based learning activities. *Computers in Human Behavior, 22*, 1027–1042. doi:10.1016/j.chb.2004.03.026

Bretts, M. (1993). She shall overcome. *Computerworld, 27*, 67–70.

Cheung, C. M. K., Lee, M. K. O., & Chen, Z. (2002). Using the Internet as a learning medium: An exploration of gender difference in the adoption of FabWeb. In *Proceedings of the 35th Hawaii International Conference on System Sciences*.

Comber, C., Colley, A., Hargreaves, D. J., & Dorn, L. (1997). The effects of age, gender and computer experience upon computer attitudes. *Educational Research, 39*(2), 123–133. doi:10.1080/0013188970390201

Dittmar, H., Long, K., & Meek, R. (2004). Buying on the Internet: Gender differences in on-line and conventional buying motivations. *Sex Roles, 50*(5/6), 423–444. doi:10.1023/B:SERS.0000018896.35251.c7

Durndell, A., Glissov, P., & Siann, G. (1995). Gender and computing: Persisting differences. *Educational Research, 37*(3), 219–227. doi:10.1080/0013188950370301

Eriksson-Zetterquist, U. (2007). Editorial: Gender and new technologies. *Gender, Work and Organization, 14*(4), 305–311. doi:10.1111/j.1468-0432.2007.00345.x

Faulkner, W., & Lie, M. (2007). Gender in the Information Society: Strategies of inclusion. *Gender, Technology and Development, 11*(2), 157–177. doi:10.1177/097185240701100202

Forson, C., & Ozbilgin, M. (2003). Dot-com women entrepreneurs in the UK. *Entrepreneurship and Innovation,* February, 13-24.

Friedman, E. J. (2005). The reality of virtual reality: The internet and gender equality advocacy in Latin America. *Latin American Politics and Society, 47*(3), 1–34. doi:10.1353/lap.2005.0034

Gefen, D., & Straub, D. W. (1997). Gender differences in the perception and use of e-mail: An extension to the Technology Acceptance Model. *Management Information Systems Quarterly, 21*(4), 389–400. doi:10.2307/249720

Gutek, B. A. (2006). Book Reviews ['Doing IT: Women working in Information Technology' by K. Scott-Dixon, Sumach Press, 2004.]. *Gender, Work and Organization, 13*(6), 621–623. doi:10.1111/j.1468-0432.2006.00325_1.x

Hattie, J., & Fitzgerald, D. (1987). Sex differences in attitudes, achievement and use of computers. *Australian Journal of Education, 31*(10), 3–26.

Hermawati, W., & Luhulima, A. S. (2000). Women in science, engineering and technology (SET): A report on the Indonesian experience. *Gender, Technology and Development, 4*(1), 87–100. doi:10.1177/097185240000400104

Hornig, S. (1992). Gender differences in responses to news anout science and technology. *Science, Technology & Human Values, 17*(4), 532–542. doi:10.1177/016224399201700406

Horrigan, J. B. (2007). A typology of information and communication technology users. *PEW Internet & American Life Project Report.* Retrieved from http://www.pewinternet.org

Igbaria, M., & Baroudi, J. J. (1995). The impact of job performance evaluations on career advancement prospects: An examination of gender differences in the IS workplace. *Management Information Systems Quarterly, 19*, 107–123. doi:10.2307/249713

Jackson, L. A., Ervin, K. S., Gardner, P. D., & Schmitt, N. (2001). Gender and the Internet: Women communicating and men searching. *Sex Roles, 44*(5/6), 363–379. doi:10.1023/A:1010937901821

Jansson, M., Mörtberg, C., & Berg, E. (2007). Old dreams, new means: An exploration of visions and situated knowledge in information technology. *Gender, Work and Organization, 14*(4), 371–387. doi:10.1111/j.1468-0432.2007.00349.x

Kling, R. (2000). Learning about information technologies and social change: The contribution of social informatics. *The Information Society, 16*(3), 217–232. doi:10.1080/01972240050133661

Knupfer, N. N. (1998). Gender di visions across technology advertisements and the www: Implications for educational equity. *Theory into Practice, 37*(1), 54–63. doi:10.1080/00405849809543786

Kraut, R., Mukhopadhyay, T., Szczypula, J., Kiesler, S., & Scherlis, B. (1999). Information and communication: Alternative uses of the internet in households. *Information Systems Research, 10*(4), 287–303. doi:10.1287/isre.10.4.287

Kusku, F., Ozbilgin, M. F., & Ozkale, L. (2007). Against the tide: gendered prejudice and disadvantage in engineering study from a comparative perspective. *Gender, Work and Organization, 14*(2), 109–129.

Lee, J. D. (2002). More than ability: Gender and personal relationships influence science and technology involvement. *Sociology of Education, 75*(4), 349–373. doi:10.2307/3090283

Little, J. C. (1999). *The role of women in the history of computing* (pp. 202–205).

Loebbecke, C., & Wareham, J. (2003). The impact of e-business and the information society on 'strategy' and 'strategic planning': An assessment of new concepts and challenges. *Information Technology Management, 4*, 165–182. doi:10.1023/A:1022946127615

Loyd, B., & Gressard, C. (1984). The effects of sex, age and computer experience on computer attitudes. *AEDS Journal, 18*(2), 67–77.

Mujtaba, B. G. (2007). *Workforce diversity management: Challenges, competencies, and strategies.* Llumina Press.

OECD. (1999). *The economic and social impact of electronic commerce* (pp. 1–168). Preliminary Findings and Research Agenda.

Oudshoorn, N., Rommes, E., & Stienstra, M. (2004). Configuring the user as everybody: Gender and design cultures in information and communication technologies. *Science, Technology & Human Values, 29*(1), 30–63. doi:10.1177/0162243903259190

Selwyn, N. (2007). Hi-tech=Guy-tech? An exploration of undergraduate students' gendered perceptions of information and communication technologies. *Sex Roles*, *56*, 525–536. doi:10.1007/s11199-007-9191-7

Shih, J. (2006). Circumventing discrimination: Gender and ethnic strategies in Silicon Valley. *Gender & Society*, *20*(2), 177–206. doi:10.1177/0891243205285474

Spotts, T. H., Bowman, M. A., & Mertz, C. (1997). Gender and use of instructional technologies: A study of university faculty. *Higher Education*, *34*(4), 421–436. doi:10.1023/A:1003035425837

Tatli, A., Ozbilgin, M. F., & Kusku, F. (2008). Gendered occupational outcomes: The case of professional training and work in Turkey. In J. Eccles & H. Watt (Eds.), *Explaining Gendered Occupational Outcomes*. Michigan: American Psychological Association (APA) Press.

Truman, G. E., & Baroudi, J. J. (1994). Gender differences in the information systems managerial ranks: An assessment of potential discriminatory practices. *Management Information Systems Quarterly*, *18*, 129–141. doi:10.2307/249761

Venkatesh, V., & Morris, M. G. (2000). Why don't men ever stop to ask for directions? Gender, social influence, and their role in technology acceptance and usage behaviour. *Management Information Systems Quarterly*, *24*(1), 115–139. doi:10.2307/3250981

Volman, M., & Eck, E. V. (2001). Gender equity and information technology in education: The second decade. *Review of Educational Research*, *71*(4), 613–634. doi:10.3102/00346543071004613

Wilson, F. (2003). Can compute, won't compute: Women's participation in the culture of computing. *New Technology, Work and Employment*, *18*(2), 127–142. doi:10.1111/1468-005X.00115

Zaucher, S., Korunka, C., Weiss, A., & Kafka-Lützow, A. (2000). Gender-related effects of information technology implementation. *Gender and Information Technology*, *7*(2), 119–132.

Chapter 7
The Impact of Gender in ICT Usage, Education and Career:
Comparisons between Greece and Germany

Bernhard Ertl
Universität der Bundeswehr München, Germany

Kathrin Helling
Leopold-Franzens-Universität Innsbruck, Germany

Kathy Kikis-Papadakis
Institute for Applied Computational Mathematics, Greece

ABSTRACT

Gender is an important issue in the context of information and communication technologies (ICT). Studies show that ICT use is subject to gender bias, e.g. in relation to ICT use and interests. This contribution describes the current situation of gender and ICT professions in Germany and Greece. Based on an empirical study, it shows particular areas in ICT education that suffer from gender inequalities in both countries. Furthermore, the chapter elaborates how gender inequalities develop from secondary to professional ICT careers based on statistics from Germany and Greece.

INTRODUCTION

Germany and Greece have different economic structures. In Germany, 72 percent of the work force relate to services, 27 percent to industry and less than one percent to agriculture. Thus, Germany has a industrial history with huge exports of industrial goods, mainly machines, vehicles and chemicals (CIA, 2010a). In contrast, Greece has a stronger agricultural tradition: 65 percent of the work force relate to services, 22 percent to industry, and 12 to agriculture. This results in comparably few exports of mainly food and beverages, and manufactured goods (CIA, 2010b). Comparing both countries, Greece's exports value about 1.5 percent of the German exports.

DOI: 10.4018/978-1-60960-759-3.ch007

As a consequence of recession-driven economic development, the information and communication technology sector has weakened in recent years in OECD countries. Still, an ICT growth of about 4 percent was observed in 2008. Although there was a decrease in 2009 due to the current financial crisis, a general upturn is expected in the long-term because of constant development of the ICT services, software, products for Internet use and communication, and infrastructure. The ICT skills of the work force contribute to the growth: the overall share of employees in ICT specialist occupations is 4 percent and increasing rapidly, and 20 percent of employment relates to occupations that use ICT extensively (OECD, 2008). The report shows that Germany is ahead of Greece with regard to the development of the ICT sector, e.g. Germany has a tertiary share of top performing ICT companies, it is spending a greater amount in ICT market segments, it has a tertiary share of ICT services in total business services, and a tertiary share of ICT employees in the business sector, and so forth.

Apart from these differences in the economic structure and ICT business sector development, the educational systems of both countries can be compared. Even though there are many differences, for example with regard to the education policy developments in both countries, or the educational attainment of the population, both countries have a similar humanistic secondary tradition with an average schooling expectancy of 16 years in Germany and 17 years in Greece. In addition, the public educational expenditures are on a similar level, 4.4 percent of the GDP in Greece and 4.6 percent in Germany (CIA, 2010a; 2010b). In this context, it is interesting to note how far gender differences reach in ICT usage, in ICT education at secondary level, in tertiary education, and in ICT-related careers in both countries. Currently, females profit less than men from the growth and employment prospects of the ICT sector, and approaches to facilitate girls at secondary level to take up ICT-related careers should be considered.

Germany and Greece have recently made progress with respect to reducing gender inequalities in using information and communication technologies. An increasing number of citizens are using ICT, and in the context of Internet usage recent studies were able to show an increase in all age groups for both countries (see Initiative D21, 2008; Observatory for the Greek IS, 2008). However, the studies also reveal that the number of male Internet users steadily exceeds the number of female users, and the gap increases with the user age.

At secondary level, self-efficacy and competencies for computer use follow the traditional western stereotype (to exemplify, see OECD, 2005): boys show more confidence than girls in performing computer tasks.

Educational trajectories from secondary to university and the uptake of careers in ICT-related professions increase gender imbalance (European Commission, 2006): Particularly in the context of ICT, the higher the level of professional development, the lower the percentage of women involved. Even if women are employed in ICT professions, they earn less for their work than their male colleagues.

In this chapter, gender differences in ICT usage behaviour, in interests and self-confidence levels at secondary, in university enrolments and professional career choice are compared between Germany and Greece and reflected in the context of existing literature and studies. Specifically, Internet usage statistics in German and Greek households are described from a gender perspective. Further, gender differences exist in the educational system especially in ICT-related subjects at secondary and university. The ICT interests and self-confidence of pupils at secondary is analysed, and tertiary education statistics of computer science and other related subjects are described. Career pathways in Germany and

Greece show an underrepresentation of women in ICT professions.

Data was retrieved from German and Greek national statistical reports, as well as from European comparative studies. Differences in ICT usage by girls and boys at secondary are also reported on the basis of a review of existing research data, and complemented with results from a survey conducted by the Universtät der Bundeswehr München and FORTH/IACM in secondary schools in Germany and Greece.

By this approach the chapter puts gender differences in ICT-related careers in an overall context, taking into account aspects of secondary education and tertiary education, as well as artefacts of a digital divide in ICT usage outside the educational system. The comparison between Germany and Greece allows the identification of similarities and differences which are considered in the context of practical implications for secondary education and the aim to increase the number of women who use ICT or choose a career in this field. This chapter intends to answer the following questions:

- What is the impact of gender on ICT usage in Germany and Greece?
- What is the impact of gender on the interest of pupils at secondary level in aspects of ICT?
- Are gender-specific usage patterns and interests in ICT at secondary level reflected in the career choices of men and women?
- What is the impact of gender on ICT-related career choice at tertiary level and in professional life?
- What can we learn from a comparison of Germany and Greece?

GENDER AND ICT USAGE

Differences in the usage of ICT – using Information- and Communication Technology (e.g. computers, the Internet) – can be found for various characteristics such as age, education, sex, and region. A study by the European Community in 2004 (Eurostat data; see Demunter, 2005) reveals that differences in ICT usage are less to do with gender than age and education. Still, the digital divide is reflected in gender differences of ICT-usage patterns in Germany and Greece. The following section describes usage statistics of men and women, and reflects on the impact of gender on the usage behaviours in both countries.

Statistical data of ICT usage is available on Germany and Greece from several sources, e.g. Eurostat data and national surveys. Yet, the fact that comparability of national data sets is limited due to differing samples, time frames, and definition of the survey items must be considered.

Furthermore, the availability of Internet access in German and Greek households, as well as broadband penetration, differs between both countries and regions within those countries. Eurostat data show that German households (with at least one member aged 16-74 years) are technically advanced in comparison with Greek households. In 2009, 79 percent of German households and 38 percent of Greek households have Internet access; the number of broadband connections per 100 inhabitants in Germany and Greece is 29.4 and 15.6 respectively (see Eurostat, 2009a, 2009b).

Data Review

European Comparison

According to Eurostat, the EU-25 average of computer usage (computer use on average once a week in the last three months) of women is 54 percent, and of men is 61.3 percent. These results differ between age groups: younger people are more active in using computers than older ones. In all age groups, however, the proportion of males exceeds that of females, although this difference is higher in the older age groups (Seibert, 2007; see Table 1).

Table 1. Proportion of women and men who used a computer at least once a week in the last three months by age, 2006

Age	16-24		25-54		55-74	
Sex	Women	Men	Women	Men	Women	Men
EU-25	81	83	60	66	21	35
Germany	91	93	77	81	33	49
Greece	65	69	33	43	3	11

Table 2. Male and female Internet users from 2005 to 2007 in percent points (Initiative D21, 2008; Observatory for the Greek IS, 2009)

Year	Germany		Greece	
	Women	Men	Women	Men
2005	47.6	63.2	12.3	23.6
2006	51.5	65.4	15.9	27.5
2007	53.8	67.1	19.0	31.5
2008	58.3	72.4	26.1	41.9

The usage data also shows the same patterns with regard to age and sex for Germany and Greece. However, in all age groups, Germany is clearly above the European average, and Greece is below it.

National Surveys

The following data from the (n)onliner-study of the Initiative D21 (2008) in Germany and the annual study (period 2005-07) on the use of Information and Communication Technologies in Greece (Observatory for the Greek IS, 2008) provide a statistically representative overview of the Internet usage in German and Greek households with a focus on the digital divide between female and male users.

The 2008 data of the (n)onliner-study is based on more than 50,000 computer-supported telephone interviews with German-speaking inhabitants aged 14 years and over, in the years 2005-2007. Internet usage was defined as using the Internet, independent of location and reason of usage, during the last 12 months.

The data on individuals' Internet usage in Greece by the Observatory for the Greek IS (2009) is also based on annual telephone interviews. In 2008, about 5,966 randomly selected households, with at least one member aged 16-74 years, distributed pro rata among the 13 Greek Regions, were contacted, and this resulted in 2,316 interviews with individuals having used the Internet at least once a week over the last 3 months.

On the basis of these studies, increased use of computers and the Internet can be observed for both men and women in both countries. Although the percentage of women using ICT is growing faster than that of men, there is still a significant gender gap. That gap seems to get smaller with time.

Furthermore, Table 2 shows that Internet use has penetrated German households much more at this time than Greek households. This result reflects the above reported Eurostat data of Internet access and broadband connection in Germany and Greece.

Table 3 presents detailed figures with respect to specific age groups in Germany and Greece. The data refers to a differing basis in each cate-

Table 3. Internet users by age groups in percent points, in 2008 (Initiative D21, 2008; Observatory for the Greek IS, 2009)

Age (DE)	Germany		Age (EL)	Greece
	Women	Men		Total
14-19	94.3	93.2	16-24	76.5
20-29	88.6	90.9	25-34	59.9
30-39	82.5	88.2	35-44	43.3
40-49	73.5	82.9	45-54	28.2
50-59	57.5	69.6	55-64	12.5
60-69	32.4	51.7	65-74	3
70+	9.2	27.2	-	-

gory for both countries. However, the comparison shows that the higher share of Internet users in Germany (see Table 1) is valid across all age groups. Furthermore, one can see that Internet use decreases with age in both countries. The figures reveal that the difference between age groups within one country is much stronger in Greece than in Germany; while Internet usage for Germans aged 50-59 is more than half, only an eighth of Greeks in the same age category uses the Internet.

It is worth mentioning that in the young cohorts, gender discrepancy in ICT use is rather small. For example, in 2007 the 10-15 years age group in the major Athens area shows almost no differences: 94 percent of the boys compared to 91 percent of the girls use computers, and 70 percent of the boys compared to 76 percent of the girls use the Internet (Observatory for the Greek IS, 2008). The situation in Germany is comparable: the (n)onliner study shows equal Internet usage statistics of female and male teenagers in the age group 14-19 years, with females being slightly ahead in 2008 (females: 94.3 percent, males: 93.2 percent; see Initiative D21, 2008; similar results were also discovered by Imhof, Vollmeyer, & Beierlein, 2007; and MPFS, 2008).

Reflection

To sum up, there is some progress with respect to gender equality in ICT usage, especially at the teenage level. Yet, Internet usage is just one aspect of ICT which needs to be considered from the perspective of gender differences. Therefore, it is important to have a closer look beyond usage statistics at skills and self-confidence with respect to particular fields of ICT. The following section will provide more detailed insights.

GENDER AND ICT IN SECONDARY

Due to the rapid development of new media and information technologies, the efforts of educating pupils in this field need to be monitored and evaluated continuously with up-to-date and relevant content. Media education and media literacy need to take into account the use of current information- and communication technology (ICT), and new digital media for teaching and learning (see e.g. BLK, 1987; European Commission, 2007; Gesellschaft für Informatik, 1999b), particularly as there are differences in the media use and competences of female and male pupils.

ICT in schools is mainly introduced by specific curricula. In *German* schools, digital literacy and ICT competence is taught in the context of media education, which also includes traditional media, and the subject of informatics. However, because of Germany's federal structure and the educational autonomy of each German state, there is no unified curriculum for media education or informatics across Germany. Informatics, for example, is part of the upper secondary curriculum in each state, but not always compulsory. Besides the differences in the respective state curricula, there is one common concept about what competencies in informatics pupils should have developed to obtain a university entrance degree (Abitur). In *Greece*, ICT has been integrated into the curriculum on all levels of education, mainly through the subject of Informatics. ICT is introduced into both levels of compulsory education (primary and lower-secondary) under the scope of providing students with opportunities to develop basic computer literacy as well as critical thinking skills, and to enhance their motivation for creative action at a personal and social level (Insight, 2007). It was recognized, however, that in everyday secondary life, ICT is present solely as a separate subject (not as a tool for other subjects). The skills intended to be developed in the frame of the Informatics lessons extent beyond what might be considered as basic computer skills as the curriculum calls for the use of standard applications and the Internet as well as programming and data base usage.

Method and Instruments

To get more insights into the issue of ICT interests and competencies at school in Greece and Germany, we have run an empirical study. In this study, the particular interests of males and females in upper secondary classes have been researched into. The instrument was a questionnaire which asked pupils about their particular preferences and interests in the context of digital literacy which was adapted from Huemer (2005). The pupils received the questionnaires in their respective mother tongue. 185 students participated in this study: 90 (48 males, 42 females) from Germany and 95 (48 males, 47 females) from Greece. The mean age of the sample was around 17 years. However, the German sample was slightly older on average than the Greek sample and therefore age was considered for all further analysis.

Results

Figure 1 shows the general interest in computer science asking students if they seek information about computers in their spare time in magazines or in the news. Results show that Greek pupils show significantly more interest. Furthermore, boys were shown to be more interested than girls in both countries. Besides, we found an effect of age indicating that older pupils were less interested than the younger ones. Considering the scale range (three items) from a minimum of 2.5 (not interested) to a maximum of 10 (very interested), one can see that the general interest was rather moderate in both countries (average of 4.61 in Germany, and 5.89 in Greece).

Besides the interest in computer science, another important aspect is the self-confidence of pupils regarding their ability in performing tasks on the computer (see Figure 2). Similar to the results with respect to interest, Greek pupils show more self-confidence than German pupils, even though the difference is not significant. Also, for this category, there was an effect that boys were more self-confident than girls in both countries. Besides, we found an effect of age indicating that older pupils were less self-confident than the younger ones. In contrast to the interest scale, pupils reported rather high level of self-confidence (average of 6.91 in Germany, and 7.90 in Greece).

In the following, we will focus on particular interests of pupils in four different aspects of computer science: hardware, programming, standard software, and Internet.

With respect to pupils' interest in hardware, we can see a stereotypical interest distribution for both countries (see Figure 3). Greek pupils show slightly more interest in hardware than German

Figure 1. Interest in computer science (N = 183; Min = 2.5; Max = 10; M = 5.25; SD = 1.75)

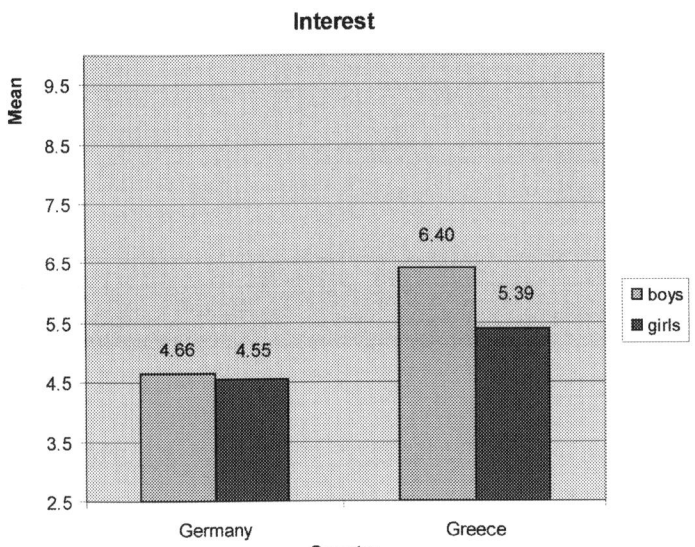

Figure 2. Self-confidence in computer science (N = 183; Min = 2.5; Max = 10; M = 7.39; SD = 1.37)

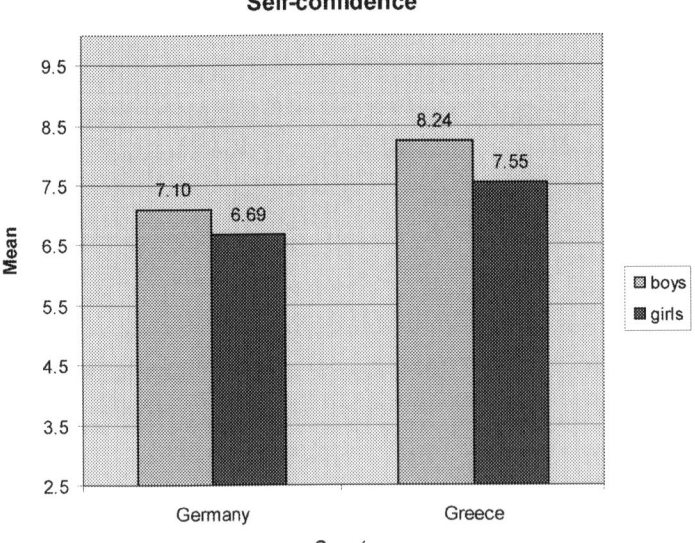

pupils. For this category, there was strong effect that boys were more interested in hardware than girls in both countries. Interest in hardware was not subject to age. In general, boys were rather or very interested in hardware, while girls were just moderately to rather interest.

Compared to hardware, pupils showed less interest in programming (see Figure 4). In this category, there were no significant differences between Germany and Greece, apart from an effect for sex: boys were more interested in programming than girls in both countries. Boys and girls were just moderately interested in program-

Figure 3. Interest in hardware (N = 183; Min = 2.5; Max = 10; M = 7.29; SD = 1.72)

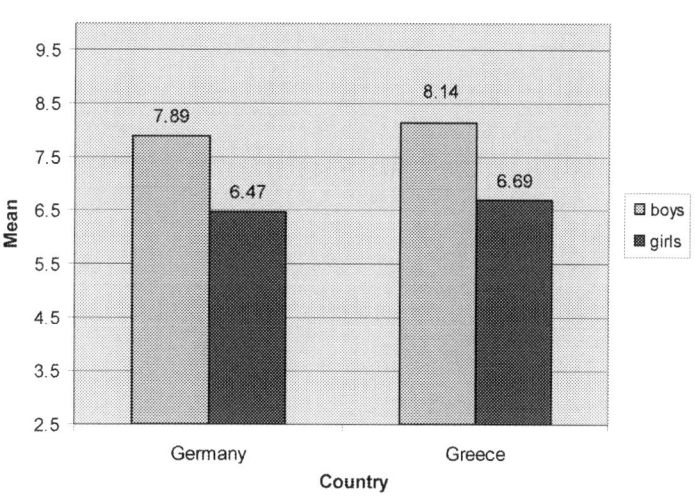

Figure 4. Interest in programming (N = 183; Min = 2.5; Max = 10; M = 6.42; SD = 1.91)

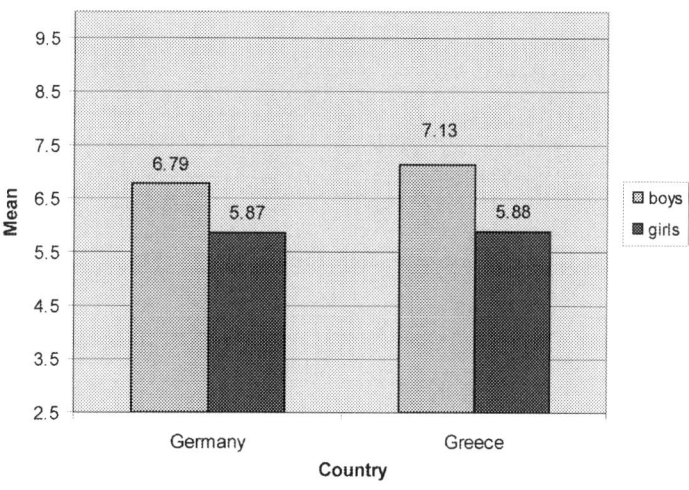

ming (average for Germany: 6.36, and Greece: 6.50); there was no effect regarding age.

With respect to standard software the gender difference has changed. Even though the difference was slight, girls showed more interest in standard software than boys (see Figure 5). Also, there were no significant differences between Germany and Greece in this category (average of Germany: 7.13, and Greece: 7.06), neither for age.

With respect to the Internet, we can descriptively see higher interest values for girls than for

Figure 5. Interest in standard software (N = 183; Min = 2.5; Max = 10; M = 7.03; SD = 1.92)

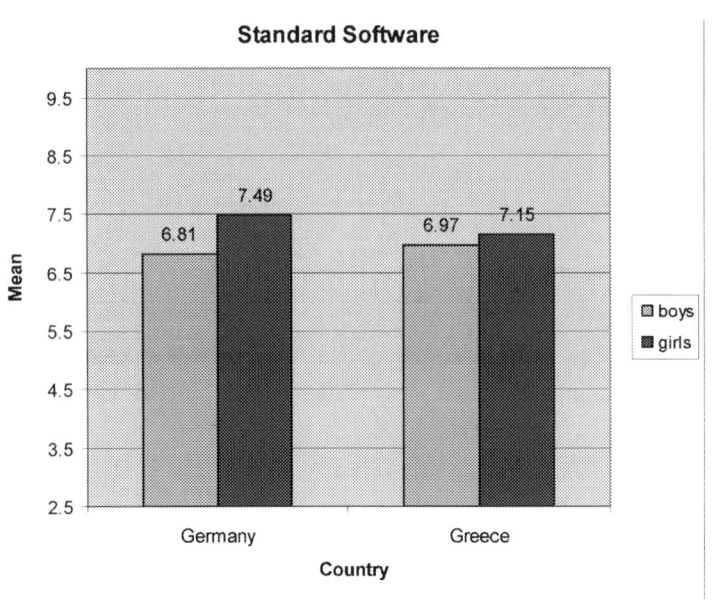

Figure 6. Interest in Internet (N = 183; Min = 2.5; Max = 10; M = 7.92; SD = 1.42)

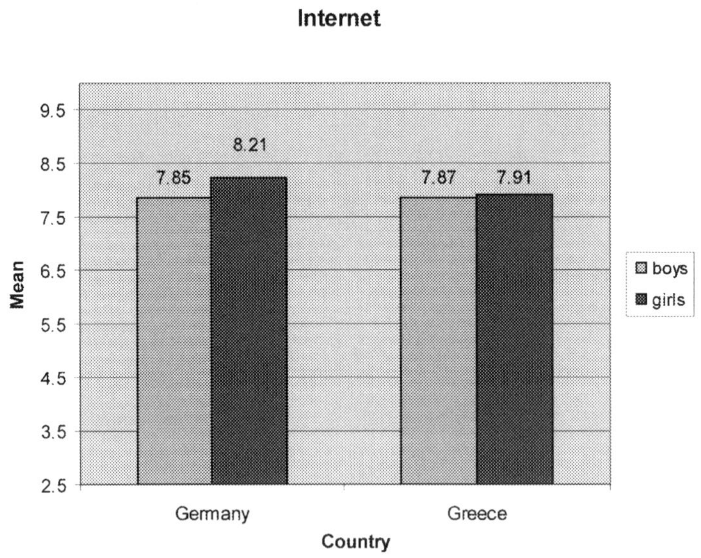

boys, and also for German pupils than for Greek pupils (see Figure 6). However, these descriptive differences are far from significance (average of Germany: 8.03 and Greece: 7.89). The pupils' age was not important for this category.

Reflection

The figures above reflect and support several other research findings from different countries. The British Educational Communications and Technology Agency (Becta, 2008) reports that implementing ICT in education improves motivation and attainment of boys and girls. However, this increase is higher for boys. This may correspond with the finding that girls use ICT at home for school work while boys use it for leisure activities (Becta, 2008; MPFS, 2008; Imhof, Vollmeyer, & Beierlein, 2007). According to Schnirch and Welzel (2004), ICT has a proportionally smaller share in girls' interests than in that of boys. This also reflects the lower interest rates of girls in our study. Furthermore, Becta (2008) emphasizes the factor of socio-economic background. In Great-Britain, this was found to be more an impact factor for girls' ICT access and use than for boys'. Considering that there is a higher ICT penetration in German households than in Greek households, one could assume that in Greece children of a comparable higher socio-economic background have access to computers than in Germany. This may be an aspect to explain the higher interest and self-confidence values in Greece.

The findings of Faulstich-Wieland and Nyssen (1998) in Germany report that parents estimate ICT skills as less important for girls. Also Becta (2008) found that girls prefer social and creative ways of using ICT (see also MPFS, 2008), collaborative work and the application of ICT for learning in formal and informal contexts (see also Munk, 2007), while boys prefer to use it for playing computer games (see MPFS, 2008). This goes along with the higher interest and self-confidence values for boys, and also with the higher interest in technical aspects like hardware and programming. In contrast, we were not able to find gender differences for the more creative and social aspects, like standard software and the Internet.

Models, e.g. the model of a student's career choice by Dick and Rallis (1991), describe how aptitudes and the cultural milieu influence the perception of socialisers and the interpretation of experiences, and thereby influence self-concepts and career values. The self-concepts may have an effect on the particular fields of interest, and therefore reduce the chance to build knowledge and skills in other areas. This means that girls' stereotypical socialization may hinder them to build the knowledge that could help to prove them digitally literate.

We were able to find out that younger pupils also had more general interest and more self-confidence in ICT than older ones. However, age had no influence in the different areas that we analyzed. This result, and the above reported usage statistics, allow for two different conclusions regarding age: younger people use computers more often and are more interested in ICT and have more self-confidence in this area. A correlation between these two aspects, high self-confidence and increased computer usage, can be presumed. According to Compeau and Higgins (1995) computer-self-efficacy is a mediator for computer usage, and findings from a study by Lopez and Manson (1997) support this presumption. However, we can also see that the interest in particular fields remains constant for age and country, and is mainly subject to gender.

To sum up, we have evidence for different ICT interest patterns for boys and girls. However, there is only little evidence for the stereotypical view of girls being less gifted than boys (see e.g. Becta, 2008) and therefore not suited to professional ICT careers. In the following sections, we will elaborate on how these differences go along with the boys' and girls' choice of ICT studies and ICT careers.

GENDER AND ICT AT UNIVERSITY

A university degree can be seen as an entrance ticket to a well-paid career in the ICT sector. In Germany, surveys of the Computerwoche magazine and Kienbaum consultancy show that the required qualification level for ICT professions is quite high and the majority of employees in the ICT sector graduated either at university or university of applied sciences (Fachhochschule "FH"; see Holzapfel, 2006). However, ICT studies (computer science, informatics, technology) at universities in Germany and Greece show again gender differences with regard to subject choices and graduation.

We consider the situation of gender and ICT at university on the basis of a review of existing surveys on national and European level. However, data of Germany and Greece is rather heterogeneous, with regard to the availability of methodological details and result analysis. Often, data is only available on an aggregated level, or on an exemplary basis, e.g. not differentiated by sex; not specifically for ICT studies but for science, technology, engineering and mathematics (STEM) in total, with a focus on specific universities only. Below, we will first describe the situation in each country, and then discuss the similarities and differences.

Data Review

European Comparison

According to Eurostat[1], Germany and Greece are among the European countries with the lowest percentage of women enrolled at university level. In 2007, 49.7 percent of the students in Germany were female, in Greece 50.4 percent of women studied. The EU-27 average is 55.2 percent. The number of maths, science, and technology (MST) graduates increased continuously in both countries, as was reported by the European Commission (2008) on the basis of Eurostat data for the years 2000-2005. However, Germany, and Greece are not among the top performing countries, and the share of women in MST graduates was 24.4 percent in Germany and 40.9 percent in Greece (EU average is 31.2 percent, in 2006).

ICT studies in Germany. According to data from the German Federal Statistics Office[2] in the winter semester 07/08 1.941.405 students were enrolled at German universities (including universities of applied science and "Verwaltungsfachhochschulen"); 47 percent of them were female. The number of enrolments of male students exceeds that of females. Nevertheless, since 2006 this proportion has changed for the total number of passed examinations. Fifty-one percent of women were among the 286.391 students who passed examinations in 2007. However, this development refers to teacher training and diploma graduation level only; other examinations such as PhD, university of applied science or Bachelor and Master degrees are still passed by a majority of men.

The number of men and women who started their first year at German universities in the winter semester 07/08 was nearly 50 percent respectively. Nevertheless, women decided on subjects according to traditional stereotypes. They enrolled for linguistic and cultural sciences, social science, or teacher training. Women are clearly underrepresented in engineering sciences and choose primarily selected subjects such as biology, architecture, interior design, or health sciences.

A report by the BMBF (2008) presents similar results about the distribution of men and women on different subjects at tertiary education level. Some subjects are clearly dominated by men, others by women (meaning that less than 30 percent of men or women study the subject). The following table (Table 4) displays the percentages of women beginning at university in STEM subjects for the year 2001 and 2002. We can see that the share of females in the selected subjects is clearly below the total of 50 percent reported for Mathematics, Information Technology, and Science.

The Hochschul Informations System (HIS) provides a study related to the whereabouts of Mathematics graduates, including mathematics teacher training, and compared to computer science and electrical engineering studies (Briedis et al., 2008). The data is based on analysis of official statistics from the Federal Statistical Office and on HIS surveys of enrolled students (for the years 1990, 1994, 1999, 2004, 2006), elementary students (for the years 1990/91, 1995/96, 2000/01, 2006/07, winter terms), graduates (for the years

Table 4. Percentages of female beginners at university and university of applied sciences (FH) in some STEM subjects in the years 2001 and 2002. Source: Federal Statistical Office (Frauen geben Technik neue Impulse e.V., 2004)

Subjects	Female beginners (in percent)			
	University		FH	
	2001	2002	2001	2002
Industrial engineering	21.5	24.5	21.5	21.9
Computer sciences	17.7	17.2	14.3	14.7
Mechanical engineering	12.0	14.1	7.6	7.2
Electrical engineering	10.2	10.0	7.0	7.7

1993, 1997, 2001, 2005), and the 18th Sozialerhebung of the Deutsche Studenten Werke (DSW).

Briedis et al. (2008) present data about the percentage of female students for the four subjects of mathematics, computer science, electrical engineering, and mathematics teacher training. Their message is quite plain that compared to humanities, which has a percentage of female students of about 70 percent, a clear underrepresentation of women can be seen for computer science and electrical engineering, both below 20 percent. In mathematics (including teacher training), the number of women is close to the number of men as well as to the total percentage of women enrolled at University (about 50 percent).

Similar results exist for the percentage of women graduating in the subjects of mathematics (including teacher training, close to 60 percent), computer science (below 20 percent), and electrical engineering (around 10 percent). Again, the results in computer science and electrical engineering reflect the aforementioned underrepresentation of female students in these subjects and their overrepresentation in humanities (Briedis et al., 2008).

IT studies in Greece. If one would like to describe the gender distribution in Greek universities in a single statement this would be "men are overrepresented in polytechnics departments and women in the humanities". Nevertheless, the percentage of female students grows almost every year, and their presence in the "technological" university departments remains among the highest in Europe. While in 1969-1970, women represented only 31.4 percent of the University student population, thirty years later they prevailed over men with 58.7 percent in 2000-2001 (Maratou-Alipranti, Dafna, Yannakopoulou, Kymperi, & Repa, 2002).

The Ministry of Education provides a large variety of data regarding the tertiary education population but no cumulative data on gender. For this reason, the data presented refer only to the academic year 2007-08. One can observe the low representation of women among the university students in ICT-related departments. 27 percent of the undergraduate in Informatics Departments are women compared to 23.4 percent to Informatics Engineering Departments (see Table 5). Mainly men pursue graduate studies and this differentiation is enhanced at PhD level. This holds even more for ICT-related departments. As can be seen from Table 5, among Master candidates of the Informatics Department (university) students, 34 percent are women compared to 28.2 percent for Informatics Engineering Departments (polytechnic). A major difference can be found between the PhD candidates: 25 percent of the Informatics Department students are women and only 15.3 percent in Polytechnics Informatics Engineering Departments.

Table 5. Percentage of females in University studies

	University Students – Academic Year 2007-2008 data					
	Undergraduate students		Graduate students			
	Total		Master		PhD	
	Male	Female	Male	Female	Male	Female
University Informatics Departments	73.7%	27.0%	65.7%	34.0%	75.4%	25.0%
Polytechnics- Informatics Engineering Departments	76.2%	23.4%	71.7%	28.2%	84.8%	15.3%

An overall picture for the distribution among gender in Greek tertiary education institutions (universities and polytechnics/technical universities) is: 60.3 percent of the undergraduate students, 56 percent of the Master students and 42 percent of the Ph.D. candidates are women – numbers much larger than those observed in ICT-related departments.

This picture is also reflected in specific surveys, e.g. performed by university departments. For example, the Department of Electrical and Computer Engineering (Aristotle University of Thessaloniki) relating gender to department enrolment, study and career orientation reports the following findings (ΑΡΙΣΤΟΤΕΛΕΙΟ ΠΑΝΕΠΙΣΤΗΜΙΟ ΘΕΣΣΑΛΟΝΙΚΗΣ, 2008). In the period 1979 (year in which the department was established) to 2008, from a total of 5,962 students that had been enrolled, only 1,123 are women (18.84 percent), and from the 2,978 students that graduated in the period 1988 to 2008, only 487 are women (16.35 percent). Moreover, the percentage of women graduates is lower than that of women enrolled. More women than men drop out during the course of their studies. In the same report it is stated that the percentage of female PhD candidates is almost the same as that of female graduates – of the 152 students that are PhD candidates 27 are women (18 percent). It is of interest that female PhD candidates need more time to obtain their degree than their male counterparts (and this is stable over the years). A possible explanation is that women work during their studies or have other family affairs to attend to. Besides, data provided by the Secondary of Civil Engineering of the National Technical University of Athens (2006) show that men outnumber women in terms of student enrolment and teaching personnel (just 19 percent are women; Mimikou, 2008).

Reflections

Explanations that have been provided in literature regarding the gender gap refer to the societal stereotypes and work conditions. Women are more attracted to professions that are offered by the State (i.e. teachers) since such positions are considered to be "stable" and easily leave space for other obligations women have (i.e. children, family obligations). This view is also supported by parents. Work in industry or in the construction sectors is considered competitive and "tough" for women.

Similarities and differences. Results of both countries show that females are underrepresented in ICT subjects but over-represented in humanities. However, there is a difference of about 10 percentage points between Germany and Greece: In all stages of the university education pathway, the share of females in ICT is at least 10 percent higher in Greece than in Germany, starting from undergraduate enrolment to PhD studies. In Germany, the share of females is relatively stable and below the EU average. In contrast, the representation of Greek women in ICT-related studies is above the EU average, and is constantly increasing. However, also in Greece there are restraining barriers (such as social stereotypes, family obligations, labour market) that have to be addressed if Greece wants to achieve the desired gender-balance of at least 30 percent of female participation in ICT.

Reasons for the underrepresentation of women may be that the above reported lower interest of female pupils in ICT has an influence on their further education choices. Besides, Engeser, Limbert, and Kehr (2008) analyzed motives for the choice of studies in the ICT context. They found that the key motivations for studying ICT are future aspects, e.g. career opportunities, the mastery of a future technology, good chances for employment, scientific challenges, and seeing the importance of computer science in society. They found action-result-expectations as another factor for ICT as career choice: Only persons who are self-confident in meeting the requirements of ICT studies choose such kinds of studies. No extra effect for gender was reported. However, the findings of our study at German and Greek schools show that females might have lower self-confidence in ICT, and thus the low number of female beginners in ICT-related subjects may be a result of their levels of self-confidence developed during school time.

GENDER AND ICT PROFESSIONS

The underrepresentation of women in ICT-related professions is even stronger than at tertiary education level. Less women than men who graduated in computer science and informatics choose a career in this field.

According to Dick and Rallis (1991), career choice is influenced by people's self-concept, their perception of their own abilities, and the subjective value of certain career pathways. These beliefs build on experiences made during school and family life, including the perceived expectations or attitudes of other people (socialisers) such as peers, parents and teachers. In this respect, stereotypical career perspectives of socialisers can influence the career choices of men and women differently. Findings of the survey conducted by Dick and Rallis show that for women, the attitude of teachers is a significant factor for choosing or not choosing a career in engineering or science; however, still other socialisers play an important role. The expected salary is less important for women, but men consider it in their career choice to a greater extent.

Employment statistics and information on the salary provide insights into on the gender gap in ICT-related professions and careers. Again, we will describe the particular situation in each country, and reflect on the differences and similarities afterwards. The heterogeneity of the data hardly provides information on the same professions for both countries, e.g. for Germany, we selected exemplary data for data processing professionals, as well as from graduates of computer science, mathematics and engineering; while for Greece the focus is on the differences of men and women working in the technical departments at universities, as well as on total employment and unemployment rates.

Data Review

ICT & Gender at Professional Level in Germany. We retrieved exemplary data from the annual survey of the German Institute for Employment Research (IAB)[3], which monitors the development of unemployment and professions subjected to social insurance contributions. The IAB provides the following data for women and men employed as data processing professional, and computer scientists (classification 774; occupation: research, development, construction, design of products, concepts and programs). The data exemplarily show an underrepresentation of women in the ICT core professions in Germany. In general, an increase in the total number of 774 employees can be seen for each year, and at the same time the proportion of women is decreasing (Table 6).

Table 6. Percentage of women employed in professions of the classification 774. (Source: IAB)

Employees (Classification 774: data professionals and computer scientists)				
Year	Total	Men	Women	Women (in %)
2005	448.383	359.630	88.780	19.8%
2006	456.534	367510	89.024	19.5%
2007	469,880	376.844	93.036	19.1%
2008	490.462	397.765	92.697	18.9%
2009	501.293	407.050	94.243	18.8%

Table 7. Percentage of unemployed women in professions of the classification 774. (Source: IAB)

Unemployed (Classification 774: data professionals and computer scientists)				
Year	Total	Men	Women	Women (in %)
2005	56.136	49.063	10.552	17.7%
2006	46.032	38.943	7.089	15.4%
2007	33.345	28.543	4.802	14.4%
2008	26.707	22.834	3.873	14.5%
2009	30.252	26.319	3.933	13.0%

Looking at the unemployment rates for 774 professionals from 2005 to 2009 an increase can be observed in total, for men and women. At the same time, the percentage of unemployed women is decreasing (Table 7).

The above mentioned HIS survey (Briedis et al., 2008) provided additional data about the progress of employment of students in the first 12 months following graduation from university. It can be concluded from the data that graduating in mathematics, and even more in computer science and electrical engineering, is associated with clearly better chances for swift entry into the job market after university than graduating in mathematics teacher training or the humanities.

It has to be noted that according to Blossfeld et al. (2009) the number of women who work as teachers clearly exceeds the number of men who work in the same profession (in total 65.3 percent teachers are female: 88 percent on primary secondary and 40 percent for grammar schools). In other professions which require a university degree or a degree from a university of applied sciences, the percentage of women is 39.6 percent.

ICT & Gender at Professional Level in Greece. In the last two decades statistics have shown that female employment has risen at faster rates than the average rate of total employment in all professional categories, and the inequality of representation in prestigious occupations between the genders has reduced. In 1961, only 3.7 percent of working women were classified under the "scientific professions" and "managerial-executive personnel" categories, whereas in 1991, 19.4 percent of working women were found in these same occupational categories. However, gender-based differentiations in payment continue to be a significant reality in specific areas, although it has been noted that in all public sector and semi-public establishments or organizations such as banks, universities etc., such an issue does not exist. Although the participation of women in the teaching personnel in primary and secondary education is high (more than 50 percent of the

Table 8. Unemployment rate by gender December 2004-2008

Gender	2004	2005	2006	2007	2008
Men	6.6	6.4	6.0	6.1	6.5
Women	16.7	14.4	14.1	13.0	12.2
Total	10.7	9.7	9.3	8.9	8.9

total personnel are women), women in tertiary education occupy posts in lower ranks. Data shows that women in the universities constitute about one third of the total staff. It seems that the higher the position in the academic hierarchy, the lower the percentage of women. Women are not as easily promoted as men, especially when they have family commitments, as there are not enough public facilities to help conciliate family and employment (Maratou-Alipranti, Dafna, Yannakopoulou, Kymperi, & Repa, 2002).

A research study conducted by the Polytechnics School in Thessalonica, namely the Department of Electrical and Computer Engineering, gives a concise picture about the graduates' employment characteristics, situation and discrepancies in relation to the gender characteristics (see ΑΡΙΣΤΟΤΕΛΕΙΟ ΠΑΝΕΠΙΣΤΗΜΙΟ ΘΕΣΣΑΛΟΝΙΚΗΣ, 2008). According to the results of this research, women face more difficulties finding a job than men because of unfair competition. For women in Greece, a very important factor according to the results from the data of this research for the choice of a professional career in general is the compatibility with the family, as well as a good salary and the perspective of progress. The majority of the female graduates in the ICT field find a job from one to six months after the completion of their studies.

According to the work done by Pouliakas and Livanos (2008), male faculty are mostly found in the more technically-oriented academic departments such as Polytechnics, Computer Science, Agricultural Studies, Physics and Mathematics, Medicine, Economics and Business and Physical Education. On the contrary, women are more heavily represented in Law, Social Sciences, Humanities, Education, Librarianship and other medical-related sciences. According to their analysis, given that the former degrees are higher-paid disciplines than the latter, it becomes obvious that the subject of degree is a potential culprit for explaining the gender wage differential of university graduates in Greece. Greek women tend to find refuge in less risky educations that consequently offer lower compensation in terms of pay.

According to the Eurostat Labour Force Survey conducted in 2008 in Greece, the unemployment rate among women with tertiary-level education was much higher than among men (exceeding the 8 percent of the total population; Eurydice, n. d.). Some research about the employment situation for women in science and technology (see Meri, 2008) shows that the largest gender disparities in terms of the absolute number of unemployed Human Resources in Science and Technology (HRST) were found in Greece (as well as in Portugal and Slovenia) where the number of female unemployed HRST was twice the figure for men. The Labour Force Survey of the General Secretariat of National Statistical Service of Greece shows that the number of unemployed men decreased from December of 2004 to 2006, and increased in 2007 and 2008. The number of unemployed women decreased from 2004 to 2008 (see Table 8, General Secretariat of the National Statistical Service of Greece, n. d.).

According to the study conducted by Protovoulia (Education and Development Initiative; Protovoulia, n. d.) from April, 2006 to June, 2008, the percentage of unemployment of the Informatics School graduates is much higher for female than male graduates (for example, 3 years post graduation, the estimated percentage of females' unemployment is 10.1 percent, in contrast to 2.1 percent for males').

Reflection

The issue of women in ICT professions, especially in leading positions, was discussed on the CeBIT 2009, a conference for professionals in the IT sector in Hannover. An initiative was started to attract more women to this sector, especially by improving the working conditions, exceeding the concept of just combining family and work but taking into account a comprehensive career planning approach. A related agenda was designed and signed by more than 100 women and men from several enterprises, including women in ICT leading positions (Kompetenzzentrum, 2009a; 2009b). The results of a short research with 125 male and female visitors of the CeBIT point out the need for action in this area. More than half of the women (59.1 percent) can imagine working in a STEM profession and even more (63.6 percent) see the STEM sector as an innovative field of work. However, only one fifth of the questioned visitors see women in management or leading positions in the ICT sector (BMBF, 2009).

Similarities and differences. To provide a comprehensive analysis of gender in ICT profession we have to deal with the issue of reliable data. Data for Germany is coded as "classification 774" which comprises some ICT jobs but which is not comprehensive for the whole ICT sector. More differentiated analysis, e. g. Apfelbaum (2009) and IGM (2008) provide more detailed information, but lack in representativeness, particularly with respect to the issue of gender. The data presented for Greece focus mostly on the educational sector due to a lack of access to specific data with categorization that will give information about employment in the IT sector. The categorization of data provided by the corresponding authorities about employment is based on general labour groups or is connected to the existing fields of studies. Also in Greece, the categorization based on labour groups does not give a clear view of IT-related employees. Yet, besides these obstacles we can see that in the ICT sector employment, the share of females is even smaller than in ICT studies at university level in both countries. And even if women are employed in ICT professions, they still earn less money for their work than their male colleagues. Furthermore, when we look at unemployment figures, we can see another imbalance. The unemployment rate of females in the German ICT sector is higher than that of males; in Greece the rate of unemployed females in this sector is about twice the rate of unemployed males. That means that there is not only an underrepresentation of females employed in the ICT sector; females also have more difficulties in getting a job in this sector in both countries.

CONCLUSION

This paper focuses on gender equality in ICT usage, in ICT education at secondary and tertiary level, and on the career level in Germany and Greece. In both countries, a digital divide between men and women can be observed: girls and women use ICT and the Internet to a smaller extent, they show less interest in ICT at school, and have lower self-confidence in performing computer tasks, enrol for humanities rather than for ICT-related subjects at university, and are clearly underrepresented in ICT careers.

If we remember the different economic structure of Germany and Greece, we can see that the results go in a similar direction. Current usage statistics however show that girls and women are catching up. In both countries, younger women (16-24 years) show nearly the same frequency in computer usage than males. We were also able to report on the Internet usage behaviour of German females which is similar to that of males'. Age factor also influenced the ICT interest and self-confidence at school: younger people are more interested in ICT and have a higher level of self confidence. Yet, there is still a long way to go. Additionally, educational trajectories from school to university, and the uptake of careers in ICT-related professions increase gender imbalance in a way that meets a leaky pipeline comparison. (European Commission, 2006): Particularly in the field of ICT, the share of women shrinks with continuing professional development, in Germany and in Greece.

Digital literacy is more than using computers and the Internet once a week; and also for young persons there are still differences in the interests and self-concepts to the advantage of boys. The findings of the study by Engeser et al. (2008) revealed that self-confidence is a crucial aspect for career choice in the ICT field; and Dick and Rallis (1991) reported that the attitudes and expectations of socialisers, e.g. teachers, can influence the career choice of women. The results of our study at German and Greek schools, as well as the PISA study (OECD, 2005), show less self-confidence of girls in many aspects of ICT usage. This is—according to Dick and Rallis—an essential aspect for future career choices. Consequently, measures for improving gender balance have to start as early as possible as our study shows that gender differences already exist in schools.

Practical Implications

The similar patterns of German and Greek ICT usage, education and careers of women and men, and the fact that women are still lagging behind, support the necessity of promoting equality in the field of ICT. A starting point would be raising the self-confidence of girls, and taking into account the facilitating role of teachers in this respect.

Such facilitation is, however, linked with school curricula and their implementation of media education, basic education in information technology, and informatics lessons at school. These subjects and related frameworks have to be an indispensable part of the curricula in Germany and Greece. However, gender support is not yet an underlying principle of the curricula of these subjects, and, if at all, specific gender support in relation to ICT is implemented on project level only (see Helling & Ertl, 2009; Kikis-Papadakis, Papanastasiou, & Margetousaki, 2009). For example, in Germany, several initiatives and projects aim to attract females to study STEM subjects, e.g. information days for pupils, information services for women at university, and special studies for women only (Studien- & Berufswahl, n. d.; www.komm-mach-mint.de). Therefore, future development of education policy at school level has to take into account both the consistent implementation of information technology education and at the same time the specific facilitation of girls and boys towards an increase and equal development of their digital literacy.

A major problem in this context is often the inadequate education of teachers. Although teacher education and training include aspects of media education and information technology education, only a small number of teachers have the competencies to adequately implement ICT in education and facilitate pupils' digital literacy (see Helling & Ertl, 2009; Kikis-Papadakis, Papanastasiou, & Margetousaki, 2009).

Future education policy development needs to deal with this challenge by ensuring the competence development of all teachers in this respect. Furthermore, the development of gender-sensitive teaching and learning materials for the facilitation of digital literacy of girls and boys needs to be advanced. Even if there are several initiatives and projects that focus on the promotion of ICT and gender issues, the described statistics and studies show that the issue of promoting equality with respect to gender and ICT will remain a critical topic in the future for both countries, and needs to be advanced from a project-based approach to a policy implementation level.

To sum up, the situation in both countries with respect to gender and ICT is still far from gender balanced. Statistics about ICT use of teenagers indicates that there may be a change. However, this has to be supported to affect the uptake of ICT careers by females positively.

ACKNOWLEDGMENT

Parts of this contribution were funded by EU (LLP-Program, Projects PREDIL 141967-2008-LLP-GR-COMENIUS-CMP and SESTEM 505437-llp-2009-GR-KA1-KA1SCR), DAAD and IKY (Project D0813016 resp. Agreement number 136 IKYDA 2009: Comparative study on gender differences in technology enhanced and computer science learning: Promoting equity).

REFERENCES

Apfelbaum, D. (2009). Wer verdient wie viel? Ergebnisse der c't-Gehaltsumfrage 2008. *c't, 6*, 92-99.

ΑΡΙΣΤΟΤΕΛΕΙΟ ΠΑΝΕΠΙΣΤΗΜΙΟ ΘΕΣΣΑΛΟΝΙΚΗΣ, (2008). *Μελέτη σπουδών και επαγγελματικής σταδιοδρομίας στο Τμήμα Ηλεκτρολόγων Μηχανικών & Μηχανικών Υπολογιστών Α.Π.Θ. σε σχέση με τα χαρακτηριστικά φύλου*. Report in Greek, 2008, Retrieved January, 29th, 2010, from http://newton.ee.auth.gr/ genderIssues/ docs/ THMMY_Gender New_vFINAL.pdf

Becta - British Educational Communications and Technology Agency. (2008). *How do boys and girls differ in their use of ICT?* Retrieved January, 29th, 2010, from http://partners.becta.org.uk/ upload-dir/ downloads/ page_documents/ research/ gender_ict_briefing.pdf

BLK – Bund-Länder-Kommission für Bildungsplanung und Forschungsförderung (1987). Gesamtkonzept für die informatische Bildung. *Materialien zur Bildungsplanung und Forschungsförderung, 16*.

Blossfeld, H.-P., Bos, W., Hannover, B., Lenzen, D., Müller-Böling, D., Prenzel, M., & Wößmann, L. (2009). *Geschlechterdifferenzen im Bildungssystem. Jahresgutachten 2009*. Wiesbaden: VS Verlag für Sozialwissenschaften. Retrieved March 23, 2009, from http://www.aktionsrat-bildung.de/ fileadmin/ Dokumente/ Geschlechterdifferenzen_im_Bildungssystem__Jahresgutachten_2009.pdf

BMBF – Bundesministerium für Bildung und Forschung (Ed.). (2008). *Studiensituation und studentische Orientierungen. 10. Studierendensurvey an Universitäten und Fachhochschulen*. Bonn. Retrieved March 11, 2009, from http://www.bmbf.de/ pub/ studiensituation_studentetische_orientierung_zehn_lang.pdf.

BMBF – Bundesministierium für Bildung und Forschung. (2009). *CeBIT-Umfrage: Frauen wollen MINT!* Retrieved March 11, 2009, from http://www.komm-mach-mint.de/ Startseite/ News/ CeBIT-Umfrage-Frauen-wollen-MINT!

Briedis, K. Egorova, T. Heublein, U. Lörz, M., Middendorff, E., Quat. H. & Spangenberg, H. (2008). *Studienaufnahme, Studium und Berufsverbleib von Mathematikern. Einige Grunddaten zum Jahr der Mathematik*. HIS: Forum Hochschule, 9. Retrieved March 17, 2009, from http://www.his.de/ pdf/ pub_fh/ fh-200809.pdf

CIA. (2010a). The world factbook: Germany. Retrieved April 4, 2010 from https://www.cia.gov/ library/ publications/ the-world-factbook/ geos/ gm.html

CIA. (2010b). The world factbook: Greece. Retrieved April 4, 2010 from https://www.cia.gov/ library/ publications/ the-world-factbook/ geos/ gr.html

Compeau, D. R., & Higgins, C. A. (1995, June). Computer Self-Efficacy: Development of a Measure and Initial Test. *Management Information Systems Quarterly, 19*(2), 189–211. doi:10.2307/249688

Demunter, C. (2005). The digital divide in Europe. Statistics in Focus, 28/2005.

Dick, T. P., & Rallis, S. F. (1991). Factors and influences on high secondary students' career choices. *Journal for Research in Mathematics Education, 22*, 281–292. doi:10.2307/749273

Engeser, S., Limbert, N., & Kehr, H. (2008). *Abschlussbericht zur Untersuchung Studienwahl Informatik.* Retrieved October 29, 2009, from http://www.psy.wi.tum.de/ Docs/ Studienwahl_ Informatik -Abschlussbericht.pdf

European Commission. (2006). *Women in Science and Technology. Creating sustainable careers.* Retrieved September 1, 2009, from http://ec.europa.eu/ research/ science-society/ document_library/ pdf_06/ wist2_sustainable-careers-report_en.pdf

European Commission. (2007) *Key Competencies for Lifelong Learning – A European Framework. (2007). Annex of the Recommendation of the European Parliament and of the Council of 18 December 2006 on key competencies for lifelong learning (2006).* Official Journal of the European Union. Retrieved January 29, 2010, from http://ec.europa.eu/ dgs/ education_culture/ publ/ pdf/ ll-learning/ keycomp_en.pdf

European Commission. (2008). *Progress towards the Lisbon objectives 2010 in education and training.* Retrieved December 2, 2009, from http://ec.europa.eu/ dgs/ education_culture/ publ/ pdf/ educ2010/ indicatorsleaflet_en.pdf? aid=14505& d=2007-11

Eurostat. (2009a) *Internet-Zugangsdichte – Haushalte. Prozent der Privathaushalte mit Internet-Zugang.* Retrieved April 5, 2010, from http://epp.eurostat.ec.europa.eu/ tgm/ table.do? tab=table& init=1& plugin=1& language=de& pcode=tsiir040

Eurostat. (2009b). *Versorgungsgrad mit Breitbandanschlüssen. Anzahl der Breitbandanschlüsse je 100 Einwohner.* Retrieved April 5, 2010, from http://epp.eurostat.ec.europa.eu/ tgm/ table.do? tab=table& init=1&plugin=1& language=de& pcode=tsiir150

Eurydice (n.d.). *Key Data on Education in Europe 2009.* Retrieved January, 29th, 2010, from http://eacea.ec.europa.eu/ education/ eurydice/ documents/ key_data_series/ 105EN.pdf

Faulstich-Wieland, H., & Nyssen, E. (1998). Geschlechterverhältnisse im Bildungssystem - Eine Zwischenbilanz. In Rolff, H.-G., Bauer, K.-O., Klemm, K., & Pfeiffer, H. (Eds.), *Jahrbuch der Schulentwicklung.* Weinheim: Juventa.

Frauen geben Technik neue Impulse e.V. (Ed.) (2004). *Studiengänge im Wettbewerb. Hochschulranking nach Studienanfängerinnen in Naturwissenschaft und Technik.* Retrieved March 30, 2009, from http://www.ranking-kompetenzz.de/ daten/ images/ Ranking%20 Broschuere.pdf

General Secretariat of the National Statistical Service of Greece. (n.d.). *Latest Statistical Data* Retrieved January, 29th, 2010, from http://www.statistics.gr/ portal/ page/ portal/ ESYE

Gesellschaft für Informatik (GI) e.V. (1999) Informatische Bildung und Medienerziehung. Empfehlung der Gesellschaft für Informatik e.V.. Beilage zu *LOG IN 19*(6).

Helling, K., & Ertl, B. (2009). *PREDIL - The National Context of Germany.* Munich, Heraklion: PREDIL Project Consortium.

Holzapfel, N. (2006). *IT Branche – Gehälter im Aufwind.* Retrieved, March 9, 2009, from http://www.sueddeutsche.de/ jobkarriere/ 309/ 300307/ text/

IGM (Ed.). (2008). *Entgelt in der ITK-Branche 2008. Eine Erhebung in der Informations- und Telekomunikationsbranche. 10. Erhebung.* Frankfurt a. M. Bund.

Imhof, M., Vollmeyer, R., & Beierlein, C. (2007). Computer use and the gender gap: The issue of access, use, motivation, and performance. *Computers in Human Behavior, 23*, 2823–2837. doi:10.1016/j.chb.2006.05.007

Initiative D21. (2008). *(N)ONLINER ATLAS 2008. Eine Topographie des digitalen Grabens durch Deutschland.* Retrieved January 29, 2010, from http://www.initiatived21.de/wp-content/uploads/2009/06/NONLINER2009.pdf

Insight (2007). *Greece, Last revised: October 2007,* Agapi Vavouraki, Hellenic Pedagogical Institute, Retrieved January, 29, 2010, from http://insight.eun.org/ww/en/pub/insight/misc/country_report.cfm?

Kikis-Papadakis, K., Papanastasiou, R., & Margetousaki, A. (2009). *PREDIL - The National Context of Greece.* Heraklion: PREDIL Project Consortium.

Komeptenzzentrum Technik Diversity Chancengleichheit, V. (2009a). *Charta für Talente der Zukunft.* Retrieved March 11, 2009, from http://www.kompetenzz.de/Features/Charta

Komeptenzzentrum Technik Diversity Chancengleichheit, V. (2009b). *Mehr Frauen in IT-Führungspositionen! Pressemitteilung.* Retrieved March 11, 2009, from http://www.komm-mach-mint.de/Startseite/Service/Presse/Pressemitteilungen/Mehr-Frauen-in-IT-Fuehrungspositionen!

Lopez, A. D., & Manson, P. D. (1997). A study of individual computer self-efficacy and perceived usefulness of the empowered desktop information system. *The Cal Poly Pomona Journal of Interdisciplinary Studies, 10,* 83–92.

Maratou-Alipranti, L., Dafna, K., Yannakopoulou, L., Kymperi, Z., & Repa, P. (2002). *WOMEN AND SCIENCE: Review of the situation in Greece.* Retrieved January, 29th, 2010, from ftp://ftp.cordis.europa.eu/pub/improving/docs/women_national_report_greece_en.pdf

Meri, T. (2008) *Women in science and technology, Statistics in focus, Science and Technology, 10/2008.* Retrieved January 29, 2010, from http://bookshop.europa.eu/eubookshop/download.action?fileName=KSSF08010ENC_002.pdf&eubphfUid=554230&catalogNbr=KS-SF-08-010-EN-C

MPFS - Medienpädagogischer Forschungsverbund Südwest (LFK/LMK). (2008). *JIM-Studie 2008. Jugend, Information, (Multi-) Media. Basisuntersuchungen zum Medienumgang 12- bis 19-Jähriger.* Retrieved January 29, 2010, from http://www.mpfs.de/fileadmin/JIM-pdf08/JIM-Studie_2008.pdf

Munk, B. (2007). LogoGo - An approach to the design of girl-specific educational software. In Zorn, I., Maas, S., Rommes, E., Schirmer, C., & Schelhowe, H. (Eds.), *Gender Designs IT. Construction and Deconstruction of Information Society Technology.* Wiesbaden: VS Verlag für Sozialwissenschaften.

Observatory for the Greek IS. (2008). *Παρουσίαση αποτελεσμάτων έρευνας για τη χρήση των νέων τεχνολογιών από τα παιδιά.* Retrieved January 29, 2010, from http://www.observatory.gr/Files/Meletes/Y8EEUR081015DOCEL_Π3%20%20Έρευνα%20Παιδιά.pdf

Observatory for the Greek IS. (2009). Measurement of eEurope/i2010. Indicators for Greece. 2008 Findings. Retrieved April 5, 2010, from http://www.observatory.gr/files/meletes/Booklet%20eEurope%202008%20en.pdf

OECD. (2005). *Are Students Ready for a Technology-Rich World? What PISA Studies Tell Us.* OECD. Retrieved January 29, 2010, from http://www.oecd.org/ document/ 31/ 0,3343,en_32252351_32236173_35995743_1_1_1_1,00.html

OECD. (2009). *Information Technology Outlook 2008.* OECD. Retrieved April 9, 2010, from http://browse.oecdbookshop.org/ oecd/ pdfs/ browseit/ 9308041E.PDF

Pouliakas, K., & Livanos, I. (2008). *The gender wage gap as a function of educational degree choices in Greece.* Retrieved January 29, 2010, from http://mpra.ub.uni-muenchen.de/ 14168, MPRA paper No. 14168, posted 24 March 2009

Protovoulia (n. d.). *Σύνοψη Μελέτης για την Απασχόληση των Πτυχιούχων Τριτοβάθμιας Εκπαίδευσης στην Ελλάδα.* Retrieved January 29, 2010, from http://studies.protovoulia.org/ files/ synopsi_meletis2.pdf

Schnirch, A., & Welzel, M. (2004). Nutzung neuer Medien im Bereich des naturwissenschaftlichen Unterrichts der Realschule. Eine Studie unter Genderperspektive. In Buchen, S., Helfferich, C., & Maier, M. S. (Eds.), *Gender methodologisch. Empirische Forschung in der Informationsgesellschaft vor neuen Herausforderungen.* Wiesbaden: VS Verlag für Sozialwissenschaften.

Seibert, H. (2007). Gender differences in the use of computers and the Internet. Statistics in focus. Population and Social Conditions, 119/2007. Retrieved April 5, 2010, from http://epp.eurostat.ec.europa.eu/ cache/ ITY_OFFPUB/ KS-EI-08-001/ EN/KS-EI-08-001-EN.PDF

Studien- & Berufswahl. (n.d.). *Frauen im Studium.* Retrieved March 10, 2009, from http://www.studienwahl.de/ print.aspx? f=2/2_3_0_0_0_0_0_content _01.aspx&id=

KEY TERMS AND DEFINITIONS

Career Choice: Intentional selection of an educational pathway leading to a specific profession.

Fachhochschule: A university of applied sciences on a technical level in Germany.

Gender: The term of gender refers to a particular perspective on females and males that goes beyond biological sex and also includes socialisation effects.

ICT: Abbreviation for Information and Communication Technologies.

Interest: A psychological concept that relates to an emotion which draws attention to something.

IS: Abbreviation for Information System.

Self-Confidence: A psychological concept describing that a person feels capable for doing something.

STEM: Abbreviation for Science, Technology, Engineering, and Mathematics.

ENDNOTES

[1] Eurostat. Country profiles. http://epp.eurostat.ec.europa.eu/guip/introAction.do?profile=cpro&theme=eurind&lang=en.

[2] Federal Statistics Office. Edution Research and Culture. Higher Education. Tables. http://www.destatis.de/jetspeed/portal/cms/Sites/destatis/Internet/DE/Navigation/Statistiken/BildungForschungKultur/Hochschulen/Tabellen.psml

[3] German Institute for Employment Research (IAB). Berufe im Spiegel der Statistik. Retrieved April 06, 2010, http://bisds.infosys.iab.de/bisds/result?region=19&beruf=BO774&qualifikation=2

Chapter 8
The Effect of Gender on Associations between Driving Forces to Adopt ICT and Benefits Derived from that Adoption in Medical Practices in Australia[1]

Rob Macgregor
University of Wollongong, Australia

Peter Hyland
University of Wollongong, Australia

Charles Harvey
University of Wollongong, Australia

ABSTRACT

Like other Small to Medium Enterprises (SMEs), medical practices can gain a great deal by adopting and using Information and Communication Technologies (ICTs). Unlike other SMEs, little is known about General Practitioners' (GPs) perceptions of the benefits of ICT use or about the differences between these perceptions by male and female GPs. This chapter reports a survey of these perceptions of the drivers for and benefits of ICT use by male and female GPs in Australia.

DOI: 10.4018/978-1-60960-759-3.ch008

INTRODUCTION

The advent of affordable Internet-based information and communications technology (ICT) has led the medical and healthcare sectors to explore the use of such technologies to improve patient care and reduce business inefficiencies within General Practice (GP). This has been recognised by the World Health Assembly, who, in 1997, saw technology as one of many parts of sustainable health systems and, in 2005, saw technology as a means of leveraging health-for-all through the interchange of information and communications (Kirigia, Seddoh, Gatwiri, Muthuri & Seddoh, 2005). From the late 90's studies began to appear detailing the design of clinical ICT systems (see for example Baldwin, Clarke & Jones, 2002; Pelletier-Fleury et al., 1999; Hsu et al. 2005), the use of such systems within medical practices (Ammenwerth, Mansmann, Iller, & Eichstadter, 2003; Catalan, 2004; Shohet & Lavy, 2004; Waring and Wainwright, 2002;) and, more recently, the decision-making behind ICT adoption (Didham, Martin, Wood, Harrison, 2004; MacGregor, Harvie, Hyland and Lee, 2007; Pan & Pokharel, 2007). Studies, for example, in New Zealand (Didham et al 2004), showed that time, costs and perceived lack of IT skill were important considerations for GPs when evaluating ICTs. Lee Cain, Chockley and Burstin (2005) found practice size and standardisation of work were of concern to many doctors, while Simon et al (2007) found practice size (both in terms of patient numbers and staff numbers) and the type of care being offered were statistically associated with the perception of both drivers for and barriers against ICT adoption.

Along with the studies examining the ICT adoption process, there have been a number of studies detailing the potential benefits derivable from ICT adoption and use in general practices. El-Sayed & Westrup (2003), for example, suggest that ICT use in medical practices improves communication within and outside the practice, makes the business side of the practice more effective and helps build new business initiatives. Baldwin et al (2002) suggest that ICTs support and enable complex interactions between GPs, consultants, patients, nurses and, in some cases, equipment. Fors & Moreno (2002) suggest that ICTs, in medical practices, alter day-to-day procedures, making the overall final product more effective, while Ray & Mukherjee (2007) note the use of ICT to promote governance and planning.

While there have been studies investigating both the driving forces behind ICT adoption and the benefits derived from that adoption, there have been no studies that have attempted to determine whether giving priority to one driving force over another leads to a perception of improvement to specific benefits. Similarly, while there have been a number of studies that have explored gender differences in the adoption and use of internet-based technology (Kolsaker & Payne, 2002; Rodgers & Harris, 2003; Oudshoorn, Rommes, & Stienstra, 2004; Yang & Lester, 2005), there have been no studies aimed at determining the relationship between priority given to driving forces and perceived benefits of ICT adoption between male and female GPs.

The purpose of this chapter is to determine whether the relationship between priority given to driving forces and perceived benefits of ICT adoption differs between male and female GPs. The chapter begins by examining the nature of ICT in medical practices, in particular the driving forces behind the adoption process and benefits derivable from their adoption and use. As medical practices in Australia are almost all specialised small businesses, the chapter examines gender differences both from a small business perspective as well as from a medical perspective. The chapter presents a study of 196 GPs (128 males, 68 females) who have adopted ICT in their practice. A series of factor analyses is applied to the driving forces behind ICT adoption to determine the groupings of driving forces and the groupings of benefits for male and female GP respondents. Using these groupings a partial least square model was developed and tested to determine whether there are gender differences in the association between perception of importance of driving forces and perception of subsequent benefits.

The Nature of ICT in Medical Practice

The nature of ICT in medical practice differs widely in the literature. At the 'cutting edge' the use of ICT involves functions such as knowledge management and knowledge translation (Ho et al 2004), video and audio components and the use of imaging equipment (Baldwin et al 2002), multiple site education (Kuruvilla, Dzenowagis, Pleasant & Dwivedi, 2004) and distance clinical treatment (Caro, 2005). At the general practice level ICT has been shown to be an effective tool in the treatment of chronic disease (Christensen & Remler, 2007) as well as a mechanism for analysing, integrating and communicating in disease management (Cherry, Moffatt, Rodriguez, & Dryden, 2002).

A number of studies (Adogbeji & Akporhonor, 2005; Kuruvilla et al., 2004; Lougheed, 2004; Ndubisi & Kahraman, 2005,) have shown that the use of ICT in medical practices is not just the province of developed economies, but is becoming more commonplace in developing economies in Africa, S.E. Asia and South America.

Studies (Ash, Gorman, Seshadri, & Hersh, 2004; Ho et al 2004; Keddie & Jones, 2005; Lougheed, 2004; Stevanovic, Stanic, & Varga, 2005) have shown that the use of ICT within medical practices has moved beyond simply being a clinical tool and now incorporates the wider role of business and practice management. At first glance these may appear to be at odds with one another; clinical aiming at quality and efficiency of medical care and business aiming towards profitability and budgetary concerns. However, recent studies (Bonneville & Pare, 2006; Lievens & Jordanova, 2004) suggest that for any medical or healthcare function to be truly viable there must be parity between medical and business efficiency. Indeed, Kuruvilla et al. (2004) suggest that ICTs in general practice not only centralise geographically dispersed resources, thus promoting flexibility and economies of scale, but they promote efficiency, enhance quality of care and encourage partnerships between practitioners as well as between patients and practitioners.

In line with these views, the benefits attributed to the use of ICT can be seen from two perspectives – medical and general business. From a medical perspective these benefits include contact with other clinicians regarding patient care (Baldwin et al., 2002; Qavi, Corley, & Kay, 2001,); elimination of redundancy in patient care (Pelletier-Fleury et al., 1999); enhancements to the effectiveness of the practice (Andersson, Vimarlund & Timpka, 2002) and improvement to patient care (Leung, Yu, Wong, Johnston & Tin, 2003). From a general business perspective benefits include the ability to strategically plan and manage the business environment (Gallagher, 1998), increase the flexibility of administration and communications (Brunn, Jensen & Skovgaard, 2002), enhance efficiency (Rees, 1998; Tetteh & Burn, 2002), and better manage costs (Nelson & Alexander 2002, Pullen, Atkinson & Tucker, 2000).

Benefits Deliverable through ICT Adoption

A number of authors (Ray & Mukherjee 2007, Pan & Pokharel 2007) suggest that the use of ICTs in any business setting improves the current ways business is transacted (improvement to efficiency); new ways that business might be transacted (improvement to effectiveness); and communication. It is appropriate to consider these separately.

Improvement to the Efficiency of the Practice

Grimson, Grimson, & Hasselbring (2000) and Baldwin et al. (2002) suggest that as medical care is information-intensive, the information coming often from a variety of sources, the use of ICTs can only enhance the overall efficiency of the process. At an observable level, these improvements are seen in such things as more timely sharing of information, knowledge and expertise (Andersson et al., 2002; Fors & Moreno 2002), improvement to medical teaching facilities (Valcke & De Wever, 2006) and the distribution of these to developing economies (Gani & Clemes, 2006).

Improvement in Effectiveness in the Practice

A number of authors suggest that the use of ICTs in healthcare substantially improves its effectiveness. As already noted, ICTs provide a more timely sharing of information. However, a number of studies (Ray & Mukherjee, 2007; Waring & Wainwright, 2002) take this further suggesting that ICTs allow frameworks to be developed that reduce complex social, organisational and political issues in the dissemination of information, while still maintaining overall governance. At a more material level, Pan & Pokharel (2007) showed that the use of ICTs is essential for the procurement, distribution, inventory management and packaging of materials necessary for good health care. They found that unlike most other businesses, healthcare has two supply chains, one external and one internal since a single product might be required by a variety of users.

In smaller practices, Pelletier-Fleury et al. (1999) found that ICTs can eliminate redundancy in patient care, while Andersson et al. (2002) and Leung et al. (2003) found ICTs added to the overall care of patients and effectiveness of the practice.

Improvement to Communications

While the link between ICTs and communications in medical practices might seem obvious, a number of studies have examined the precise role of ICTs. Qavi et al. (2001) and Baldwin et al. (2002) suggest that ICTs can improve patient care through communication with other clinicians. Brunn et al. (2002) suggest that increased communications can bring with it flexibility in administration and general procedure, while El-Sayed & Westrup (2003) take this even further suggesting that the removal of temporal an spatial limits allows new models of communications, administration and procedure to be developed. Compton, Lang, Richardson & Hess (2007) note that while there is some obvious crossover from general business to medical practice clearly a number of drivers have come from research undertaken within the medical sector. He counts among these medical imaging, diagnosis and treatment at a distance and medical information management. Sandberg (2003) and Stevanovic et al. (2005) have suggested that ICT is the nucleus of chronic disease management. A number of studies have examined ICT in specific medical roles – Hubner & Selleman (2005) suggest ICT use in acute patient care, Torp, Hanson, Ulstein, & Magnusson (2008) have shown that ICT contributes to the health promotion of elderly spousal carers, while Hagglund Scandurra, & Koch (2007) and Meijer & Ragetlie (2007) have shown that ICT can empower patients.

Drivers for ICT Adoption

As might be expected, the driving forces can be aligned with the perceived benefits. However, one additional driving force needs to be added – pressure to adopt ICTs.

Pressure to Adopt ICTs

Fors & Moreno (2002) suggest that where once ICT adoption and use might have been considered an optional addition to practice, today, access to and use of information is essential to patient care, adding that at the wider business level, the use of ICTs is a cornerstone providing both information and empowerment.

Studies by Andersson et al (2002), Ray & Mukherjee (2007) take this a stage further suggesting that the adoption, design and use of ICTs must be based on supporting internal and external stakeholder pressure and should logically lead to an e-governance framework, through which other tasks can be achieved.

Gender Differences from a General Business Perspective

As most general practices in Australia are specialised small businesses, it is appropriate to examine the findings from this 'wider' perspective, before examining medical practices or the use of ICT. It is also appropriate to note the observations of Baker Aldrich, & Liou (1997) and Carter (2000) who found that research into gender differences in the ownership/management of small to medium enterprises (SMEs) is scarce by comparison to research that has examined SMEs in general. That having been said, however, there are a number of interesting findings in the literature that compare various facets of gender differences in the ownership/management of SMEs. These facets include comparisons of ownership/management statistics and reasons for the movement into the SME sector, finance availability, management style, networking, business types and success or failure of the business. These will now be considered separately.

Reasons for Movement into the SME Sector

A number of studies (Carter, 2000; Brooksbank, 2000; Reynolds, Savage, & Williams 1994,) have suggested that the primary motivation for moving into the SME sector is the desire to become self-employed. Studies by Brush & Hisrich (1999), Nillson (1997) and Sandberg (2003) have provided similar findings in Europe, US and Scandinavia. While it is possible to argue that these figures are a little dated, more recent studies (Carrington, 2006; Leung, 2006; Walker & Webster 2007) would suggest that little has changed.

It is interesting to note that while early studies (Goffee & Skase, 1985; Hisrich & Brush, 1986; Watkins & Watkins, 1984) concentrated on the motivational similarities (Male – Female), more recent studies found that females saw becoming self-employed within the SME sector as a means of circumventing the 'glass ceiling'.

Finance and Finance Availability

Studies of finance and finance availability have provided very different findings, depending on whether they were undertaken in Europe or North America. European studies undertaken by Carter (2000) and Sandberg (2003) found that males were more likely to make use of bank loans and overdrafts than females. In fact females were less likely to use or rely on financial institutional arrangements including cheaper sources of finance such as extended supplier credit than were their male counterparts. The UK study also showed that, on average, the capital expenditure on actual start-up by female owner/managers was 33% of that spent by males.

Not only do males and females differ in their use of finance, but Carter (2000) and Sandberg (2003) have shown that access to finance differs between male and female owner/managers. The studies have shown that while financial institutions may have a non-discriminatory policy, the application of those policies stereotypes and prejudices women.

The increased difficulty of obtaining finance by female owner/managers have affected four areas of financing:

- The ability to raise start-up finance (Carter, 2000)
- Differences in guarantees required to attract financing (Carter, 2000; Sandberg, 2003)
- Attraction of ongoing finance through females failing to penetrate the informal financial networks (Sandberg, 2003)
- Sexual stereotyping (Carter, 2000)

By comparison, studies carried out in North America (Bowlin & Renner, 2008; Carrington, 2006; Orser, Riding, & Manley, 2006,) suggest that female business owners were as likely as men to seek most types of external finance. These studies also found that male female and executives receive comparable salaries after allowing for differences between companies e.g. performance, size and pay philosophy.

Management Style

The literature surrounding gender differences and management style in the SME sector provides disparate results. While early studies (Maupin, 1990; Powell, 1993) suggest that there are few real differences in leadership styles between men and women, others have identified a number of important differences. Carter & Cannon (1992), for example, found that female owner managers were less confident, less aggressive and lacking in problem-solving abilities than males, while Waldstrom & Madson (2007) found that females adopt the same approaches to management as males. Verheul, Risseeuw, & Bartelse (2002) and Mcgregor & Tweed (2001) found that female managers of SMEs were more relaxed with giving instructions to staff through informal conversation than their male counterparts. While the male managers stressed the role and use of power, female managers stressed the importance of interpersonal communication. Both studies also found that female owner/managers were more likely to hire external expertise and were more inclined to develop business strategies that were specific to their particular business than their male counterparts. One interesting finding was reported by Mohr & Wolfram (2008). This was that listeners showed less irritation to male communication than female communication.

Studies in Sweden (Sandberg 2003) and New Zealand (Mcgregor & Tweed, 2001) showed that female owner managers:

- paid more attention to business-to-business links than males
- paid more attention to strategic alliances than males
- were more mindful of both their customers and their staff than were male managers
- were significantly better at dealing with the details of the day-to-day business
- were far more aware and capable of managing budgets than their male counterparts

Networking

The role of business networking has become an important area of research and government strategy within the SME sector, and is equally important for GPs. Describing them as 'self designing' partnerships, Eccles & Crane (1998 cited in Dennis, 2000), observe that business networks, within the SME sector are dynamic arrangements evolving and adjusting to accommodate changes in the business environment. Member organisations have interconnected linkages that allow more efficient movement towards predetermined objectives than would be the case if they operated as single, separate entities. In addition to providing much needed information alliances, business networks often provide legitimacy to their members. This legitimacy is particularly useful for businesses that provide a service and whose products are intangible; company image and reputation become crucial, since customers can rarely test or inspect the service beforehand. Not only do networks provide assistance and legitimacy, studies by Veradarajan & Cunningham (1995), O'Donnell, Gilmore, Cummins, & Carson (2002), Johannisson, Ramirez-Pasillas, and Karlsson, (2002) suggest that they enhance business structure, business process and a distribution of power. Again, the results of studies vary. Where early studies (Brush, 1997; Carter, 2000; Carter & Rosa, 1998; DeWine & Casbolt, 1983; Sandberg, 2003; Smeltzer & Fann, 1989) suggest that male networks were unwelcoming of females, reducing the females' ability to use network partners to gain finance or attract technical or marketing assistance, more recent studies (Klyver & Terjesen, 2007; Vinokurova, 2007; Waldstrom, 2007) suggest that females are often treated more formally than their male counterparts.

Gender Differences Among GP's

There have been a number of studies examining gender differences among GPs. Studies in Australia (Britt et al. 1996) and the Netherlands (van den Brink-Muinen et al. 1998) found, not unexpectedly, that female GPs tended to handle more gynaecological patients than male GPs. These studies also found that female GPs were more often sought out for psycho-social problems. Chambers & Campbell (1996) and Boerma & van den Brink-Muinen (2000) found that female GPs apply fewer technical procedures than males, while van den Brink-Muinen et al. (1998) found that female GPs were more attentive with problems from staff then were their male counterparts.

A recent study of 112 GPs (56 male and 56 female) (Adams et al 2008) found

'Female doctors recall more patient cues overall, particularly about history presentation' (Adams et al 2008, pp 1)

The study also showed that female doctors took more account of male age when considering diseases that did male doctors.

One study (Boerma & van den Brink-Muinen 2000) has shown there are differences between males and females in terms of preferred remuneration, with female GPs more often preferring wages than business share. This same study of 8183 GPs across Europe also found a number of other gender differences including, female GPs:

- tended to work fewer hours
- tended to allocate more time to patients
- had less technical equipment at their disposal
- made fewer house calls
- were less involved in services outside the practice (with the exception of providing training)
- preferred group practices to solo practices.

A recent study (Ribeiro 2008) found that in France and Portugal there was a vast difference in remuneration between non-national male GPs and non-national female GPs.

Gender Differences in Internet Use

There have been a number of studies conducted over the years that have examined gender differences in Internet use. Studies by Jackson et al. (2001) and Odell et al. (2000) found that females tended to use the Internet more as a communication tool, while males searched for information. Schoon & Cafolla (2002) found that females have a more difficult time navigating arbitrary web-sites than males, while Singh (2001), Oudshoorn et al. (2004) and Yang & Lester (2005) suggest that the Internet has a stronger masculine 'feel' than a feminine one. Studies by Yang & Lester (2005) found that most purchases through the Internet are subsequent to navigation and information seeking. As such, males were more likely to purchase on the Internet than females (Yang & Lester 2005, Akhter 2003).

Although our knowledge of gender differences in relation to Internet adoption as users and consumers is broad, our understanding of gender differences in relation to ICT adoption as business owners is scant and inadequate. One exception is a study of e-commerce and tele-working in 112 Spanish small businesses by Perez et al. (2002). The authors found that small businesses with female managers were significantly more concerned with the difficulty of using the technology than those with male managers. The study also cited

Figure 1. Drivers for ICT adoption in general practices

How would you rate the importance of the following factors in influencing your adoption of ICT?

Factors	Un-important	Somewhat un-important	Neither Un-important nor Important	Somewhat Important	Very Important
Pressure from patients	☐	☐	☐	☐	☐
Pressure from suppliers	☐	☐	☐	☐	☐
Pressure from competing GPs	☐	☐	☐	☐	☐
Pressure from medical authorities	☐	☐	☐	☐	☐
Improve information storage & retrieval	☐	☐	☐	☐	☐
Improve communication	☐	☐	☐	☐	☐
Reduce business costs	☐	☐	☐	☐	☐
Improve business efficiency	☐	☐	☐	☐	☐
Improve patient care/contact	☐	☐	☐	☐	☐
Improve capacity to support a systematic approach to disease management	☐	☐	☐	☐	☐
Streamlining of billing & accounting functions	☐	☐	☐	☐	☐
Strengthen relations with business related partners	☐	☐	☐	☐	☐
Facilitates e-Commerce*	☐	☐	☐	☐	☐
Keeping in touch with medical & other developments	☐	☐	☐	☐	☐
Generating prescriptions	☐	☐	☐	☐	☐
Contact with hospitals	☐	☐	☐	☐	☐

cost of technology and changes to work procedures as being of more concern to female managers.

The findings of all of these studies formed the basis for a set of driving forces (see Figure 1) and a set of benefits (see Figure 2).

Methodology

A series of interviews was undertaken with GPs to determine whether the sets of drivers and benefits listed in figures 1 and 2 were applicable and complete. The set of drivers and benefits were found to be appropriate, complete and without duplication.

It should be noted that all GPs in Australia practise within a Division of General Practice, the specific division dependent on the geographic location of the practice. There are many divisions of general practice within Australia, each containing approximately 200 – 300 GPs, each having a head of division and each staffed by GPs and non-medical staff. The role of the division is to disseminate material to GPs, to deal with disputes etc. and to keep members in touch with the latest developments (medical, legal, financial etc.).

10 divisions were contacted and asked to participate in the study. 5 (Illawarra, Hunter, Ballarat, South East NSW and Rockhampton) agreed. Questionnaires were sent to each division for distribution to members. Each questionnaire had a reply-paid envelope back to the researchers. Consequently, detail of divisional membership was not disclosed. Respondents were asked, amongst other things, to rate the importance of drivers and benefits of ICT adoption across a 5 point Likert scale (1 very unimportant, 5 very important). Respondents were also asked their gender. 890 surveys were distributed across 5 locations in Australia (Illawarra, Hunter, Ballarat, South East NSW and Rockhampton).

Figure 2. Benefits of ICT adoption in general practices

In what ways has ICT contributed to the functioning/viability of your general practice?					
Contribution	Not at all important	Un-important	Neither important nor un-important	Important	Very important
Expanding the patient/ customer base by broadening the area of coverage	☐	☐	☐	☐	☐
Improvement to business efficiency (time saving/patient care)	☐	☐	☐	☐	☐
Reduction of the overall workload and increased leisure time	☐	☐	☐	☐	☐
Enabling more time to be spend on patient care	☐	☐	☐	☐	☐
Reduction of business operating costs	☐	☐	☐	☐	☐
Improvement to the way the business is operated	☐	☐	☐	☐	☐
Allowing the business to expand	☐	☐	☐	☐	☐
Information storage and retrieval	☐	☐	☐	☐	☐
Communication with fellow GPs	☐	☐	☐	☐	☐
Communication with other medical organisations	☐	☐	☐	☐	☐
Disease Management	☐	☐	☐	☐	☐
Streamlining of billing and accounting functions	☐	☐	☐	☐	☐
Adding to the Skills of the practice	☐	☐	☐	☐	☐
Communication with hospitals	☐	☐	☐	☐	☐
Ordering drugs	☐	☐	☐	☐	☐
Other communication with general practice business suppliers	☐	☐	☐	☐	☐
Reducing the importance of distance (remoteness) in the provision of high quality medical care	☐	☐	☐	☐	☐

Results

Of the 890 surveys distributed, 198 (129 male, 69 female) were returned giving a response rate of 20.2%.

Before applying any statistical examination it was imperative to determine whether sufficient sample size had been achieved.

A formula for sample size was used

$$E = z_{\alpha/2}\, \sigma/\sqrt{n}$$

or

$$n = (z_{\alpha/2}\, \sigma/E)^2$$

At 99.9% degree of confidence $z_{\alpha/2}$ was determined to be 2.59.

The highest value for σ was 2.71
The margin of error was 1.

The minimum sample size was 50 (rounded up). A series of Levene tests was carried out to determine homogeneity of variance. The Levene's tests provided a significance of <.001 for all questions being examined, indicating that data was sufficiently robust to apply t-tests, linear regressions and chi-square tests.

For the purpose of clarity, the drivers and benefits will be considered separately.

Drivers

The results of Kaiser-Meyer-Olkin MSA (.914) and Bartlett's Test of Sphericity ($\chi^2 = 1987$, $p = .000$) indicated that the data set satisfied the assumptions for factorability. Principle Components Analysis was chosen as the method of extraction in order to account for maximum variance in the data using a minimum number of factors. A three-factor solution was extracted with eigenvalues

Table 1. A 3 Factor Model for ICT Drivers

Total Variance Explained Rotation Sum of Squared Loadings			
Component	Eigenvalue	% Variance	Cumulative %
Improvement to Medical & Business Efficiency & Effectiveness	7.567	47.295	47.295
Pressure	2.271	14.195	61.490
Improvement to Communications	1.022	6.388	67.878

Table 2. Rotated component matrix

Driver	Improvement to Medical & Business Efficiency & Effectiveness	Pressure	Improvement to Communications
Pressure from patients		.798	
Pressure from suppliers		.807	
Pressure from competing GPs		.820	
Pressure from medical authorities		.761	
Improve information storage & retrieval	.825		
Improve communication	.794		
Reduce business costs	.645		
Improve business efficiency	.831		
Improve patient care/contact	.855		
Improve capacity to support a systematic approach to disease management	.812		
Streamlining of billing & accounting functions	.748		
Strengthen relations with business related partners		.455	
Facilitates e-Commerce*			.790
Keeping in touch with medical & other developments			.653
Generating prescriptions	.730		
Contact with hospitals			.582

7.567, 2.271 and 1.002 and was supported by an inspection of the Scree plot. These 3 factors accounted for 67.878% of the variance as shown in Table 1.

The 3 components were rotated using the Varimax procedure and a simple structure was achieved as shown in the Rotated Component Matrix in Table 2.

The 3 factors are independent and uncorrelated, as an orthogonal rotation procedure was used.

Benefits

Again, the results of Kaiser-Meyer-Olkin MSA (.937) and Bartlett's Test of Sphericity ($\chi^2 = 2384,, p = .000$) indicated that the data set satis-

Table 3. A 3 Factor Model of ICT Benefits

Total Variance Explained Rotation Sum of Squared Loadings			
Component	Eigenvalue	% Variance	Cumulative %
Improved Communications	9.335	54.912	54.912
Improved Medical & Business Efficiency	1.286	7.565	62.477
Improved Medical & Business Effectiveness	1.146	6.744	69.220

Table 4. Rotated Component Matrix

Benefit	Improved Communications	Improved Medical & Business Efficiency	Improved Medical & Business Effectiveness
Expanding the patient/ customer base by broadening the area of coverage			.674
Improvement to business efficiency (time saving/ patient care)			.529
Reduction of the overall workload and increased leisure time			.722
Enabling more time to be spend on patient care			.783
Reduction of business operating costs			.693
Improvement to the way the business is operated		.689	
Allowing the business to expand			.530
Information storage and retrieval		.826	
Communication with fellow GPs	.705		
Communication with other medical organisations	.706		
Disease Management		.582	
Streamlining of billing and accounting functions		.804	
Adding to the Skills of the practice		.595	
Communication with hospitals	.742		
Ordering drugs	.646		
Other communication with general practice business suppliers	.761		
Reducing the importance of distance (remoteness) in the provision of high quality medical care			.678

fied the assumptions for factorability. Principle Components Analysis was chosen as the method of extraction in order to account for maximum variance in the data using a minimum number of factors. A three-factor solution was extracted with eigenvalues 9.335, 1.286 and 1.146 and was supported by an inspection of the Scree plot. These 3 factors accounted for 69.220% of the variance as shown in Table 3.

The 3 components were rotated using the Varimax procedure and a simple structure was achieved as shown in the Rotated Component Matrix in Table 4.

Figure 3.

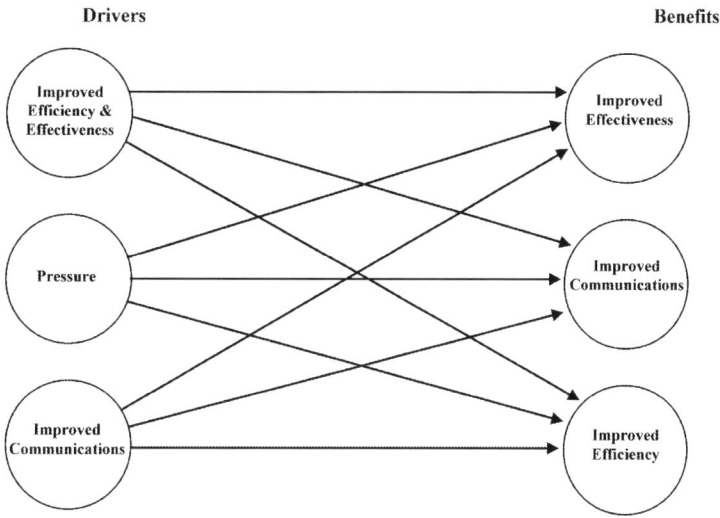

The results of the factor analysis on the drivers produced 3 underlying factors – improvement to medical and business efficiency and effectiveness, pressure and improvement to communications (see Table 2). Similarly the results of the factor analysis on the benefits produced 3 underlying factors – improved medical and business efficiency, improved communications and improved medical and business effectiveness (see Table 4). A simple model was developed using the combined male and female data (see Figure 3). Figure 3 provided all possible associations between driving force factors and benefit factors for ICT adoption. It was now appropriate to determine whether those associations differed between male and female GPs. The data was subdivided into male and female GP responses and the two sets of data formed the basis for testing the model. The model was tested using partial least squares (PLS) with PLSGraph. PLS is a combination of principal components analysis, path analysis and regression. PLS offers a number of advantages. It is suitable for exploratory studies (Chin 1998, Gefen et al 2000), it has minimal requirements on sample size and residual distribution (Gefen et al 2000) and it is an appropriate procedure for small response levels, other methods requiring greater than 200 responses (Lai 2004). The results can be seen in Figures 4 and 5 and Tables 5 & 6.

An examination of Figure 4 and Table 5 shows that those respondents that placed a higher priority on improvement to medical and business efficiency and effectiveness noted a higher level of benefit in terms of improvement to efficiency, improvement to effectiveness and improvement to communications. The data in Figure 4 shows that placing a higher priority on improvement to communications or pressure to adopt ICTs did not significantly alter the perception of any of the three groups of benefits.

In the PLS analysis, the square roots of the Average Variance Analysis (AVE) values for all constructs are higher than the correlations between constructs and the composite reliability values are above 0.70 (Gefen et al 2000). These results indicate good convergent and discriminant validity and reliability.

An examination of Figure 5 and Table 6 shows that those female respondents that placed a higher priority on improvement to medical and business efficiency and effectiveness or improvement to communications saw no significant differences in the perception of any of the three groups of

Figure 4. Partial Least Squares Model of Drivers and Benefits (Males)

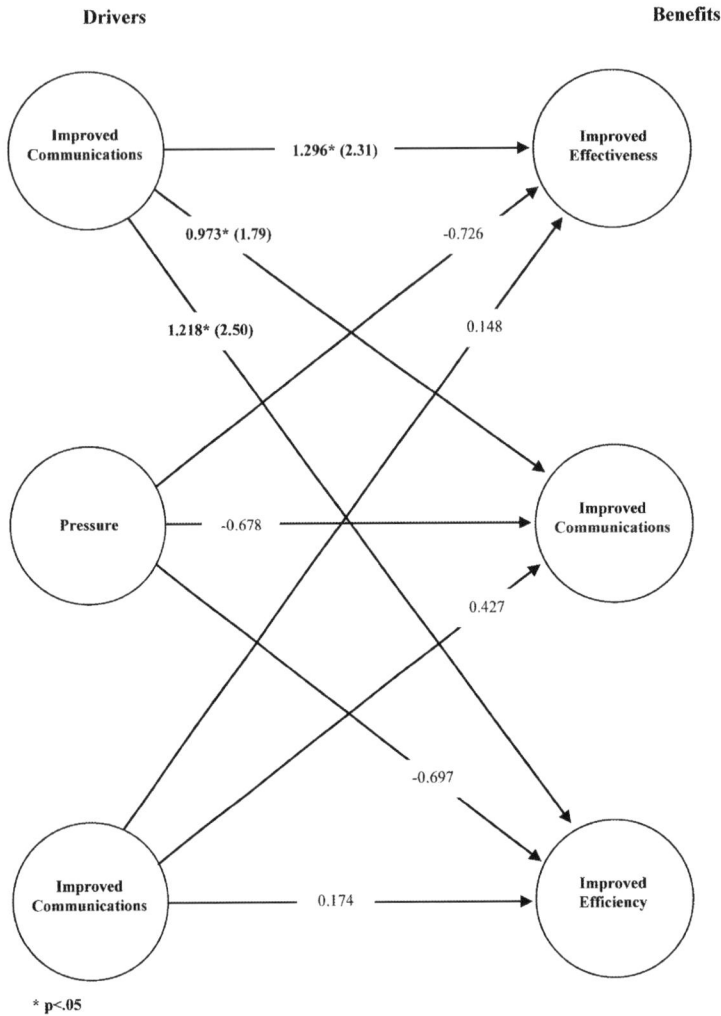

benefits. However, those respondents that had adopted ICT primarily through pressure did show a significantly positive difference in the perception of all three benefit groups.

Again, in the PLS analysis, the square roots of the Average Variance Analysis (AVE) values for all constructs are higher than the correlations between constructs and the composite reliability values are above 0.70 (Gefen et al 2000). These results indicate good convergent and discriminant validity and reliability.

DISCUSSION

An examination of Table 1 shows that there are 3 factors underlying the 16 drivers for ICT adoption. The data also shows that the highest priority given by the respondent GPs was to those drivers that loaded onto the medical and business efficiency factor (eigenvalue 7.567, % variance accounted for 47.295). An examination of Table 2 shows that 8 of the 16 drivers loaded onto the improved medical and business efficiency and effectiveness factor; 5 drivers loaded onto the pressure factor

Figure 5. Partial Least Squares Model of Drivers and Benefits (Females)

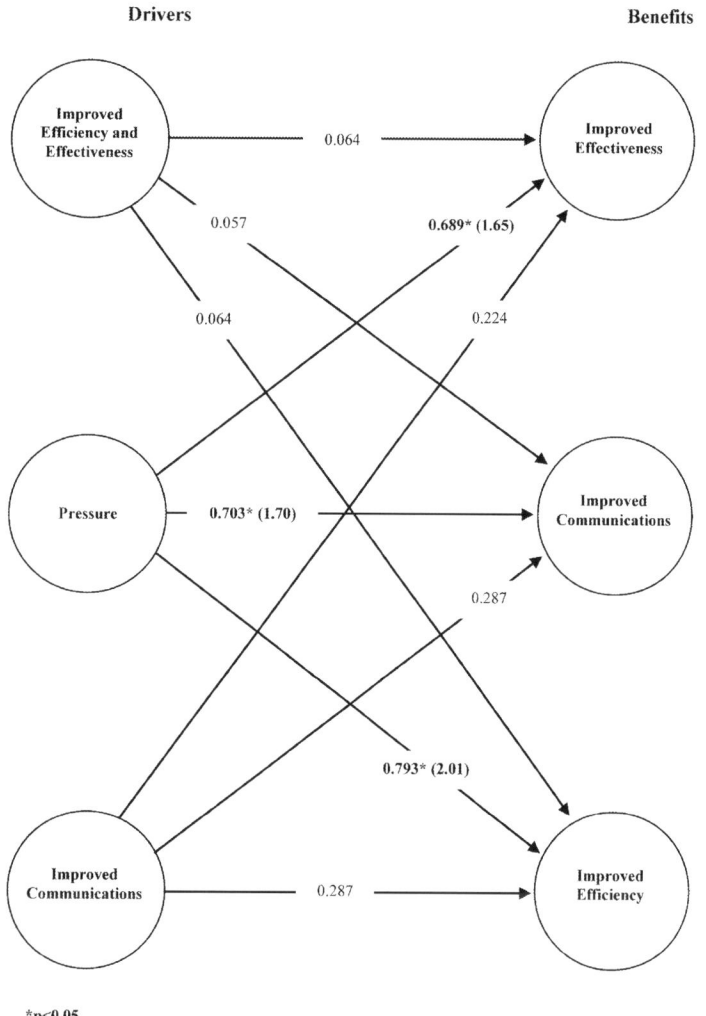

and 3 drivers loaded onto the improved communications factor.

The first result of significance can be seen by comparing Tables 1 and 3. In Table 1 we see that contrary to the literature, when GPs were asked to consider drivers for ICT adoption they grouped improvement to efficiency and improvement to effectiveness together as a single group. However, when they were asked to consider the benefits of ICT adoption, improved efficiency and improved effectiveness were considered to be 2 unique and uncorrelated groups.

An examination of Table 3 shows, again, that there are 3 factors underlying the 17 benefits derived from ICT adoption. The data shows that respondents considered the most important benefits were those that loaded onto the improvement to communication factor (eigenvalue 9.335, % variance accounted for 54.912). An examination of Table 4 shows that there are 5 benefits loaded onto the communications factor, 5 benefits loaded onto the efficiency factor and 7 benefits loaded onto the effectiveness factor.

Table 5. Square Root of AVE (diagonal numbers) Correlations and Composite Reliabilities of Constructs

NB Square Root of AVE (diagonal numbers): Males						
	Medical & Business Efficiency & Effectiveness-	Pressure	Comm's Driver	Comm's Benefit	Medical & Business Efficiency	Medical & Business Effectiveness
Medical & Business Efficiency\ Effectiveness Driver	.871					
Pressure		.964				
Comm's Driver			.831			
Comm's Benefit	.937	-.678	.427	.871		
Medical & Business Efficiency Benefit	1.218	-.697	.174		.909	
Medical & Business Effectiveness	1.296	-.726	.148			.888
Composite Reliability	.962	.985	.870	.956	.960	.949

Table 6. Square Root of AVE (diagonal numbers) Correlations and Composite Reliabilities of Constructs

NB Square Root of AVE (diagonal numbers): Females						
	Medical & Business Efficiency & Effectiveness-	Pressure	Comm's Driver	Comm's Benefit	Medical & Business Efficiency	Medical & Business Effectiveness
Medical & Business Efficiency\ Effectiveness Driver	.906					
Pressure		.913				
Comm's Driver			.915			
Comm's Benefit	.057	.707	.198	.806		
Medical & Business Efficiency Benefit	-.098	.793	.287		.922	
Medical & Business Effectiveness	.064	.689	.224			.888
Composite Reliability	.973	.974	.939	.926	.966	.948

As indicated, a simple model (see Figure 3) was developed using the combined male and female data such that it provided for all possible associations between driving forces and benefits. This model was tested using male and female data to determine which associations were significant.

An examination of Figure 4 shows that male GPs who placed greater importance on improvement to efficiency and effectiveness through ICT adoption, saw, as benefits, improvement to effectiveness (Beta = 1.296, t = 2.31), improvement to communications (Beta =.973, t = 1.79) and improvement to efficiency (Beta = 1.218, t = 2.50). The data showed that those that placed a greater importance on improvement to communications saw no significant difference in any of the three groups of benefits. It is interesting to note that those male respondents that felt their primary reason for ICT adoption was the pressure placed upon them to adopt ICTs saw a negative 'benefit' (denoted by the negative Beta values -.697, -.726 and -.678 respectively). These results support similar findings in the small business community (MacGregor & Vrazalic 2007) and while not significant, suggest that pressure from governments, medical authorities or other medical practitioners appears to lead to a loss of efficiency, effectiveness and communications within the practice.

An examination of Figure 5 shows that female GPs who placed a greater importance on improvement to effectiveness and efficiency saw no significant difference in and of the three groups of benefits. As with their male counterparts, female GPs who placed a greater importance on improvement to communications saw no significant difference in any of the three groups of benefits. Perhaps most surprising was the result for pressure to adopt ICTs. Unlike their male counterparts, female GPs reacted 'positively' to pressure being placed on them to adopt ICT in their practices. The results show that female GPs who placed greater importance on having adopted ICTs through pressure, saw, as benefits, improvement to effectiveness (Beta =.689, t = 1.65), improvement to communications (Beta =.703, t = 1.70) and improvement to efficiency (Beta =.793, t = 2.01).

There are a number of possibilities to explain the marked differences between the results in Figures 4 and 5. One possibility is that those female GP respondents who placed greater emphasis on improvement to communications or improvement to effectiveness and efficiency expected far greater changes in their day-to-day activities than was forthcoming from the introduction of ICT. In other words, male GP respondents were far easier pleased with the changes brought about by ICT adoption (particularly if that adoption had been at their instigation) than were female GPs. If this was the case, only those female GPs who had been pressured, and presumably expected little positive outcome from the adoption, would have seen positive changes to their day-to-day activities.

An alternative explanation to the findings might be that male GPs tend to be more independent in their choices of technology than female GPs. Thus, males under pressure to adopt ICT would react negatively, compared to female GPs. It is interesting to note that a European study, comparing male and female GPs (Boerma & van den Brink-Muinen 2000) found that female GPs had less technology than males and that they undertook fewer technical procedures than their male counterparts. On the surface, the current study would tend to support those earlier findings.

The results of this study are significant in several ways. The analysis has shown that 16 of the most common driving forces to ICT adoption and 17 of the most common benefits can each be grouped in relation to 3 factors each. This gives researchers a powerful explanatory tool because it reduces the "noise" in the data. Instead of accounting for 16 drivers or 17 benefits, adoption can be explained in terms of 3 factors. The Rotated Component Matrices also enable the prediction of the scores of each individual driver or benefit based on the score of the 3 factors. Whereas before researchers identified various drivers or benefits, this study shows that they are logically correlated to 3 factors. This makes it simpler not only to explain, but also predict ICT adoption in general practice.

Secondly, the results show that emphasis on certain driving forces for ICT adoption produces different perceived benefits, these differing markedly between male GPs and female GPs. While male GPs who emphasised improvement to efficiency and effectiveness as the most important reason for adopting ICT perceived a significant improvement in all three groups of benefits, this was not the case for female GPs. Similarly as male GPs saw an erosion of benefits when adopting ICT under pressure, female GPs perceived a significant improvement in the perception of benefits.

Limitations of the Study

It should be noted that this study has several limitations. The data for the study was collected from several areas in Australia. Therefore, although conclusions can be drawn, the results may not be generalisable to other countries. Also, this is a quantitative study, and further qualitative research is required to gain a better understanding of the key issues.

CONCLUSION

The results show that ICT drivers and benefits can be grouped simplifying explanation. The results show that as GPs appear to separate effectiveness and efficiency in terms of benefits, they do not separate them as drivers for ICT adoption are concerned. The study has also shown that there is a significant and important link between driving forces for ICT adoption and perceived benefits of that adoption.

The results also show that there are differences between male and female GPs where it comes to ICT adoption and use. Clearly further studies need to be undertaken to determine why these differences do occur and to determine measures to minimise the effect of these differences.

REFERENCES

Adams, A., Buckingham, C. D., Lindenmayer, A., & McKinlay, J. B. (2008). The influence of patient and doctor gender on diagnosing coronary heart disease. *Sociology of Health & Illness*, *30*(1), 1. doi:10.1111/j.1467-9566.2007.01025.x

Adogbeji, B., & Akporhonor, B. A. (2005). The impact of ICT (Internet) on research and studies: The experience of Delta State University students in Abraka, Nigeria. *Library Hi Tech News*, *10*, 17–21. doi:10.1108/07419050510644347

Akhter, S. H. (2003). Digital divide and purchase intention: Why demographic psychology matters? *Journal of Economic Psychology*, *24*, 321–327. doi:10.1016/S0167-4870(02)00171-X

Ali, K.A.M., Jemain, A.A., Yusoff, R.Z., & Abas, Z. (2007). efficient cost management through excellence quality management practices among local authorities in Malaysia. *Top Quality Management and Business Excellence, 18*(1-2, 99.

Amarsaikhan, D., Lkhagvasuren, T., Oyun, S., & Batchuluun, B. (2007). online medical diagnosis and training in rural Mongolia. *Distance Education*, *28*(2), 195–211. doi:10.1080/01587910701439241

Ammenwerth, E., Mansmann, U., Iller, C., & Eichstadter, R. (2003). factors affecting and affected by user acceptance of computer-based nursing documentation: Results of a two-year study. *Journal of the American Medical Informatics Association*, *10*(1), 69–84. doi:10.1197/jamia.M1118

Andersson, A., Vimarlund, V., & Timpka, T. (2002). Management demands on information and communication technology in process-oriented health-care organisations. *Journal of Management in Medicine*, *16*(2/3), 159–169. doi:10.1108/02689230210434907

Anvari, M. (2007). Impact of information technology on human resources in healthcare. *Healthcare Quarterly (Toronto, Ont.)*, *10*(4), 84–88.

Ash, S., Gorman, P. N., Seshadri, V., & Hersh, W. R. (2004). Perspectives on CPOE and patient care. *Journal of the American Medical Informatics Association*, *11*(2), 95–99. doi:10.1197/jamia.M1427

Baker, T., Aldrich, H. E., & Liou, N. (1997). Invisible entrepreneurs: The neglect of women business owners by mass media and scholarly journals in the USA. *Entrepreneurship and Regional Development*, *9*(3), 221–238. doi:10.1080/08985629700000013

Baldwin, L. P., Clarke, M., & Jones, R. (2002). Clinical ICT systems: Augmenting case management. *Journal of Management in Medicine, 16*(2/3), 188–198. doi:10.1108/02689230210434925

Boerma, W.G.W., & van den Brink-Muinen. (2000). Gender-related differences in the organisation and provision of services among general practitioners in Europe: A signal to Health Care Planners. *Medical Care, 38*(10), 993–1002. doi:10.1097/00005650-200010000-00003

Bonneville, L., & Pare, D. J. (2006). Socioeconomic stakes in the development of telemedicine. *Journal of Telemedicine and Telecare, 12*(5), 217–219. doi:10.1258/135763306777889073

Bowlin, W. F., & Renner, C. J. (2008). Assessing gender and top-management-team pay in the mid-cap and small-cap companies using data envelopment analysis. *European Journal of Operational Research, 185*(1), 430. doi:10.1016/j.ejor.2007.04.022

Braet, O., & Ballon P (2007). Business model scenarios for remote management. *Journal of Theoretical and Applied Electronic Commerce research, 2*(3), 62–79.

Britt, H., Bhasale, A., & Miles, D. A. (1996). The sex of the general practitioner: A comparison of characteristics, patients and our conditions managed. *Medical Care, 34*(1), 403–415. doi:10.1097/00005650-199605000-00003

Brooksbank, D. (2000). Self employment and small firms. In Carter, S., & Jones-Evans, D. (Eds.), *Enterprise and Small Business: Principles, Policy and Practice*. London: Prentice Hall.

Brunn, P., & Jensen, M., & Skovgaard. (2002). E-marketplaces: Crafting a winning strategy. *European Management Journal, 20*(3), 286–298. doi:10.1016/S0263-2373(02)00045-2

Brush, C. G. (1997). Women's entrepreneurship. In *Proceedings of the OECD Conference on Women Entrepreneurs in Small and Medium Enterprises*. Paris: OECD.

Brush, C. G., & Hisrich, R. (1999). Women owned businesses: Why do they matter? In Acs, Z. (Ed.), *Are Small Firms Important? Their Role and Impact*. Boston, MA: Kluwer Academic Publishers.

Butt, G., & Lance, A. (2005). Secondary teacher workload and job satisfaction: Do successful strategies for change exist? *Educational Management Administration & Leadership, 33*(4), 401. doi:10.1177/1741143205056304

Cabral, R., Kwong, H., & Tang, W. (2007). Managing ICT resources for the improvement of health quality in China. *International Journal of Healthcare Technology and Management, 8*(1-2), 5. doi:10.1504/IJHTM.2007.012107

Caro, D. H. J. (2005). The axis and nexus of e-health alliances in 2020. *Canadian Journal of Public Health, 96*(4), 1–3.

Carrington, C. (2006). Small business financing profiles: Women entrepreneurs. *Journal of Small Business and Entrepreneurship, 19*(2), 83.

Carter, S. (2000). Improving the numbers and performance of women-owned businesses: some implications for training and advisory services. *Education + Training, 42*(4/5), 326–333. doi:10.1108/00400910010373732

Carter, S., & Cannon, T. (1992). *Women as entrepreneurs*. London: Academic Press.

Carter, S., & Rosa, P. (1998). The financing of male and female owned businesses. *Entrepreneurship and Regional Development, 10*(3), 225–241. doi:10.1080/08985629800000013

Catalan, J. (2004). Internet medicine sales and the need for homogeneous regulation. *International Journal of Medical Marketing, 4*(4), 342–349. doi:10.1057/palgrave.jmm.5040185

Chambers, R., & Campbell, I. (1996). Gender differences in general practitioners at work. *The British Journal of General Practice, 46*, 291–293.

Cherry, J. C., Moffatt, T. P., Rodriguez, C., & Dryden, K. (2002). Diabetes disease management program for an indigent population empowered by telemedicine technology. *Diabetes Technology & Therapeutics, 11*(6), 783–791. doi:10.1089/152091502321118801

Chowdhury, S. K. (2006). Investments in ICT-capital and economic performance of small and medium scale enterprises in East Africa. *Journal of International Development, 18*(4), 533. doi:10.1002/jid.1250

Christensen, M. C., & Remler, D. (2007). Information and Communications technology in Chronic Disease Care. *Medical Care Research and Review, 64*(2), 123–147. doi:10.1177/1077558706298288

Compton, S., Lang, E., Richardson, T. M., & Hess, E. (2007). Knowledge translation consensus conference: Research methods. *Academic Emergency Medicine, 14*(11), 991.

Davies, I., Mason, R., & Lalwani, C. (2007). Assessing the impact of ICT on UK general haulage companies. *International Journal of Production Economics, 106*(1), 12. doi:10.1016/j.ijpe.2006.04.007

Dennis, C. (2000). Networking for marketing advantage. *Management Decision, 38*(4), 287–292. doi:10.1108/00251740010371757

DeWine, S., & Casbolt, D. (1983). Networking: External communication systems for female organisational members. *Journal of Business Communication, 20*, 57–67. doi:10.1177/002194368302000205

Didham, R., Martin, I., Wood, R., & Harrison, K. (2004). Information technology systems in general practice medicine in New Zealand. *The New Zealand Medical Journal, 117*, 1198.

El Sayed, H., & Westrup, C. (2003). Egypt and ICTs: How ICTs bring national initiatives, global organizations and local companies together. *Information Technology & People, 16*(1), 76–92. doi:10.1108/09593840310463041

Elbeltagi, I. (2007). E-commerce and globalization: An exploratory study of Egypt. *Cross Cultural Management, 14*(3), 196. doi:10.1108/13527600710775748

Fors, M., & Moreno, A. (2002). The benefits and obstacles of implementing ICTs strategies for development from a bottom-up approach. *Aslib Proceedings, 54*(3), 198–206. doi:10.1108/00012530210441746

Fuchs, C. (2008). The implications of new information and communication technologies for sustainability. *Environment, Development and Sustainability, 10*(3), 291–309. doi:10.1007/s10668-006-9065-0

Gallagher, M. (1998). *Evolution of facilities management in the healthcare sector Construction paper no. 86, The Chartered Institute of Builders* (pp. 1–8). Ascot.

Gani, A., & Clemes, M. D. (2006). Information and communications technology: A non-income influence on economic well being. *Information and Communications Technology, 33*(9), 649–663.

Gefen, D., Straub, D. W., & Boudreau, M. C. (2000). Structural equation modeling and regression: Guidelines for research practice. *Communications of the AIS, 4*(7), 1–78.

Goffee, R., & Skase, R. (1985). *Women in charge: The experience of female entrepreneurs*. London: Allen & Unwin.

Grimson, J., Grimson, W., & Hasselbring, W. (2000). The IS challenge in health care. *Communications of the ACM, 43*(6), 49–55. doi:10.1145/336460.336474

Hagglund, M., Scandurra, I., & Koch, S. (2007). Using scenarios to capture work processes in shared home care. *Studies in Health Technology and Informatics, 130*, 233–239.

Harrigan, P. O., Boyd, M. M., Ramsey, E., Ibbotson, P., & Bright, M. (2008). The development of e-procurement within the ICT manufacturing industry in Ireland. *Management Decision, 46*(3), 481. doi:10.1108/00251740810863906

Hisrich, R., & Brush, C. G. (1986). *The woman entrepreneur: Starting, financing and managing a successful new business*. Lexington, MA: Lexington Books.

Ho, K., Lauscher, H. N., Best, A., Walsh, G., Jarvis-Selinger, S., Fedeles, M., & Chockalingam, A. (2004). Dissecting technology-enabled knowledge translation: Essential challenges, unprecedented opportunities. *Clinical and Investigative Medicine. Medecine Clinique et Experimentale, 27*(2), 70–78.

Hsu, J., Huang, J., Kinsman, J., Fireman, B., Miller, R., Selby, J., & Ortiz, E. (2005). Use of E-Health Services Between 1999 and 2002: A Growing Digital Divide. *Journal of the American Medical Informatics Association, 12*(2), 164–171. doi:10.1197/jamia.M1672

Jackson, L. A., Ervin, K. S., Gardner, P. D., & Schmitt, N. (2001). Gender and the Internet: Women communicating and men searching. *Sex Roles, 44*, 363–379. doi:10.1023/A:1010937901821

Johannisson, B., Ramirez-Pasillas, M., & Karlsson, G. (2002, August). Theoretical and Methodological Challenges Bridging Firm Strategies and Contextual Networking. *Entrepreneurship and Innovation*, 165-174.

Johansson, U. (2003). Regional Development in Sweden: October 2003, Svenska Kommunförbundet. Retrieved from http://www.lf.svekom.se/tru/RSO/Regional_development_in_Sweden.pdf

Keddie, Z., & Jones, R. (2005). Information Communications Technology in General Practice: A Cross-Sectional Survey in London Informatics. *Primary Care, 13*(2), 113–123.

Kirigia, J. M., Seddoh, A., Gatwiri, D., Muthuri, L. H. K., & Seddoh, J. (2005). E-Health: Determinants, Opportunities, Challenges and the Way Forward. *WHO African Region BMC Public Health, 5*, 1–11.

Kolsaker, A., & Payne, C. (2002). Engendering trust in e-commerce: a study of gender-based concerns. *Marketing Intelligence & Planning, 20*(4/5), 206–214. doi:10.1108/02634500210431595

Kuruvilla, S., Dzenowagis, J., Pleasant, A., & Dwivedi, R. (2004). Digital Bridges Need Concrete Foundations: Lessons from the Health InterNetwork India. *British Medical Journal, 328*(7449), 1193. doi:10.1136/bmj.328.7449.1193

Lee, J., Cain, C., Chockley, N., & Burstin, H. (2005). The Adoption Gap: Health Information Technology in Small Physician Practices. *Health Affairs, 24*(5), 1364–1366. doi:10.1377/hlthaff.24.5.1364

Leung, D. (2006). The male/female earnings gap and female self employment. *Journal of Socio-Economics, 35*(5), 759. doi:10.1016/j.socec.2005.11.034

Leung, G. M., Yu, P. L. H., Wong, I. O. L., Johnston, J. M., & Tin, K. Y. K. (2003). Incentives and Barriers that Influence Clinical Computerisation in Hong Kong: A Population Based Physician Survey. *Journal of the American Medical Informatics Association, 10*(2), 201–212. doi:10.1197/jamia.M1202

Lievens, F., & Jordanova, M. (2004). Is There a Contradiction between Telemedicine and Business? *Journal of Telemedicine and Telecare, 10*(1), 71–74. doi:10.1258/1357633042614393

Lougheed, T. (2004). Wireless Points the Way in Africa. *Appropriate Technology, 31*(3), 50.

Lucas, H. (2008). Information and communications technology for future health systems in developing countries. *Social Science & Medicine, 66*(10), 21–22. doi:10.1016/j.socscimed.2008.01.033

MacGregor, R. C., Harvie, C., Hyland, P. N., & Lee, B. C. (2007). Benefits Derived from ICT Adoption in Regional Medical Practices: Perceptual Differences between Male and Female General Practitioners. *International Journal of Healthcare Information Systems and Informatics, 2*(1), 1–13. doi:10.4018/jhisi.2007010101

MacGregor, R. C., & Vrazalic, L. (2007). *E-commerce in Regional Small to Medium Enterprises.* Hershey, PA: IGI Global. doi:10.4018/978-1-59904-123-0

Matthews, P. (2007). ICT assimilation and SME expansion. *Journal of International Development, 19*(6), 817–827. doi:10.1002/jid.1401

Maupin, R. (1990). Sex Role Identity and Career Success of Certified Public Accountants. *Advances in Public Interest Accounting*, 97-105.

McGregor, J., & Tweed, D. (2001). Gender and Managerial Competence: Support for Theories of Androgyny. *Women in Management Review, 16*(6), 279–286. doi:10.1108/09649420110401540

Meijer, W. J., & Ragetlie, P. L. (2007). Empowering the patient with ICT tools: The unfulfilled promise. *Studies in Health Technology and Informatics, 127*, 199–218.

Mohr, G., & Wolfram, H.-J. (2008). Leadership and Effectiveness in the Context of Gender: The Role of Leaders' Verbal Behaviour. *British Journal of Management, 19*(1), 4. doi:10.1111/j.1467-8551.2007.00521.x

Ndubisi, N. O., & Kahraman, C. (2005). Malaysian Women Entrepreneurs: Understanding the ICT Usage Behaviours and Drivers. *Journal of Enterprise Information Management, 18*(6), 721–739. doi:10.1108/17410390510628418

Nelson, M. L., & Alexander, K. (2002). The emergence of supply chain management as a strategic facilities management tool. In K. Alexander (Ed.), *Proceedings of the Euro FM Research Symposium in Facilities Management.* The University of Salford

Nillson, P. (1997). Business Counselling Services Directed Towards Female Entrepreneurs – Some Legitimacy Dilemmas. *Entrepreneurship and Regional Development, 9*(3), 239–258. doi:10.1080/08985629700000014

O'Donnell, A., Gilmore, A., Cummins, D., & Carson, D. (2001). The Network Construct in Entrepreneurship Research: A Review and Critique. *Management Decision, 39*(9), 749–760. doi:10.1108/EUM0000000006220

O'Dowd, T. C., McNamara, K., Kelly, A., & O'Kelly, F. (2006). Out-of-hours co-operatives: general practice satisfaction with governance and working arrangements. *The European Journal of General Practice, 12*(1), 15–18. doi:10.1080/13814780600757195

Odell, P. M., Korgen, K. O., Schumacher, P., & Delucchi, M. (2000). Internet use among female and male college students. *Cyberpsychology & Behavior, 3*, 855–862. doi:10.1089/10949310050191836

Orser, B. J., Riding, A. L., & Manley, K. (2006). Women Entrepreneurs and Financial Capital. *Entrepreneurship Theory and Practice, 30*(5), 643. doi:10.1111/j.1540-6520.2006.00140.x

Oudshoorn, N., Rommes, E., & Stienstra, M. (2004). Configuring the User as Everybody: Gender and Design Cultures in Information and Communication Technologies. *Science, Technology & Human Values, 29*(1), 30–63. doi:10.1177/0162243903259190

Pan, Z. X. T., & Pokharel, S. (2007). Logistics in Hospitals: A Case Study of some Singapore Hospitals. *Leadership in Health Services, 20*(3), 195–207. doi:10.1108/17511870710764041

Pelletier-Fleury, N., Lanoe, J. L., Philippe, C., Gagnadoux, F., Rakotonanahary, D., & Fleury, B. (1999). Economic Studies and Technical Evaluation of Telemedicine: The Case of Telemonitored Polysomnography. *Health Policy (Amsterdam), 49*, 179–194. doi:10.1016/S0168-8510(99)00054-8

Perez, M. P., Carnicer, M. P. L., & Sanchez, A. M. (2002). Differential Effects of Gender Perceptions of Teleworking by Human Resources Managers. *Women in Management Review, 17*(6), 262–275. doi:10.1108/09649420210441914

Powell, G. N. (1993). *Women and Men in Management* (2nd ed.). Newbury Park, CA: Sage Publications.

Pullen, S., Atkinson, D., & Tucker, S. (2000). Improvements in Benchmarking the Asset Management of Medical Facilities. In *Proceedings of the International Symposium on Facilities Management and Maintenance Brisbane* (pp 265–271).

Qavi, T., Corley, L., & Kay, S. (2001). Nursing Staff Requirements for Telemedicine in the Neonatal Intensive Care Unit. *Journal of End User Computing, 13*(3), 5–13. doi:10.4018/joeuc.2001070101

Ray, S., & Mukherjee, A. (2007). Development of a Framework Towards Successful Implementation of E-governance Initiatives in Health Sector in India. *International Journal of Health Care, 20*(6), 464–483. doi:10.1108/09526860710819413

Rees, D. (1998). Management Structures of Facility Management in the National Health Service in England: A review of Trends 1995-1997. *Facilities, 15*(3 /4), 254–261. doi:10.1108/02632779810229075

Reynolds, W., Savage, W., & Williams, A. (1994). *Your own business: A Practical guide to success.* ITP.

Ribeiro, J. S. (2008). Gendering Migration Flows. *Physicians and Nurses in Portugal Equal Opportunities International, 27*(1), 77–87. doi:10.1108/02610150810844956

Rigby, M. (2006). Evaluation – The Cinderella Science of ICT in Health. *Yearbook of Medical Informatics*, 114–120.

Rodgers, S., & Harris, M. A. (2003). Gender and e-commerce: An exploratory study. *Journal of Advertising Research, 43*(3), 322. doi:10.1017/S0021849903030307

Sandberg, K. W. (2003). An Exploratory Study of Women in Micro Enterprises: Gender Related Difficulties. *Journal of Small Business and Enterprise Development, 10*(4), 408–417. doi:10.1108/14626000310504710

Schoon, P., & Cafolla, R. (2002). World Wide Web Hypertext Linkage Patterns. *Journal of Educational Multimedia and Hypermedia, 11*, 117–139.

Selwood, I., & Pilkington, R. (2005). Teacher Workload: Using ICT to release time to teach. *Educational Review, 57*(2), 163. doi:10.1080/0013191042000308341

Shohet, I. M., & Lavy, S. (2004). Healthcare Facilities Management. *State of the Art Review Facilities, 22*(7/8), 210–220. doi:10.1108/02632770410547570

Simon, S. R., Kaushal, R., Cleary, P. D., Jenter, C. A., Volk, L. A., & Poon, E. G. (2007). Correlates of Electronic Health Record Adoption in Office Practices: A Statewide Survey. *Journal of the American Medical Informatics Association, 14*(1), 110–117. doi:10.1197/jamia.M2187

Singh, S. (2001). Gender and use of the Internet at home. *New Media & Society, 3*(4), 395–415.

Smeltzer, L. R., & Fann, G. L. (1989). Gender Differences in External Networks of Small Business Owner/Managers. *Journal of Small Business Management*, 25–32.

Smith, R. (2004). Access to Healthcare via Telehealth: Experiences from the Pacific. *Perspectives on Global Development and Technology*, *3*(1), 197. doi:10.1163/1569150042036693

Stevanovic, R., Stanic, A., & Varga, S. (2005). Information Systems in Primary Health Care. *Acta Medica Croatica*, *59*(3), 209–212.

Tetteh, E., & Burn, J. (2001). Global Strategies for SME-business: Applying the SMALL Framework. *Logistics Information Management*, *14*(1-2), 171–180. doi:10.1108/09576050110363202

Torp, S., Hanson, E., Ulstein, I., & Magnusson, I. (2008). A pilot study of how information and communication technology may contribute to health promotion among elderly spousal carers in Norway. *Health & Social Care in the Community*, *16*(1), 75–85. doi:10.1111/j.1365-2524.2007.00725.x

Valcke, M., & De Wever, B. (2006). Information and Communication Technologies in Higher Education: Evidence-based Practices in Medical Education. *Medical Teacher*, *28*(1), 40–48. doi:10.1080/01421590500441927

Van den Brink-Muinen, A., Bensing, J. M., & Kerssens, J. J. (1998). Gender and Communication Style in general practice: Differences between Women's Health care and regular health care. *Medical Care*, *36*, 100–106. doi:10.1097/00005650-199801000-00012

Varadarajan, P. R., & Cunningham, M. (1995). Strategic Alliances: A Synthesis of Conceptual Foundations. *Journal of the Academy of Marketing Science*, *23*(4), 282–296. doi:10.1177/009207039502300408

Verheul, I., Risseeuw, P., & Bartelse, G. (2002). Gender Differences in Strategy and Human Resource Management. *International Small Business Journal*, *20*(4), 443–476. doi:10.1177/0266242602204004

Waldstrom, C., & Madsen, H. (2007). Social relations among managers: Old boys and young women's networks. *Women in Management Review*, *22*(2), 136. doi:10.1108/09649420710732097

Walker, E. A., & Webster, B. J. (2007). Gender, age and self-employment: Some things change, some stay the same. *Women in Management Review*, *22*(2), 122. doi:10.1108/09649420710732088

Waring, T., & Wainwright, D. (2002). Enhancing Clinical and Management Discourse in ICT Implementation. *Journal of Management in Medicine*, *16*(2/3), 133–149. doi:10.1108/02689230210434880

Watkins, J., & Watkins, D. (1984). The Female Entrepreneur: Backgrounds and Determinants of Business Choice – Some British Data. *International Small Business Journal*, *2*(4), 21–31.

Yang, B., & Lester, D. (2005). Gender Differences in e-commerce. *Applied Economics*, *37*, 2077–2089. doi:10.1080/00036840500293292

ENDNOTE

[1] A previous version of this chapter was published in the International Journal of E-Politics, Volume 2, Issue 1, edited by Celia Romm, pp. 68-85, copyright 2011 by IGI Publishing (an imprint of IGI Global).

Section 2
Gender and Computing in Cyberspace

Chapter 9
The Not So Level Playing Field:
Disability Identity and Gender Representation in Second Life

Abbe E. Forman
Temple University, USA

Paul M. A. Baker
Georgia Institute of Technology, USA

Jessica Pater
Georgia Institute of Technology, USA

Kel Smith
Anikto LLC, USA

ABSTRACT

The study reported in this chapter examined gender and disability identity representation in the virtual environment, Second Life. In Second Life, identity representation is the choice of the user and is a matter of convenience, style or whim, rather than a fixed characteristic. A survey of groups that identify as disabled or having a disability, especially focusing on gender, was conducted in Second Life. The distinct categories analyzed in this study included: groups associated with disability/being disabled, race/ethnicity, gender, aging, and sexuality. In the virtual world, the visual cues that exist in the "real world" are removed. However, in the "real world", those visual cues serve to activate schemas that may help explain the stigmas and ensuing isolation often felt by people with disabilities. Interestingly, in Second Life even when the visual cues are removed, users with disabilities still associate with others who identify as having disabilities. The study specifically explored groups (i.e. "communities") found in Second Life that jointly identify by gender and a disability identities. Regardless of binary gender framework, the differences between the groups that are externally classified as having some degree of disability, and those who choose to self identify or affiliate with disability related groups, have rich import for the sociology of online communities as well as for the design and characteristics of games.

DOI: 10.4018/978-1-60960-759-3.ch009

INTRODUCTION

The famous caption of a cartoon by Peter Steiner published by *The New Yorker*, "On the Internet, nobody knows you're a dog", speaks to the fact that a degree of anonymity and choice in self-representation exists for users of the Internet. The mutability of identity (gender, race, and portrayal of physical identity) especially in the virtual milieu is considered by many a liberating condition. A great deal has been written on the "inflation of desirability" in the world of online interactions, especially in synthetic, simulated environments such as Second Life where the actual identity of an individual (or if indeed, it is actually a "real" individual, rather than a software simulation) is not intuitively apparent. Other online environments such as Facebook, while not comparable simulated environments, present similar opportunities to shape or manage the representation of actual identity albeit in less detailed dimensions (that is two dimensional rather than virtual 3D) than those afforded to participants in Second Life.

Any discussion of identity and portrayal, online or otherwise, presupposes that there is an "other" – the audience, community, or observer who perceives the individual. While physical communities have typically been geographically constructed, online communities are generally communities of self selection rather than ones in which membership is automatically attributed by virtue of locale (Baker & Ward, 2002). This paper examined the nature of disability, gender and identity portrayal in virtual environments, in which representation is a matter of convenience, style or whim. To explore this idea, a survey was conducted of self-identified groups (i.e. communities), in the virtual space, Second Life. The Second Life platform was chosen because it offers unique opportunities for self expression and identity development within an immersive environment. In Second Life environments personal and group identity are related mainly, but not solely, to the 'avatar', embodiment with interactive and immersive characteristics (Bortoluzzi and Trevisan, 2009). For this study, immersivity is crucial because "digital environments allow us to transform our self-representations dramatically, easily, and in ways that are not possible in the physical world," which is of even greater significance because "Collaborative Virtual Environments allow [for] geographically-separated individuals to interact via networking technology, oftentimes with graphical avatars" (Yee, 2007).

The importance of this work is both theoretical, as well as practical. First, the development of Disability Schema Theory, an extension of Schema Theory, will provide researchers a potential framework for continuing the discussion regarding disability in both the virtual and real world. This continuation could lead to what we refer to as the "level playing field" in the real world, similar to that which exists in the virtual world. Secondly, any movement toward understanding what causes stigmatization for people with disabilities including gendered stigmatization, can be used to help reduce that stigma therefore creating a world where people with disabilities are no longer marginalized and will be considered equal members of the society in which they live. Additionally, while this shifting kaleidoscope of identity is itself of interest, a more interesting phenomenon is one of individuals who choose to make apparent and explicit, visible manifestations of their disability. This study explored these alternative expressions of gender and disability that occur in virtual environments.

IDENTITY, DISABILITY, AND GENDER IN VIRTUAL ENVIRONMENTS

"Cyberspace has been cast as a post gender, post-human world, where the 'lived' body or 'meat' is be [sic] left behind in the real social world. It has been suggested that there is an absence of a physical body in cyberspace, and instead, a disembodied free floating electronic/cyber self or cyber-persona manifests itself in netspace." (Ward, 2001, p. 189).

Previous studies have typically focused on either gender representations or disability representation in virtual environments, but rarely both. The extension of the concept of the digital divide to encompass the idea of a disability divide is further hindered by the minimal representation of women with disabilities. Yet in one aspect of digital "life" more than twenty percent of gamers are believed to have some degree of functional limitation (Ingham, 2008).

Research involving virtual environments has often included the study of individuals with disabilities, typically of individuals who participate in chatrooms and other venues of discourse constructed to address the interests of the disabled community. Little attention, however, has been paid to the inclusion of users with disabilities, specifically women; in general non-targeted virtual environments open to everyone, such as Second Life. Additionally, the question of how individuals identify themselves in these virtual environments has gotten little attention in the literature.

This chapter reports on an exploration of the representation of gender, disability and identity in Second Life in order to develop a framework for understanding nuanced communities of communication in virtual environments. The framework draws upon Schema Theory, coupled with literature on online identity and representation to provide the groundwork for our empirical study. Specifically, Gender Schema Theory, a prior extension to Schema Theory is discussed as well as a new extension, Disability Schema Theory.

SCHEMA THEORY

Schema theory articulates the process of creating a perception built on the preexisting schema (i.e. cognitive structure) coupled with new incoming information (Bem, 1981). According to Bem, a schema helps to shape an individual's perception allowing individuals to attach meaning onto vast amounts of incoming information. A schema develops after an individual repeatedly observes similar events (Fiske & Taylor, 1984). Eventually the observed characteristics and behaviors will be incorporated into a preexisting schema or a new schema will be developed (Perry, Davis-Blake, & Kulik, 1994). Park and Hastie (1987) found that schemas can be learned through explicit instruction as well as vicarious observation. Perry et al. noted that if an individual repeatedly observes exceptions to existing schemas, a new *subschema* may be developed. Subschemas allow an individual to retain their original schema, while allowing for a newer schema incorporating the observed variations (Perry et al.). Once schemas are formed, they are stored in long term memory and may be used in conjunction with other schemas, or alone, as the basis for judgments and decisions based on judgment (Perry et al.).

Bem (1993) defined gender schemas as schemas developed within individual role expectations based on the biological sex of the individual under observation. Creation of a gender schema usually occurs when an individual observes or is taught that certain behaviors are attributed to one specific gender (Perry et al., 1994). Although the formation of gender schemas begins at a young age, as individuals mature cultural differences may challenge the attributes of those schemas (Lemons & Parzinger, 2007). In addition to cultural differences, an individual may challenge or reject traditional gender schemas and thus become gender non-conformists who form and use non-traditional gender schemas when processing information based on observed behaviors, thus rejecting traditional gender roles (Bem).

Bem (1993) noted that there are two antecedents to gender schema theory; social learning theory and cognitive-development theory. Social learning theory informs gender schema theory by noting that discourse and the social structures of a culture help determine the gender schemas that are formed by a developing child. Bem added that cognitive-development theory helps to explain an individual's self-identity constructed using the gender schemas formed as a child.

The following quote from Latour illustrates the construct of a schema based on the observation of a person with a gun:

You are a different person with a gun in your hand. Essence is existence and existence is action. If I define you by what you have (the gun), and by the series of associations that you enter when you use what you have (when you fire the gun), then you are modified by the gun – more or less so, depending on the weight of the other associations that you carry. This translation is wholly symmetrical. You are different with a gun in hand; the gun is different with you holding it. You are another subject because you hold the gun; the gun is another subject because it has entered into a relationship with you (Latour, 1994, p. 33).

If in place of the gun, we substitute the example of a wheelchair, a person becomes associated with the wheelchair therefore informing the construction of a schema based on a perceived condition or disability, extending the general schema theory to construct a disability schema theory. If gender schema theory lends itself to the internalization of gender expectations based on the biological sex, then disability schema theory could lead to the development of internalized notions derived from the observation of individuals with one or more (visible) disabilities. This becomes more complicated when dealing with non-visible disabilities, or when the observer is aware that conditions of disability exist, but they are not readily apparent. For example consider the dissonance that can occur when observing an individual who parks a car in a "handicapped parking" slot, has plates or a tag identifying them as having a disability, yet apparently walks "normally" into a store.

ONLINE GENDER IDENTITY

The notion that online identity can be more fluid than "real life" identity is pervasive in the literature and takes into account the different ways in which attributes of identity including gender, age, religion, cultural heritage, etc., manifest in online as opposed to physical settings (Bowker & Tuffin, 2007; Whitley, 1997; Suler, 2002; Nowak & Rauh, 2005) Because of the relative absence of objective identifying information or cues, and the use of text based communication, individuals can depict themselves in whatever form they choose, allowing freedom from embodied identities (Coates, 2001). Furthermore, the lack of cues can create an environment where individuals can explore alternative aspects of identity, including gender (Roberts & Parks, 2001). In computer mediated environments, decisions regarding gender presentation are an important element in an individual's overall self-presentation (Samp, Wittenberg, & Gillett, 2003).

Online gender identity does not necessarily map onto the biological "sex" or gender of the user. Roberts and Park (1999) reported that users may choose to gender swap when representing themselves in the computer mediated environment. However, research regarding the degree or frequency of online gender swapping has been inconclusive. While Roberts and Parks found that between forty and sixty percent of users admitted gender-swapping, Samp et al. (2003) reported that only 28 percent of their respondents had admitted to portraying themselves as the "opposite" gender. Samp et al. had hypothesized that individual's with strong gender schemas (either masculine or feminine) would be more likely to gender swap in a computer mediated environment than those who were gender aschematic (more androgynous). Kacen (2000) reported that women have a greater tendency to mask their true gender identity in a computer mediated environment. Kacen added that since physical appearance can be so significant in the formation of gender schemas, online gender identity may be superfluous. Computer mediated environments free the user from the necessity of adopting normative binary gender rules and presents the user with fluidity regarding the presentation of one's online gender identity.

ONLINE DISABILITY IDENTITY

"Models of disability provide a framework for understanding the way in which people with impairments experience disability. They also provide a reference for society as laws, regulations and structures are developed that impact on the lives of disabled people. There are two main models that have influenced modern thinking about disability: the medical model and the social model." (Open University, 2006) According to Thoreau (2006), the often criticized medical model defines disability as sickness or impairment and a deviation from normality. Thoreau stated that the social model is defined by the barriers, mental and social, imposed by a nondisabled society on people with disability. These barriers limit or remove the opportunity for community and often lead to oppression (Bowker & Tuffin, 2002), particularly for people with disabilities for whom the presence of a disability can be largely undetectable. People with visible disabilities (those that can be seen), are often marginalized and undervalued in society (Bowker & Tuffin, 2007), and those who live with autism and other cognitive disorders suffer additionally from a lack of awareness on the part of people unfamiliar with such disorders. Individuals develop schemas through observation which can result in social stigmatization for people with disabilities (Goffman, 1986). However, in computer mediated environments people with disabilities have the opportunity to "level the playing field" by controlling the image that is "viewed" (Bowker & Tuffin, 2003), potentially nullifying the impact of the schema. Participating in these environments can offer people with disabilities the opportunity to escape from the isolation and stigma frequently associated with disability (Dobransky & Hargittai, 2006). By operating in a medium where visual perceptions are not the primary element, physical disabilities can be masked leading to what Bowker and Tuffin described as a "more positive, socially valued identity" (p. 64). However, prejudices are not removed by computer mediated environments. In fact, the "level playing field" provided by the lack of visual cues creates an option for people with visible disabilities that they are not afforded in "real world" situations. Instead, in the computer mediated environment, people with disabilities have the option or choice of whether or not to disclose their disability. This choice can either mitigate much of the prejudice or can put situational control into the hands of people with disabilities. This led us to our first research question:

RQ1. *Do people with disabilities identify as disabled in Second Life?*

Computer mediated environments offer people with disability the capacity to represent their identity based on personal choice, independent of any actual physical characteristics, and thus the ability to counter the potentially negative schemas they face. In a discourse analysis study, Thoreau (2006) found that people with disability tended to represent themselves as disabled through the use of humor and irony using both nonmedical and generic descriptors. Therefore the decision to disclose the presence or absence of disability is controlled by the individual and is only considered relevant when it is specific to the context of the conversation (Bowker & Tuffin, 2002). Bowker and Tuffin added that removal of the conceptual presence of disability was seen as positive by respondents who indicated a greater level of acceptance after disclosing a disability, likely due to the absence of visual cues in computer mediated environments. By operating in an anonymous environment, these respondents are able to experience an identity that offered a level of "able bodied-ness" and acceptance not experienced outside of computer mediated environments.

Although there is some literature on the characteristics of participation of people with disabilities in online settings (see for instance: Curran, et. al. (2007)) very little extant literature explores the actual characteristics of individuals with disabilities online. In attempting to provide context and some degree of linkage with the "real" (i.e. "non-virtual world") we draw upon Cardinali & Gordon (2001) for analogous statistics. They note that much of the prior research regarding gender and disability has focused on males in both disability studies as well as feminist studies (Cardinali & Gordon). Cardinali and Gordon also provided the following statistics:

- Men with disabilities are nearly twice as likely to have jobs as women with disabilities
- 42% of men with disabilities are in the labor force but only 24% of women with disabilities
- 12% of women with disabilities have full time employment as opposed to 30% of men with disabilities
- Women with disabilities who are employed full time earn 56% of what full time employed men with disabilities earn.

In the absence of a robust "census" of people with disabilities online, we suggest the statistics above as a surrogate for online participation. This gap in the research only serves to highlight the need to level the playing field through computer mediated environments. Further research is clearly needed to not only understand the role of computer mediated environments as the great equalizer, but to find ways to translate that into opportunities for women with disabilities in the physical world.

CHARACTERISTICS OF THE SECOND LIFE PLATFORM

At first glance, it would appear that online virtual worlds with their emphasis on 3D graphics and complex interface controls have little to offer people with disabilities. On the contrary, virtual worlds serve as a form of augmented reality where users transcend physiological or cognitive challenges to great social and therapeutic benefit. A number of intriguing developments exist within the accessibility sector, particularly for users of Linden Lab's Second Life framework: haptic input devices for the blind, virtual regions developed according to universal design principles, communities dedicated to people with cognitive disorders, the use of the avatar as counselor, applications in higher education, and customizable personae that either transcend or represent a disabled person's self-identity. Therefore with the increased ability to participate on a "level playing field," it is possible to engage in interactions with other inhabitants in a way in which specific physical characteristics are a matter of preference rather than default of visually apparent as they are in the real world. As noted above, the fact that people choose to portray themselves as "augmented" or enhanced is not surprising, but that individuals would intentionally portray themselves as disabled, potentially facing the consequential results of negative disability schemas, is of interest, which led us to our second research question:

RQ2. *Given the lack of visual cues, are there groups created in Second Life that are specific to people with disabilities? Are any of these groups additionally identified by gender?*

When Second Life achieved mainstream attention at the close of 2006, collective attitudes regarding virtual worlds were still evolving. These graphical landscapes, built from bytes of computer code and populated by self-constructed identities, called avatars, offered virtual residents the opportunity to engage in new forms of social and physical interaction. A virtual world is a simulated three-dimensional (3D) environment accessed through a computer. More than a flat (2 dimensional) website and utilizing the same technological aspects as games; virtual worlds are typically accessed via the Internet. Participants then interact with their environment, and with other users of the environment through a presentation layer consisting of such features as animated avatars, customizable objects, instant chat messaging and voice-activation. The use of virtual worlds has been explored for many purposes spanning entertainment, socialization, education and commerce industries.

Virtual worlds are sometimes considered an extrapolation of "serious games" – a software or hardware player application developed with gaming technology or design principles, intended for use beyond pure entertainment (Hinton, 2006). These programs have been developed and deployed for such purposes as education, marketing, advertisement, workplace training, or health awareness. The main difference between a "game" and a virtual world is in the objective: game players expect to be confronted with obstacles that are intentionally built into the software, while users of virtual worlds seek to engage and navigate their way through an environment empathetic to achieving user-centered goals (Smith, 2009).

Second Life, the virtual world platform developed by Linden Lab, has attracted the most widespread attention and name recognition outside of specialized gaming communities. While other applications such as Blizzard's World of Warcraft, boast greater numbers of simultaneous users, they tend to focus more on goal-driven gameplay rather than on social interaction (Hinton, 2006). There are other multi-user virtual environments, for example the IMVU platform, that provide such tools for peer-to-peer collaboration as 3D messaging, personalized profile pages, developer tools and group forums (Caoili, 2008).

From an accessibility standpoint, it would appear that gaming interfaces and virtual worlds have little to offer people with disabilities. The experience is largely visual in nature, with multiple interaction paradigms offering deep levels of customization. User inputs often require extensive hand/eye coordination to precisely control an avatar's movements. Some applications use non-persistent sound and fading messages to deliver information; for users who are unaccustomed to this level of multitasking, the resulting cognitive load can be severe.

It is interesting, then, that a new form of social literacy has begun to take shape. There exists a vital demographic of virtual world participants with a wide range of functional limitations (i.e. "disabilities"): visual impairments, motor skill disorders, degenerative illness, limited mobility, and cognitive difficulties, among others. Many of these users utilize virtual technology to great social and therapeutic benefit. For these users, avatar-driven 3D environments serve as more than a game (Deeley, 2008). Virtual worlds operate as a form of augmented reality, one where it's possible to transcend a user's physiological or cognitive challenges into something extraordinary. In fact, the choices users make when creating and maintaining their avatars, can lead to the creation and maintenance of new environments and cultures that exist within Second Life (Diehl & Prins, 2008).

TYPES OF VIRTUAL WORLD USERS

To fully understand this emerging paradigm in assistive technology, it's important to recognize how people use virtual world gaming software. Users can fall into any of three categories: *augmentationists, immersionists, or experimentalists* (Duranske, 2008). All have applications of relevance to people with disabilities. Augmentationists view the virtual world as a means to enhance their real life existence. They view their virtual personae as extensions of their identities, and they are more willing to disclose their real life identities to others in-world. Many who conduct business online, such as attorneys who practice aspects of virtual law, feel comfortable representing themselves with an avatar that closely resembles their real life appearance (Duranske, 2008).

Another practical example of an augmentationist would be someone with a visible disability who chooses to represent him- or herself as authentically as possible. This user will go so far as to outfit his or her avatar with a wheelchair, dark glasses, a guide dog, or other visual attributes representing their disability.

First-time users tend to start off as pure augmentationists, but they do not remain that way for long (Duranske, 2008). Within a short time, it's possible for a person to become proficient at making choices regarding her or his avatar's appearance and functionality. Some users with disabilities will take advantage of this feature by making the experience easier to navigate. For example, a visually-impaired resident of Second Life may dress her avatar in light colors to help visually track her location on the screen.

It is at this stage that augmentationists become immersionists – people who view virtual worlds as an alternative parallel to their real life existence. These types of users generally keep their real life identities separate from that of their avatars, with the idea that the two streams will never cross paths (Duranske, 2008). An example of an immersionist might be someone with Aspergers syndrome who exploits the anonymity of virtual worlds to practice social interaction skills.

Some avatars employ radical means to differentiate their virtual experience from real life, choosing to discard any attributes common to disability. Rather than depict themselves as "broken" with wheelchairs and canes, they make themselves available for such enjoyable activities as walking, running, surfing, dancing and riding horseback. For people with disabilities that prevent them from engaging in these real-life physical activities, virtual worlds offer a unique opportunity for users to simulate the experience at an immersive level. Interestingly, virtual reality applications have been used to augment rehabilitation therapy for patients struggling with the loss of a limb. Research demonstrates that the brain's perception to pain can be reduced when it is "tricked" into operating a replicative appendage (Ramachandran, 2006).

A third group of virtual world users are the experimentalists, who use virtual worlds as a controlled laboratory to conduct training or educational sessions. Experimentalists usually take the form of educators and trainers, or perhaps a counselor working with patients dealing with substance abuse. Another example of an experimentalist might be someone who seeks to gain empathy by undergoing a simulated experience. The Sacramento Mental Health Center in Second Life, for example, provides a virtual replica of their real-world facility, including an authentic representation of a schizophrenic episode. With visual hallucinations and subliminal voices providing an accurate depiction, the site provides visitors an opportunity to directly experience what someone with schizophrenia may go through (Deeley, 2008).

Given that so many different types of users exist in Second Life, our third research question seeks to understand whether or not people in Second Life (avatars), group together/associate with other similar "bodied" avatars. Therefore, our third research question asks:

RQ3. *Based on search term identifiers, do people appear to associate with others who identify with specific characteristics (i.e. disabilities) or genders when the cues that reinforce schemas are not apparent or observable?*

METHODOLOGY

In order to explore the identity representation of disability as well as gender identity in a virtual world environment, a survey was conducted of user groups in Second Life, the virtual world chosen for this study. Given the size (in terms of number of participants), malleability, and the presence of formal agencies and groups associated with specific groups/causes currently using the platform, it was the most representative research environment. Specifically, the survey was designed to answer the following research questions:

RQ1. *Do people with disabilities identify as disabled in Second Life?*

RQ2. *Given the lack of visual cues, are there groups created in Second Life that are specific to people with disabilities? Are any of these groups additionally identified by gender?*

RQ3. *Based on search term identifiers, do people appear to associate with others who identify with specific characteristics (i.e. disabilities) or genders when the cues that reinforce schemas are not apparent or observable?*

The Second Life platform is particularly useful as it offers the freedom of representational choices that range from near correspondence to the real world to complete abstraction. This range of use cases provides a variety of unique perspectives spanning multiple types of disability: vision, motor skill, mobility and cognition. In essence, Second Life chose to mirror the real world in many important aspects in order to provide a place that feels familiar and comfortable, while granting freedoms not possible in the real world (Ondrejka, 2004). Interestingly, while Second Life was designed to mirror real life, users can choose a completely different experience. Creation of an avatar as representation of one's self can take on dimensions not possible in the real world. For instance, people with visible disabilities may choose to create an avatar with no disabilities. The question then becomes, if the visual cues that reinforce schemas are removed, how do people identify themselves with respect to gender or disability? Do people choose to associate with others who are similarly abled or gendered when the cues that reinforce schemas are not visible? Given the lack of visual cues, are there groups created in Second Life that are intended to attract people that identify as disabled? Are the groups specifically gendered?

The data for this analysis was collected through using the survey approach noted above between March 30th and April 15th 2009, using an avatar account maintained by one of the study's authors.

Methodologically, concerns have been raised in the literature about the validity and complexity of conducting behavioral research solely within virtual spaces. Yet, many virtual worlds like Second Life present evolving cultures with their own social institutions that are becoming more significant to society at large (Noveck, 2004). A growing number of researchers have demonstrated the feasibility and importance of developing research methodologies that keep up with the realities of technological change (Boellstorff, 2008). Therefore, ethnographic and other empirical research and analysis are critical to understanding how group identities are different in immersive virtual settings than they are in the traditional "real" world. This study surveyed groups that identified themselves as having a disability and focusing on gender identity, in addition to referencing minority and sexual orientation identity groups for comparison. One of the issues present in this study relates, iteratively, to the focus of the study - self-*disclosure*. For instance, in the subject platform one of the rationales for the "game" of second life is the ability or rather, the characteristics of malleability of representation of character. We are interested here in capturing evidence of the phenomena of representing oneself in a manner, which, in the real world, disability, is characterized as a functional limitation. The fact that individuals choose to represent themselves in a realist representation rather than in a fantasy manner can be somewhat counter intuitive relative to the typical practice. Having established that, indeed people choose to express themselves as they *are* rather than as what would be considered "typical" is a first and necessary step in understanding the phenomena. We would expect that subsequent research would involve in-world ethnographic interviews individuals to develop robust understanding from subjective perspective.

There are an estimated 185,000 groups in the Second Life platform, which has a population/user base of approximately 16.5 million registered users (Linden Labs, 2009). It should be noted that this figure potentially inflates the number of unique users that log into the platform every 60 days (1,353,522 users within 60 days of the April 17th log-in) (Linden Labs, 2009). For context, the *Encyclopedia of Associations* has information on over 135,000 nonprofit organizations worldwide (Encyclopedia of Associations, 2009), although obviously an identity group and a recognized organization are not by any means equivalent.

The groups selected for this study were identified using the search feature designed into the Second Life architecture. The tool allows a user to search on specifically designated entities or keywords. The keyword/search engine approach has been used successfully in a number of different applications, especially in exploratory designs. (Beard, et. al., 2009; Fang & Lee, 2009; Norris, 2009). Through the search function one can conduct searches for services, groups, individuals, places, events, etc. Group database records in Second Life contain standard information: a group charter (can vary in length); identification of group owners and visible members; a log of all notices distributed by the group in the last 14 day time frame, and any land associated with the group. This latter condition reflects the fact that in order to purchase land from Linden Labs, a group must be formed to which the land is then deeded. Once the search results were returned, the mission statement of each group was read to verify that the group's main focus pertained to the keywords that were searched. For a group to be included (indicating that there was a degree of relationship to one of the target variables, "disability" or "gender" or "sexuality") several criteria had to be met:

1. Within the name and/or charter of the group, does the group use the terms "disabilities," "disabled," or suggest the presence of "handicap or impairment"?
2. Did the group have more than one member; and is the group English based (non-English terms were not used in the search).
3. Is gender specifically noted by group members?

DATA ANALYSIS

The data collected for this study was strictly descriptive in nature. For this foundational study, the individual cases were categorically sorted within and analyzed for patterns within the groupings of the keywords used for the search. Four distinctive groupings were analyzed in this study: groups associated with disabilities or having a disability, race/ethnicity, gender, and sexuality. Table 1 shows the search terms used for each of these group designations.

Basic average and summation tests were conducted to analyze for patterns that would signify that deeper analysis is warranted. Simple comparison tests were also conducted on membership statistics within these groups (both the groups that were returned during the search and the groups that actually met the criteria).

RESULTS

Comparison of Groups

The total number of groups that were returned within the search varied greatly. The numbers ranged from the Race/Ethnicity grouping returning 101 keyword matches to 1283 matches based

Table 1. Key search terms by group

Gender	Race/Ethnicity	Disability	Aging	Sexuality
Female	African American	Disability	Retired	Homosexual
Woman	Hispanic	Disabled	Aging	Heterosexual
Male	Latino	Handicap	Geriatric	Bisexual
Masculine	Islamic	Impaired	Geezer	LGBT
Transgender	Muslim	Muscular Dystrophy	Old	Gay
		Aspergers		Lesbian
		Blind		Queer
		Learning disorder		
		Aging		
		Wheelchair		

Figure 1. Comparison of search returns vs. actual number of groups that met the search criteria

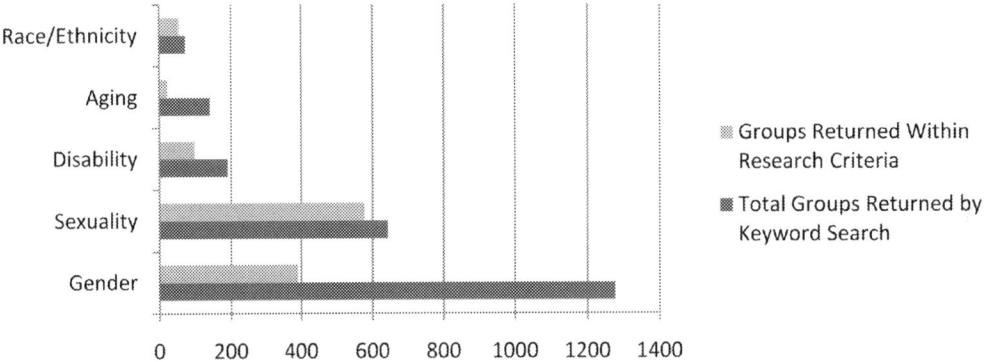

on the Gender keywords. The percentage of these groups that actually passed the search criteria discussed above also varied greatly. The Sexuality grouping had the highest passage rate with 87.8% of groups originally found to represent the criteria specified in the questions above. The Aging group had the lowest passage rate of 18.9%. Figure 1 depicts this variance.

Figure 2 highlights that the Gender and Sexuality groupings have a greater representation within the Second Life architecture. The Disability groups represented only 5% of the total number of groups that passed the search criteria filter, while the Aging group represented an even smaller amount, 2% of the total number of groups returned in the search.

Figure 3 depicts the comparison of the average size of the groups within the Second Life architecture, taking into account that groups with only one member were not counted in this analysis.

Disability Grouping

The disability-identified groups had a great deal of in-group variance. Compared to the whole group, the disability group was significantly less represented as compared to the gender and sexual orientation groups. The tables highlight these trends. Table 2 highlights the specifics of the disability associated groups analyzed for this chapter. Table 3 gives a more nuanced description of the specific groups that represent the disability

Figure 2. Percentages of groups found within search criteria based on keyword

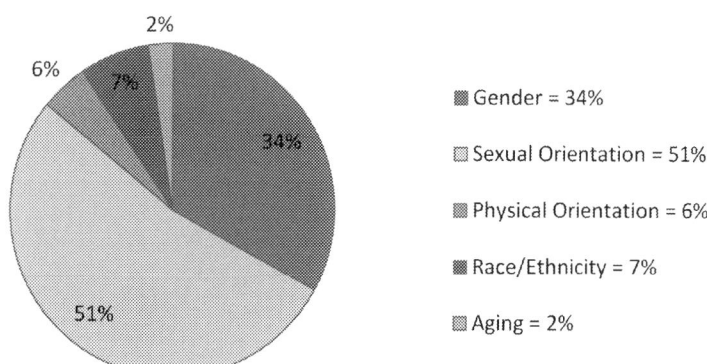

Figure 3. Average number of avatars in associated groups

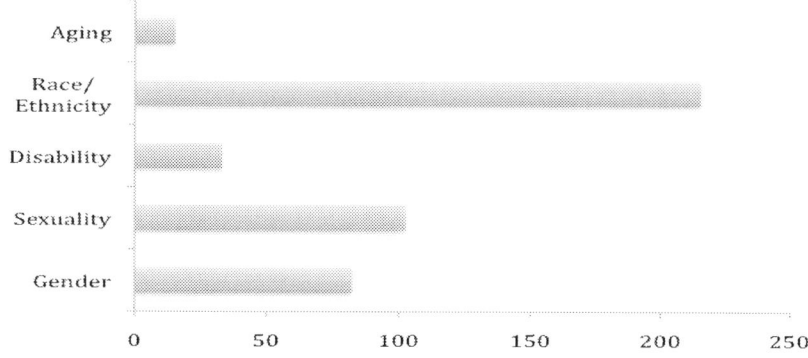

grouping: a total of 10 keywords were used for the data collection. Of the 65 groups that met the search criteria, the average group size was 30.75 members. There was a difference of 53.3 members between the lowest and highest average number of group members.

For context, an estimated 19.4% of non-institutionalized civilians in the US (approx. 48.9 million people) have a disability, with about half of these individuals being considered to have a severe disability (Kraus, 1996). According to Mobility International and the United States Access Board, there are at least 53 formal disability groups (real life groups) operating on a national scale in the United States. When comparing estimated populations to numbers of estimated disability groups, the numbers are striking. There is one national "real life" group for every 5,000,000

Table 2. Disability group statistics

Total number of groups returned	166
Total number of groups meeting criteria	65
Average population of group	30.75
Std Deviation	57.998

U.S. citizens, while there is one group for every 215,000 Second Life citizens, who may reflect the comparative ease of identification, and formation of communication based groups in a virtual setting.

For comparison, the same analysis for disability groups was applied to one of the comparator groups. In the physical world, we find that in the U.S. there are estimated to be around 135 groups representing the Lesbian, Gay, Bisexual, and Transgender (LGBT) individuals (rough estimate

Table 3. Disability Group specific data

Key Term	Total number of hits returned	Total number of hits meeting the specified criteria	Average group size
Aspergers	2	2	18.50
Disability	67	14	63.71
Disabled	5	5	50.40
Handicap	23	21	54.57
Impaired	5	5	10.40
Muscular Dystrophy	7	4	12.25
Blind	35	2	25.00
Learning Disorder	11	6	42.20
Aging	7	2	5.50
Wheelchair	4	4	25.00

of formal groups by the US Small Business Association and the National Gay and Lesbian Task Force). When comparing estimated populations to the number of estimated LGBT groups, the numbers are even more polarizing than the disability numbers. There is one group for every 2,000,000 U.S. citizens while there is one group for every 29,000 Second Life citizens. One possible explanation for this phenomenon is that in Second Life groups, group names, and group keywords are selected solely by users (Diehl & Prins, 2008). Noting that these group names may reflect "real life" references, it is important to acknowledge that these labels are chosen, not applied. While this pattern seems enlightening, further data collection and analysis needs to be obtained, specifically in relation to groups sizes of the real-world groups, so that populations can be normalized in order for a formal comparison to be made.

Given that this study is exploratory, future research might probe this condition further. The use of a multimodal system of in-world surveys, and interviews, in addition to potential social and cultural modeling and potential software analysis tools appears to be an appropriate next step. The use of purely descriptive data and statistics, while highlighting patterns, does not give us enough to claim there is a statistically significant correlation that exists.

CONCLUSION

Not surprisingly, if virtual places are any reflection of underlying social construction the number of individuals gendered or otherwise who choose to represent themselves as having a disability is relatively small. The meaning of this observation is somewhat complex to interpret as it is uncertain as to 1) whether this suggests that people with disabilities choose to alter their identity based on characteristics, or abilities (or lack thereof), or alternatively, 2) whether people with disabilities are underrepresented in the virtual/gaming population, or both. Statistics on income and information and communication technologies (ICT) use, suggest that people with disabilities have a much lower rate of Internet use, and it would be a valuable exercise to conduct exploratory research to examine the relationship between income and other accessibility issues and the lower Internet usage rates among people with disabilities (Dobransky & Hargittai, 2006).

Of particular interest, this work developed a foundation for future disability research by extending schema theory to include the issue of disability identity. This new extension, disability schema theory, has both online as well as "real life" application. Application of disability schema theory to interpretation of Second Life research seems to suggest similar perceptual impacts that also occur in the physical world. In the "real world", the visual cues that activate schemas can serve as an explanation for the stigmas and ensuing isolation often felt by people with visibly apparent disabilities.

In Second Life the visual cues are removed unless a conscious effort is made to create an avatar with characteristics that mirror real life disability. In the case of an avatar created without disability, study results suggest that Second Life users with disabilities are still associating with others who identify as having disabilities. This finding helps us to probe research question number three: Based on search term identifiers, do people appear to associate with others who identify with specific characteristics (i.e. disabilities) or genders when the cues that reinforce schemas are not apparent or observable? This is an area that is ripe for future research. In world interviews might probe users (avatars) regarding their actual disability status/identity in order to answer questions regarding the choice to either identify or not identify as a person with a disability. Additionally, in world interviews might seek to answer questions regarding decisions to associate with others who identify with similar disability status, especially in an environment where visual cues are removed. Because of the exploratory nature of this study, this was in fact a limitation of the current study.

In response to research question number two: Given the lack of visual cues, are there groups created in Second Life that are specific to people with disabilities? Are any of these groups gendered? Gender appears to play a role in many groups (i.e. "communities") found in Second Life. This is evident in the number of groups that were identified using the keyword 'gender'. Interestingly, groups such as "Gimp Girl" are specific to people who identify as both female and disabled. In fact, avatar wheelchairs are available for purchase at the Gimp Girl site and a number of avatars exist that are depicted as wheelchair users. Future research might delve further into refining both gender and disability schema theory in an effort to determine their ability to explain the choices an individual makes when deciding whether or not to identify (online) as a person with a disability as well as within the traditional binary gender paradigm. It is especially interesting to note that groups exist as a "safe space" for those who identify within the criteria of the group under review. Groups such as Gimp Girl exist not only in Second Life, but also have a presence on the Web as well as on the social networking site, Facebook. Another area for future research would be a replication of this study within social networking sites such as Facebook and MySpace.

As noted above, research suggests some 20% of gamers have some degree of disability (Faylor, 2008; Ingham, 2008) – in contrast the results of our empirical examination of membership of disability related groups shows that < 1% of total groups in-world self-describe or affiliate with disability. This is an interesting answer to research question number one: Do people with disabilities identify as disabled in Second Life? A reasonable response to these relatively stark observations might be to conduct additional in-world survey and interview research, in Second Life as well as other virtual platforms, with both avatars that self-identify as disabled, as well as with individuals who choose not to so identify but who may in fact be disabled. Another interesting question for future research is 'Why do some people with disabilities choose to gather within gender specific Second Life sites?' Additionally, it may be of interest to open a query regarding the stigma of gender in virtual environments. Future research might seek to discover why individuals choose a more fluid representation of gender, or no representation of gender at all.

We believe that both within and outside of the binary gender framework, the differences between the two disability identified groups, those externally classified as having some degree of disability, and those who choose to self identify, or more accurately affiliate "visibly" with disability related groups, have rich import for the sociology of online communities as well as for the design and characteristics of games. More practically, the applied value of this inquiry is in the potential for the development of new employment and community engagement applications as well as providing suggestions about new approaches to developing virtual and online environments to facilitate enhanced participation in society. Additionally, any research that serves to lessen the stigma felt by those who identify as disabled and those with visible disabilities will help to "level the playing field" in the real world, similar to that which exists in virtual environments such as Second Life.

ACKNOWLEDGMENT

Thank you to Celia Romm-Livermore and to the reviewers whose comments improved the manuscript substantially. We wish to acknowledge the support of Temple University, School of Communication and Theatre and the College of Science and Technology, Verizon, The Georgia Tech Research Institute, the Center for Advanced Communications Policy, and the Rehabilitation Engineering Research Center on Workplace Accommodations, funded by the National Institute on Disability and Rehabilitation Research (NIDRR) of the U.S. Department of Education under grant number H133E070026. The opinions contained in this publication are those of the authors and do not necessarily reflect those of the U.S. Department of Education.

REFERENCES

Baker, P. M. A., & Ward, A. C. (2002). Bridging temporal and spatial "gaps": The role of information and communication technologies in defining communities. *Information Communication and Society*, *5*(2), 207–224. doi:10.1080/13691180210130789

Beard, L., Wilson, K., Morra, D., & Keelan, J. (2009). A survey of health-related activities on second life. *Journal of Medical Internet Research*. *11*(2). Retrieved on March 19, 2010 from http://www.jmir.org/2009/2/e17/HTML

Bem, S. L. (1981). Gender schema theory: A cognitive account of sex typing. *Psychological Review*, *88*(4), 354–364. doi:10.1037/0033-295X.88.4.354

Bem, S. L. (1993). *The lenses of gender: Transforming the debate on sexual inequality*. New Haven: Yale University Press.

Boellstorff, T. (2008). *Coming of age in Second Life an anthropologist explores the virtually human*. New York: Princeton UP.

Bortoluzzi, M., & Piergiorgio, P. (2009). Multimodal Analysis of Virtual Learning Environments: A University Campus in Second Life. In Sapio, B. (Eds.), *The Good, the Bad and the Challenging. The User and the Future of Information and Communication Technologies* (*Vol. 1*, pp. 443–453). Koper, Slovenia: ABS-Center.

Bowker, N., & Tuffin, K. (2002). Disability discourses for online identities. *Disability & Society*, *17*(3), 327–344. doi:10.1080/09687590220139883

Bowker, N., & Tuffin, K. (2003). Dicing with deception: People with disabilities' strategies for managing safety and identity online. *Journal of Computer-Mediated Communication*, *8*(2). Retrieved on January 15, 2009 from http://jcmc.indiana.edu/vol8/issue2/bowker.html

Bowker, N. I., & Tuffin, K. (2007). Understanding positive subjectivities made possible online for disabled people. *New Zealand Journal of Psychology, 36*(2), 63–71.

Caoili, E. (2008). IMVU reaches 20 million registered users, largest virtual goods catalog. *Worldsinmotion.biz*. Retrieved on April 24, 2008 from http://www.worldsinmotion.biz/2008/06/imvu_reaches_20_million_regist.php

Cardinali, R., & Gordon, Z. (2001). Cliff jumping: Empowering actions for disabled women. *Equal Opportunities International, 20*(8), 17–24. doi:10.1108/02610150110786651

Coates, G. (2001). Disembodied cyber co-presence: The art of being there while being somewhere else. In Watson, N., & Cunningham, S. (Eds.), *Reframing the Body*. New York: Palgrave.

Curran, K., Walters, N., & Robinson, D. (2007). Investigating the problems faced by older adults and people with disabilities in online environments. *Behaviour & Information Technology, 26*(6), 447–453. doi:10.1080/01449290600740868

Diehl, W. C., & Prins, E. (2008). Unintended outcomes in Second Life: Intercultural literacy and cultural identity in a virtual world. *Language and Intercultural Communication, 8*(2), 101–118. doi:10.1080/14708470802139619

Dobransky, K., & Hargittai, E. (2006). The disability divide in Internet access and use. *Information Communication and Society, 9*(3), 313–334. doi:10.1080/13691180600751298

Duranske, B. (2008). *Virtual law: Navigating the legal landscape of virtual worlds*. American Bar Association.

2009*Encyclopedia of associations*. Dialog Online. Proquest. Retrieved on May 14, 2009.

Fang, Y., & Lee, L. (2009). A review and synthesis of recent research in Second Life. *Interactive Technology and Smart Education, 6*(4), 261–267. doi:10.1108/17415650911009236

Faylor, C. (2008). 1 in 5 casual gamers have disability, survey says. *Shacknews.com*. Retrieved on March 19, 2009 from http://www.shacknews.com/onearticle.x/53088

Fiske, S. T., & Taylor, S. E. (1984). *Social cognition*. New York: Random House.

Goffman, E. (1986). *Stigma: Notes on the management of spoiled identity*. New York: Simon & Schuster.

Hinton, A. (2006). Clues to the future. *Inkblurt.com*. Retrieved 24 April 2008 from http://www.inkblurt.com/2006/01/23/ia-summit-2006-clues-to-the-future/

Ingham, T. (2008). 20% of casual gamers are disabled. *CasualGamong.biz*. Retrieved February 7, 2009 from http://www.casualgaming.biz/news/27527/20-of-casual-gamers-are-disabled

Kacen, J. J. (2000). Girrrl power and boyyy nature: The past, present, and paradisal future of consumer gender identity. *Marketing Intelligence & Planning, 18*(6/7), 345–355. doi:10.1108/02634500010348932

2009Key US Disability Organizations. *MIUSA*. Mobility International USA. Retrieved on May 14, 2009 from http://www.miusa.org/idd/IDDresourcecenter/intldevelopment/keyorganizations/.

Kraus, L. E., Stoddard, S., & Gilmartin, D. (1996). *Chartbook on disability in the United States: An info use report*. Washington, DC: U.S. National Institute on Disability and Rehabilitation Research.

Latour, B. (1993). *We Have Never Been Modern*. London: Harvester Wheatsheaf.

Lemons, M. A., & Parzinger, M. (2007). Gender schemas: A cognitive explanation of discrimination of women in technology. *Journal of Business and Psychology*, *22*(1), 91–98. doi:10.1007/s10869-007-9050-0

Linden Labs. (2008). Economic statistics: Raw data files. *Secondlife.com.* Retrieved April 17, 2008 from http://secondlife.com/statistics/economy-data.php

2009*National Gay and Lesbian Task Force*. Retrieved on May 14, 2009 from http://www.thetaskforce.org/.

Norris, J. (2009). The Growth and Direction of Healthcare Support Groups in Virtual Worlds. *Journal of Virtual Worlds Research, 2*(2). Retrieved on March 19, 2010 from https://journals.tdl.org/jvwr/article/view/658/500.

Noveck, B. S. (2004). Introduction: The state of play. *New York Law School Law Review. New York Law School*, *49*(1), 1–18.

Nowak, K. L., & Rauh, C. (2005). The influence of the avatar on online perceptions of anthropomorphism, androgyny, credibility, homophily, and attraction. *Journal of Computer-Mediated Communication*, *11*(1), 153–178. doi:10.1111/j.1083-6101.2006.tb00308.x

Ondrejka, C. (2004). A piece of place: Modeling the digital on the real in Second Life. *SSRN.com.* Retrieved on April 28, 2008 from http://ssrn.com/abstract=555883.

Open University. (2006). Making your teaching inclusive. *The Open University, Walton Hall Milton Keynes, UK.* Retrieved January 23, 2010 from http://www.open.ac.uk/inclusiveteaching/pages/understanding-and-awareness/models-of-disability.php.

Park, B., & Hastie, R. (1987). Perception of variability in category development: Instance versus abstraction-based stereotypes. *Journal of Personality and Social Psychology*, *53*(4), 621–635. doi:10.1037/0022-3514.53.4.621

Perry, E. L., Davis-Blake, A., & Kulik, C. T. (1994). Explaining gender-based selection decisions: A synthesis of contextual and cognitive approaches. *Academy of Management Review*, *19*(4), 786–820.

Roberts, L. D., & Parks, M. R. (2001). The social geography of gender-switching in the virtual environments on the Internet. In Green, E., & Adam, A. (Eds.), *Virtual Gender*. London: Routledge.

Samp, J. A., Wittenberg, E. M., & Gillett, D. L. (2003). Presenting and monitoring a gender-defined self on the Internet. *Communication Research Reports*, *20*(1), 1–12.

Smith, K. (2009). *The use of virtual worlds among people with disabilities.* Paper presented at the California State University, Northridge, Center on Disabilities' 24th Annual International Technology and Persons with Disabilities Conference, Los Angeles, March 16-21, 2009.

Suler, J. R. (2002). Identity management in cyberspace. *Journal of Applied Psychoanalytic Studies*, *4*(4), 455–459. doi:10.1023/A:1020392231924

Thoreau, E. (2006). Ouch!: An examination of the self-representation of disabled people on the Internet. *Journal of Computer-Mediated Communication*, *11*(2), 442–468. doi:10.1111/j.1083-6101.2006.00021.x

2009United States Access Board. Retrieved on May 14, 2009 from http://www.access-board.gov/.

2009*U.S. Gay/Lesbian Business Groups*. Small Business Association. Retrieved on May 14, 2009 from http://www.smallbusinessnotes.com/interests/usgaybuslks.html.

Ward, K. (2001). Crossing cyber boundaries: Where is the body located in the online community? In Watson, N., & Cunningham, S. (Eds.), *Reframing the Body*. New York: Palgrave.

Whitley, E. A. (1997). In cyberspace all they see is your words. A review of the relationship between body, behavior and identity drawn from the sociology of knowledge. *Information Technology & People*, *10*(7), 147–163. doi:10.1108/09593849710174995

Yee, N. (2007). *The Proteus Effect: Behavioral Modifications via Transformations of Digital-Self Representations.* Unpublished Ph.D. Thesis, Department of Communications, Stanford University.

A previous version of this chapter was published in the International Journal of E-Politics, Volume 2, Issue 2, edited by Celia Romm, pp. 1-17, copyright 2011 by IGI Publishing (an imprint of IGI Global).

Chapter 10
Overcoming the Segregation/Stereotyping Dilemma:
Computer Mediated Communication for Business Women and Professionals

Natalie Anne Sappleton
Manchester Metropolitan University, UK

INTRODUCTION

Since the 1970s, there has been a swift and sizeable uptake of business ownership amongst women in the United States (U.S. Census Bureau 2002; Lowrey 2005). Women now own around 40 percent of all non-agricultural businesses – amounting to an increase of more than 25-fold since records began in 1972 (Center for Women's Business Research 2008). Moreover, the fastest recent growth in women-owned firms has been in traditionally male industries like telecommunications and construction. According to the National Women's Business Council, the number of privately held women-owned firms in non-traditional industries[1] grew by 17.5% between 1997 and 2002, outstripping the 10.4% growth in the number of women-owned firms in traditional industries 2004 (CWBR, 2008). Growth has been strongest in the construction industry; between 1997 and 2002, the number of women-owned firms grew by 35.5% (*ibid*).

All business owners must identify, access, and mobilize resources to put to use in their ventures, and the need for resources is highest in capital intensive, traditionally male industries such as high-technology and construction. Setting up such businesses is no easy task, and network members may be able to provide direct or indirect access to the required assistance, support, information and tangible resources (Kim and Aldrich 2005). This means that entrepreneurship cannot happen unless a business owner establishes the right connections to the right others (Witt 2004). Entrepreneurship should therefore be viewed as an inherently social activity; "embedded in

DOI: 10.4018/978-1-60960-759-3.ch010

a social context, channeled and facilitated or constrained and inhibited by people's positions in social networks" (Aldrich and Zimmer 1986: 4). Little wonder, then, that having well-structured social networks is linked to entrepreneurial intentions (Ljunggren and Kolvereid 1996; Aldrich, Elam et al. 1997; Renzulli, Aldrich et al. 2000; Liñán and Santos 2007), start-up completion (Hansen 2000), odds of survival (Srinvasan, Woo et al. 1994; Uzzi 1996), opportunity recognition (Ozgen and Baron 2007), idea gathering (Birley 1985), ability to secure venture capital (Baum and Silverman 2004) and finance before floatation (Florin, Lubatkin et al. 2003), internationalization (Coviello and Munro 1995), and growth (Brown and Butler 1995; Hansen 1995; Liao and Welsch 2001; Roomi 2007).

There is evidence that women operating their firms in male-dominated industries encounter greater levels of gender stereotyping, sex discrimination and difficulty accessing established networks than women in sectors like retail and services (Weiler and Bernasek 2001; Coyle and Flannery 2005). These women are thus provided with a clearer motivation for forging homophilious networks – that is, networks comprised primarily of other women. The problem is that relying on same-sex networks for leads and information denies women access to the privileged resources held by men in the industry. The resources, information, knowledge and expertise held by men operating in these fields is likely to be of a better quality due to their longer establishment and better entrenched positions in network hierarchies. As argued twenty years ago "women must break into the 'old boys' network by deliberately invading male turf however possible. A 'new girls" network will create strong ties and promote social support but with… most of the major corporate and financial centers of power controlled by men, sex-segregated separate networks are a decided handicap for women (Aldrich 1989: 128).

It is important that women in male-dominated industries are able to network effectively with their male colleagues because the ability to secure resources is a necessary prerequisite for increasing the numbers of women-owned firms in nontraditional sectors. There is evidence that *entrepreneurial segregation* contributes to levels of gender inequality in similar ways to sex segregation in employment (Marlow et al, 2008). Women earn less than men in self-employment and business ownership, and segregation makes a significant contribution to earnings disparities (Hundley, 2001; Lowrey, 2005). The sectors of the economy in which women's businesses are concentrated generate lower levels of turnover than typically male sectors such as manufacturing and high technology (Loscocco et al, 1991).

The proposition advanced in this chapter is that virtual networking via web pages, email, chat rooms and networking sites is one way that women in nontraditional industry locations might attempt to 'invade male turf'. As well as providing women who may have domestic responsibilities with a more flexible means of networking (Blisson and Kaur Rana 2001), virtual communities may help women to reduce the social and spatial barriers preventing them from establishing connections with prestigious or powerful others, whilst simultaneously breaking down gender stereotypes (Nohria 1992). As Sproull and Kiseler (1991: 13) predicted almost twenty years ago, social computing allows individuals to "cross barriers of space, time and social category to share expertise, opinions and ideas". Additionally, virtual communities can provide access to a very large number of diverse others and network maintenance is substantially easier than face-to-face interaction.

This chapter is organized as follows. The chapter begins with a discussion of the theoretical underpinnings of the work. The following section discusses the phenomenon of *homophily* in social networks, including entrepreneurial networks. Homophily – or the tendency to associate with similar others – appears to occur naturally in entrepreneurial networks because of structural and dispositional factors. In particular, women working in sectors unaccustomed to a female presence may find greater acceptance by associating with other women. However, as discussed in the following section, same-sex networks disadvantage female entrepreneurs operating in male-dominated contexts because men in those industries are privileged with more voluminous and better quality resources. A review of the literature on gender and virtual networking then follows. The literature points to the promise of virtual networks in helping women to cross gender barriers. The chapter concludes with a general appeal for empirical research on the potential of virtual networking for overcoming the "segregation/stereotyping bind".

THEORETICAL FOUNDATION

Gender, Organisations and Networks

Institutions are not gender neutral. Sociologists have long recognized that business organizations, and the environments in which they are based are gendered, and the gender is generally male. As Acker (1991: 166) has argued: "Individual men and particular groups of men do not always win in these processes, but masculinity always seems to symbolize self-respect for men at the bottom and power for men at the top, while confirming for both their gender's superiority". Since entrepreneurship is about organizations, and organizations which have, until now, been led by men, entrepreneurship too is gendered (Green and Cohen, 1995; Mirchandani, 1999; Ahl, 2006; Bird and Brush, 2002; Lewis, 2006). That is, the popular notions of the 'entrepreneur' – as hero, captain, adventurer, explorer – are undoubtedly masculine, and the features of entrepreneurship as an activity: risk-taking, innovation, emotional detachment, initiative, ambition align closer with the stereotyped notions of men, rather than women (Schein, 2001; Carli, 2006). Thus, to study women entrepreneurs without examining the gender structuring of entrepreneurship legitimizes the gender-blindness which renders masculinity invisible and thereby turns masculinity into "the universal parameter of entrepreneurial actions" (Bruni et al, 2004: 410).

While entrepreneurship is undoubtedly gendered, and recognized as so, the *contexts* in which entrepreneurship is practiced have been universalized and stripped of gender. The meaning and salience of gender varies according to entrepreneurial context because these settings denote particular occupations that "carry characteristic images of the kinds of people that should occupy them" (Kanter, 1977: 250). Imagine a firefighter, and you will probably conjure up an image of a strong, muscular male; think of a nurse and the image is likely to be of a caring female. Thus, gender is inherent not only in the organization, but is also inscribed in the jobs that take place within it, and the service or product that the organization produces. Business ownership is male-typed indeed, but in a society where women own 40 per cent of all firms, the association between actual numbers of women in entrepreneurship and gender stereotypes may be weakening. It is no longer unusual to see a woman in a management or ownership position (Powell et al, 2002). Instead, the general stereotype regarding women in business may be being replaced by more specific stereotypes regarding the *type* of business that men and women should own and operate.

Entrepreneurial settings are also important because they influence the shape and nature of working relations – determining the sex ratios of workforces, the likelihood of cross-sex or same-sex supervisor-subordinate and colleague linkages and interactions, and thereby the ways in which gender is "performed" on the job (West and Zimmerman, 1987). People perform their maleness or femaleness in different ways depending on the expectations and assumptions of their interactants and these expectations are in turn determined by the activity in which they are engaged. Since membership in a sex category may be invoked to discredit or legitimize their performance, the context in which gender is 'done', is highly relevant (West and Fenstermaker, 1993). Women entrepreneurs working in male-dominated sectors must deal with a greater proportion of opposite-sex colleagues, employees and associates than women based in typically female sectors. This means that it "is unquestionable that it [gender] will not intrude into the experience of self-employment" (Marlow et al, 2008: 337) for these *nontraditional* women. This is not to say that gender performance only occurs in cross-sex interactions (West and Fenstermaker, 1993). As Hanson and Blake (2008: 137) explain: "…Networks play an active role in the ongoing development of individual and group identities. Because networks are always about social interaction and the positioning of individuals relative to each other, they are also always about gender". Thus, since gender is an identity that is negotiated and constructed in routine social interaction, examination of the challenges that nontraditional entrepreneurs may face in their daily associations with others is essential. This chapter is about those interactions.

HOMOPHILY IN SOCIAL NETWORKS

Social capital/social networks is one of the most heavily studied topics in the field of entrepreneurship. This is because personal relations are increasingly being viewed as the key to the construction, functioning and effectiveness of a viable business enterprise (Granovetter, 1973; 1985). It is increasingly recognized that success in business is linked to not *what* you know, but *who* you know. When considering who people know, it is certainly true that "birds of a feather flock together". Homophily, or the 'like-me' principle, dictates that individuals deliberately choose to associate with others on the basis of similar ascribed or achieved characteristics like race, gender or cultural values (McPherson, Smith-Lovin et al. 2001) Social similarity is said to increase levels of liking, trust, understanding and attraction between associates (Aldrich, Carter et al. 2002) and may lead to better communication amongst interactants and greater predictability of behavior (Ibarra 1992). One of the earlier observers of the phenomenon, Almack (1922: 52) remarked:

"The clerk, the artisan, the street laborer work by and with those that are assigned to places by them, and perhaps it never occurs to them, and never to the employer that the inefficiency and discontent that pervades the establishment may be due to mal-adjustment of co-operating or contiguous individuals and groups. Homogeneity is one of the requisites for efficiency in socialization"

Thus, demographic and social similarity increases liking and influence amongst interactants, and increased liking can influence the flow of resources. Studies show that resources are more likely to be distributed within-group when members of a group share a common identity or a sense of regard, are friends, perform interdependent tasks or share demographic attributes (McKnight, Cummings et al. 1998; Anderson and Miller 2003) This can have beneficial effects on task outcomes. For example, gender-based homophily of entrepreneurial teams has been found to raise sales volumes (Fertala 2005).

Homophilious tendencies therefore lead men to associate with other men and women with other women. Gender itself clearly shapes patterns of interaction (e.g. who associates with whom, and in what situations), but so too does context. A crucial point is that homophily may occur as a result of structural as well as agential forces (Feld 1982; McPherson, Smith-Lovin et al. 2001). McPherson and Smith Lovin (1987) distinguished between 'choice homophily' – the result of personal preferences - and 'induced homophily', arising from the availability of similar others.[2] In other words, in certain contexts (an all-female high school, for example), homophily is almost inevitable. In their study of voluntary association members, although they found evidence of both types, same-sex relationships developed largely because of induced homophily. But, women and minorities tend to have less homophilious networks than white males. In her study of the employees of one advertising firm, Ibarra (1992: 425) concluded that "preferences for homophily and status will tend to coincide for men and exist in competition for women". That is, men's tendency toward gender homophily simultaneously allows them access to high status others, but rational women may view interaction with men as a more effective strategy in the pursuit of power and influence. As a result, women carved out differentiated, gendered networks: pursuing friendship and support from their relations with females, but instrumental resources and influence from their relations with men.

The problem is that in contexts in which there are few high status women, homophilious strategies are harmful to women and detrimental in their efforts to access resources (Cabrera and Thomas-Hunt 2007). This means that nontraditional women entrepreneurs may be constrained towards interactions with those best positioned to enhance their access to resources – men – despite preferences with same-sex associations. Mehra and colleagues (1998) provide some evidence for this hypothesis. In their study of the friendship networks of a prestigious MBA program, they found significant levels of gender homophily, but for women, homophily was induced by their lower levels of attractiveness as associates to males. The authors concluded: "the marginalization of women resulted more from exclusionary pressures than from their preferences for women friends" (Mehra, Kilduff et al. 1998: 447).

WOMEN BUSINESS OWNERS IN MALE-DOMINATED SECTORS AND THE SEGREGATION/STEREOTYPING BIND

Business ownership is considerably segregated by sex (Sappleton 2009). This fact is aptly illustrated in figure 1. This shows US business ownership by sex and three-digit industry classification in 2002 (the latest available data). While most businesses are male-owned, the differences according to sector are quite evident. Women are overrepresented in Healthcare and Social Assistance and Educational Services and underrepresented in Construction and the primary sectors. Men are underrepresented in the sectors in which women are most commonly found, but in most industries, men form a majority.

In spite of these observations, the absolute numbers as well as proportions of women-owned firms in nontraditional industries have seen a spurt in growth in recent years. According to the National Women's Business Council, the number of women-owned firms in non-traditional industries grew by 17.5% between 1997 and 2002, outstripping the 10.4% growth in the number of women-owned firms in traditional industries (NWBC, 2004). Growth has been strongest in the Construction category; where, between 1997 and 2002, the number of women-owned firms grew by 35.5% (*ibid*). This increase in women-owned nontraditional business increases the extent to which women business owners must network across gender-boundaries. That is, the growing number of women in traditionally male industries poses a challenge to women's preferences for homophilious relations. This challenge is outlined below.

The Stereotyping Problem

In an industry that is predominately populated by men, structural forces will induce women to possess networks that resemble the industry pro-

Figure 1. Percent Business Ownership by Sex and 3-Digit Industry Classification, 2002 (Source: US Census Bureau, 2002. Publicly owned businesses excluded.)

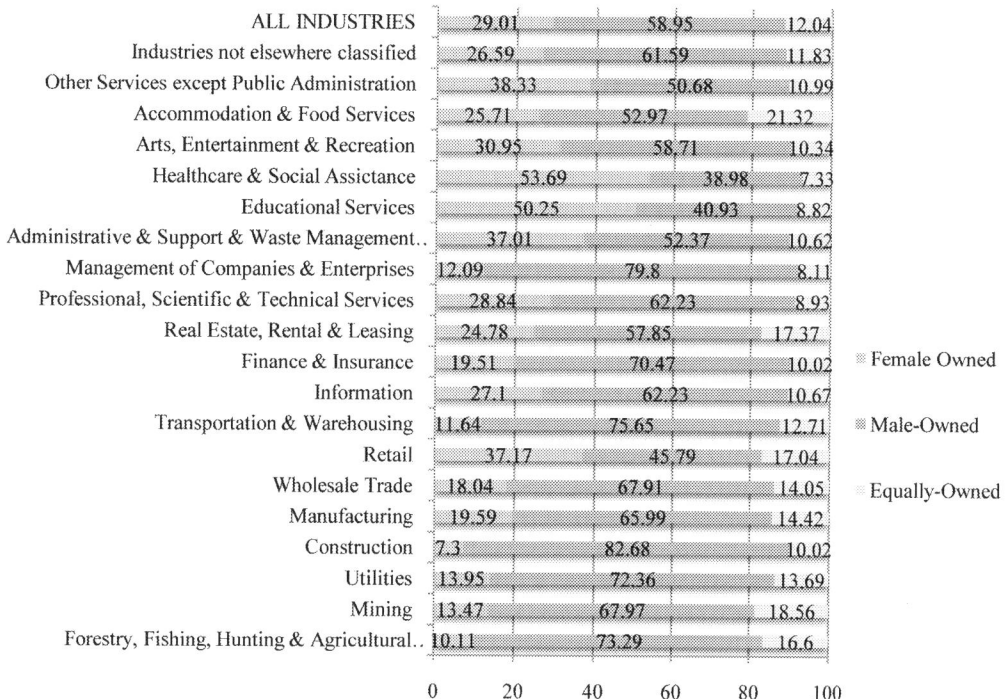

file. That is, the vast proportion of nontraditional women business owners' daily interactions will be with male staff, male colleagues, male clients, male suppliers, male financiers, accountants, associates, advisors and so on. There is some evidence that these interactions may be colored by incidents of gender stereotyping which impact on the volume and quality of resources exchanged between interactants. Gender stereotyping is the act of unnecessarily categorizing and evaluating a person according their sex, evaluating their credentials along dimensions relevant to their group's stereotype and selectively interpreting their traits (Cejka and Eagly 1999; Glick and Fiske 1999). In the past, researchers have suggested that women entrepreneurs have suffered from a general attitude that they have no place running a business (Heilman and Block 1989; Morrison and von Glinow 1990; Owen and Todor 1993; Schein 1994; 2001). Yet, now that women own 40 per cent of all firms, and this figure is increasing, the association between actual numbers of women in entrepreneurship and gender stereotypes may be weakening. It is no longer unusual to see a woman in a management or ownership position. Instead, the general stereotype regarding women in business may be being replaced by more specific stereotypes regarding the *type* of business that men and women should own and operate. Stereotyping occurs most readily when three conditions are satisfied: 1) when a person moves into a nontraditional position; 2) when that person is isolated or a few-of-a-kind and 3) where information about the person is ambiguous to perceivers (Kelly, Young et al. 1993). The migration of women into business ownership in sectors previously unaccustomed to a female presence meets each of these conditions.

Empirical evidence of stereotyping against nontraditional women can be found in literature studying businesspersons' associations with consumers, financiers and employees. Traditional women-owned enterprises serve a household clientele, but larger male-typed firms operating in global markets are far more likely to sell their products and services to other corporations and government agencies. Bates (2002) has contended that women-owned business enterprises have encountered buyer discrimination when they seek out markets beyond the (female) household clientele.[3] Using the US Census Bureau Characteristics of Business Owners data, Bates finds that although women business owners have less capacity for serving business and government clients (in terms of their size, age and industry location), even when these dimensions are statistically controlled, women-owned businesses are less likely to sell to other firms and government agencies than male-owned firms. This finding is redolent of buyer discrimination: even in skilled services, construction and goods industries (the industries most easily able to penetrate business-government markets), and "capacity notwithstanding, owner gender by itself is a major factor shaping market access" (Bates, 2002: 321).

Coyle and Flannery (2005) also found evidence of customer discrimination in their interviews with women business owners in male dominated industries. They suggested that discrimination was related to the way that business was drummed up through insider networks and word of mouth. A comment from one interviewee is illustrative: "…And it develops this whole network and often it's a network that women and minorities are just not included in. And so it takes twice as much effort and work to get past that. It really does. You have to be better at what you do because if you're equal it's [the work] going to go to someone else" (Coyle and Flannery, 2005 p. 8). Some women in that study also reported encountering flirting, sexual innuendos and inappropriate touching from male clients. Finally, the authors concluded that it was the women based in the most densely male dominated fields that experienced the greater number of gender-related barriers.

Raising finance has repeatedly been identified as problematic for women entrepreneurs, particularly so in the capital-intensive industries. Brophy (1989: 73) has argued that women's financing difficulties have "been due to attitudes held by representatives of male-dominated institutions – and often reinforced by businesswomen themselves – regarding the proper role of women in business. That role has been seen as staff or part-time employee or business hobbyist, and – if an entrepreneur at all – one confined to businesses traditionally run by women: retail and service businesses for the most part". The attitude to which Brophy refers is summed up nicely in a quote from a venture capitalist who specializes in financing computing firms: "I would never invest in a women-led business. Don't get me wrong, women are great for day care centers and have done a lot for customer service, but as an investor, you can't take a chance that they might leave to get married or pregnant" (Brush, Carter et al. 2004: 72). One-quarter of the women in Borooha et al's (1997: 86) study of Northern Irish business owners described their gender as hindering their access to external finance; many complained that banks "took them seriously only when the chosen business was in 'women's' area". Marlow and Strange (1994: 181) contended that "bank managers are still reluctant to fund female ventures, particularly those which stray beyond traditional feminized occupations". And, in interviews with male and female bankers, Blake (2006) found that certain sectors (such as construction) were seen as more appropriate for male owners while others were deemed suitable for women. A male loan officer in that study cited the example of a female who began a cleaning firm with a loan from his bank. Despite the applicant's previous job as an auto mechanic, the banker admitted that he would have had "a greater degree of difficulty granting the loan to her if she had wanted to start a business as a mechanic" (Blake, 2006: 195).

With regard to employee relations, Herrick's (1999) case study comparing the experiences of two upper level female managers in one manufacturing firm is illuminating because it highlights the importance of both sex segregation and the gender typicality of roles on staff attitudes to managers. This manufacturing firm, known as Phoenix, is described as highly sex segregated – all forklift drivers, warehouse foremen and supervisors are men, whereas women are found in clerical roles. One of the subjects of the case study, Rose, is a new senior level manager who is responsible for all-male staff. Rose is recognized as competent but disliked and viewed as a "bitch". Because of this, her subordinates refuse to take orders and will not work under her. Finally, she is replaced by "a strapping 6'2", no-nonsense kind of guy, someone tried-and-true from the old days. He too barks directions, and he too is soon disliked, but orders begin to go out on time. Now Rose's only duties are the distribution computer work. Rose seems relieved. She begins to smile more. She begins to wear her hair in a ponytail and chat with other employees. Some of the guys in the warehouse – much to their own amazement – begin to find her attractive. They claim she 'looks different somehow - softer'. They even flirt with her and dare to consider asking her out" (Herrick, 1999: 287). It is interesting that despite the male replacement's use of the same "no-nonsense" style of management as Rose, his authority was accepted. This is because the male employees found those qualities natural and acceptable for a man. To emphasize the point, Herrick also describes the experience of another female senior manager. This manager, Kathy, is well liked and powerful in the organization, which is partly attributed to her role as personnel manager (a stereotypically female role), and partly to her effort to behave "in accordance with the local norms at Phoenix for feminine behavior – baking cookies for people's birthdays, circulating birthday cards, arranging for cakes and parties to commemorate employee anniversaries, organizing baby showers for the women in the plant, providing candy for every team meeting, listening to complaints and problems, and never openly confronting anyone" (*ibid:* 291).

Identifying and appropriating business resources may mean aggressive networking activity. And, endorsing one's own talents, skills, accomplishments and strengths is essential in situations where an individual is in a minority. However, tactics of impression management are much more difficult – and dangerous - for women to employ successfully, particularly to male audiences (Carli 2001). In one study, Rudman (1998) found that women who behave confidently and assertively are evaluated more negatively and are socially less popular than males following the same strategy. This is because in youth, males are socialized to be competitive and outspoken, whereas females are taught to be community-oriented and modest (Reskin 1993; Heilman and Chen 2003). Individuals that capitulate to perceivers' expectations, whilst reconfirming gender stereotypes, create an environment that is socioculturally comfortable to all parties. Thus, both men and women exert greater influence when acing in a gender stereotypical manner (Carli 2006). In the workplace, women in managerial roles are trapped in a bind where they are perceived as "either likeable but weak and ineffective, or as unlikeable but competent and professional" (Herrick, 1999: 275). This explains why women who show communality (for example, by nodding and smiling) exert more influence than women who do not (Carli, 2001).

The literature examined above suggests that women in male dominated business contexts face a greater degree of stereotyping from male associates because of commonly held gendered beliefs about their abilities in the field. These observations explain why the "old boys' network" which supplies its members with crucial insider information on business trends has been identified as a major hindrance to the operation and growth of women owned, traditionally male firms (Weiler & Bernasek, 2001). On the other hand, "in industries where men-owned businesses historically have had strong support from customers, business networks, and families, they probably continue to receive it" (Bird and Sapp 2004: 10). The answer for many women is to carve out same-sex networks. For example, there is evidence that women nontraditional business owners hire a disproportionate number of female staff than men in the same industry (Smith, Smits et al. 1992; Verheul, Risseeuw et al. 2002). However, as the studies below show, segregation may be no more beneficial in helping these women to secure business resources.

The Segregation Problem

Faced with "the actual and too negative experiences of dealing with men, or more often, the perception that men would not be helpful" (McGowan and Hampton 2007: 122), female entrepreneurs in male dominated contexts may opt to carve out female-only or mostly female associations, and, certainly, there are increasing numbers of formalized female-only networks worldwide (Travers, Pemberton et al. 1997; Wood 1999). Rational individuals join networks only where the costs of membership are lower than the benefits (Welter and Trettin 2006). Sex segregated networks guarantee at least some women positions of power in the group, and it also links women to potential mentors and role models (Popielarz 1999). The homogeneity of such associations also fosters dense and emotionally close enclaves, leading to shared understandings, mutual favour exchange and reciprocation, moral support and encouraging the business owner to persevere in difficult times. In all, these groups can act as both a cheering squad and a sounding board to the nontraditional female entrepreneur. A recent comment from a member of an all-women's business network in Northern Ireland sums this up: "It's all women there and you know that you won't be laughed at" (McGowan & Hampton, 2007: 122). Moreover, "the social capital that arises from this strong bond consequently enables the building of one's reputation because information will travel to many nodes within the network" (Suseno 2008: 154).

In spite of the advantages, single-sex networks are disadvantageous because they partition the available pool of human, social, cultural and financial capital. Dense associations like these offer minimal benefits where the cluster is isolated from others (i.e. clusters of men). Depending on the context, men and women hold disproportionate amounts of resources. In male-dominated fields, it is men that hold the power, status, political influence and industrial resources.[4] This uneven distribution of resources means that, in male-dominated settings, men become more interesting to women, women are less useful to other women and women are unnecessary or even burdensome to men (Lipman-Blumen 1976). As Lipman-Blumen (1976: 16) explains:

"The pragmatic recognition that males controlled economic, political, educational, occupational, legal and social resources created a situation in which men identified with and sought help from other men. Women, recognizing the existential validity of the situation, also turned to men for help and protection. By now, it is practically a psychological truism that individuals identify with other individuals whom they perceive to be the controllers of resources in any given situation"

Homophily is thought to be one of the major causes of the unequal distribution of social capital (Cross and Lin 2008). When high status individuals share resources amongst themselves, low status and other excluded groups are prevented from accessing the stocks of the privileged. This could be harmful to women entrepreneurs, and especially those in non-traditional industries because of their out-group status (Cabrera et al, 2007). Although there is little direct evidence specifically pertaining to entrepreneurs, there is much evidence to support this theory in the literature on employees. In his well-cited network analysis of employees in one newspaper publishing company, Brass

(1985) found that although women occupied more central positions in the organizational structure, they were excluded from the male dominant coalition because of their low levels of interaction with male colleagues. This negatively impacted on their levels of influence and opportunities for promotion. Cross and Lin (2008) too observed an overwhelming amount of co-ethnic homophily in social networks, and found that this was associated with significantly lower access to social capital.

Belliveau (2005) argued that educational segregation allows employers to offer women lower starting salaries because a gender wage gap is less likely to be detected where women have all-female reference groups. She compared the number and value of salary offers of 83 graduating seniors from three elite liberal arts colleges, only one of which was co-ed, and collected data on the size, heterophily of the seniors' "job advice networks", as well as the sex composition and reputation of the school. Interestingly, women graduating from all-female colleges, despite similar Grade Point Averages, received significantly lower salary offers, even when controlling for demographics, human capital and school reputation. Women from mixed sex schools received higher salary offers. This finding was attributed to the lower network heterophily of women's networks, but also to network homogeneity that employers perceived women from same-sex schools would have. Put another way, employers believe that women from same sex school will consult only with other women and therefore offer them lower starting salaries. Belliveau estimated that actual and perceived network homogeneity translated into a 9 per cent wage penalty for women-only college students.

Based on these studies, it is proposed here that all-female associations are likely to be more useful to women based in traditionally female industries, than to the non-traditional female entrepreneur. Women operating in male-dominated sectors require bridges to similarly placed or more senior males if they are to attain the crucial resources of their sector. "Women in men-dominated environments might obtain a greater variety of information if they networked with some men rather than all women" (Smeltzer and Fann 1989: 26). Such a strategy is deemed vital if these firms seek growth. "A dedication to all-female networks would be a myopic strategy and one to be avoided… there is a clear need for a commitment by female entrepreneurs to building quality into their networking efforts if their enterprises are to grow, especially on a continuous, and indeed, deliberate basis" (*ibid:* 124). But, as discussed earlier, attempting to forge networks with men has its own pitfalls. This amounts to a kind of segregation/stereotyping bind – a damned-if-they-do-damned-if-they-don't situation that might help to explain the relatively low numbers of women operating in male-dominated business contexts. How, then can nontraditional women overcome this bind? Recent literature suggests that computer mediated networking might provide a solution for women facing this dilemma, and these studies are discussed in the next section.

VIRTUAL NETWORKING

Virtual or computer mediated networking allows users to contact and develop ties with individuals via Internet methods, including, but not limited to Web pages, email, chat rooms and online communities. There already exist many women's organizations with specialized websites geared towards supporting entrepreneurs and professionals. Two prominent examples are Working Women's Network and Women's Forum. Through such means, women business owners can look to others for social, support, strategic, supply and learning purposes (Wood 1999).

Virtual networking offers businesswomen many advantages over traditional face-to-face networking, and this perhaps explains their proliferation worldwide (Travers et al, 1997). Firstly, virtual means of networking provide instant access to individuals on a global basis, and are available 24 hours a day (Knouse and Webb 2001). This means that women with domestic or other responsibilities are provided with greater flexibility in how and when they develop and maintain networks. Whilst asynchronous, computer mediated communication approximates real time interaction, allowing for fast, convenient communications at low costs (Sproull and Kiesler, 1991). Additionally, virtual communities can provide access to a very large number of diverse others and network maintenance is substantially easier and less costly than face-to-face interaction (Lind, 1999).

Because of the benefits of virtual networking, several commentators have suggested that businesswomen use them to access "larger numbers of same sex and same race contacts, thus increasing the similarity or homophily of the network" (Knouse and Webb, 2001: 227). But, as argued above, homophily is not a desired trait in the networks of the nontraditional female entrepreneur. There is no reason why these women should not utilize Internet groups as a way of overcoming the challenges of cross-sex networking. Indeed, the features of virtual networks – the anonymous, decontextualized, ephemeral, asynchronous communication - appear to present to nontraditional women a convenient means of overcoming problems associated with their gender. Below, the very small, but growing body of studies that offer empirical support for this suggestion are reviewed.

Almost twenty years ago, Sproull and Kiesler (1991: 13) predicted that new technological ways of working would have equalizing effects, allowing people to "cross barriers of space, time and social category to share expertise, opinions and ideas". This is because "when communication lacks the dynamic personal information of face-to-face communication or even of telephone communication, people focus their attention more on the words in the message than on each other. Communicators feel a greater sense of anonymity and detect less individuality in others. They feel less empathy, less guilt, less concern over how they compare with others and are less influenced by social conventions" (*ibid*: 40). In particular, the removal of nonverbal cues and the ephemerality and plainness of text-speak reduces people's fears of appearing foolish or inferior to others and increases self-confidence. In face-to-face interaction, high-status people (often men) tend to dominate group discussions (Carli 2006), but research shows that online interaction gives peripheral individuals a voice, allowing them more of a chance to contribute equally (Sproull and Kiesler, 1991).

Martin and Wright (2005) argue that the internet-based technology is a "great equalizer" which allows women nontraditional entrepreneurs to compete on a level playing field with their male counterparts. Their discussions with ten women operating small ICT firms revealed a great degree of appreciation for a technology that allowed its users to appear 'invisible'. The women respondents were said to be able to run high technology firms regardless of the usual barriers of gender because "people don't realize you're women and take you more seriously – they judge you as a business no as an individual" (Martin and Wright, 2005: 170). For example, women were more easily able to find suppliers for key services using internet based methods, and links that began as virtual contacts, over time became stronger, face-to-face interaction. As one respondent (a software developer) to that study stated, "In space no-one can hear you scream? In cyberspace no-one can tell if you're male or female – you are another web presence" (*ibid*: 172).

Popular tomes on communication like John Gray's *Men are from Mars, Women are From Venus* or Deborah Tannen's *You Just Don't Understand: Women and Men in Conversation* maintain that women and men have distinctive communication styles that create barriers of understanding between them. The anonymity and fact that communication does not take place in real time may allow virtual networkers to experiment with different forms of self-representation. Thomson (2006a;

2006b) argues both men and women use gender typical language patterns when interacting with networks partners of the same sex, and this flexibility smoothes communication. Additionally, in lab-based and real-life experiments, Thomson (2006a) found that when discussing gendered topics, participants accommodated to an expected language style in order to gain acceptance or approval. There was a higher frequency of female preferential features (e.g. personal features, emotion, question asking, agreement, apology, compliment, self-derogatory statements, solidarity with others) in discussions about female stereotypical topics and a higher frequency of male-typed features in discussions about male stereotypical topics (directives, disagreement, insult, adjectives, statements emphasizing differences between group members). Thus, women appear 'male' when discussing male-typed topics with male network partners in online environments. This is because when gender markers are hidden, the topic of conversation, rather than individual gender, becomes the salient group norm. This finding may explain why women in male-dominated companies communicate with colleagues via email to a significantly greater extent than do men (Lind 2001; Brosnan 2006).

O'Brien (1999) too argues that online interactions offer individuals the opportunity to dislocate gender from its corporeal markers. In the absence of embodied characteristics, the effects of social categories of difference (sex, race, social status) on interaction are effectively erased. Thus, cyberspace allows people to cross the boundaries conventionally imposed by bodily physicality. By disguising their gender online, women business owners can conduct business activity in ways typically associated with male assertiveness, without paying the penalties that social psychologists predict in face-to-face interaction. Lind's (1999) comparison of mixed sex virtual and face-to-face work groups showed that women were more satisfied with the virtual work group, and perceived levels of cohesion, support and inclusiveness to be higher. Lind attributed these findings to the more equal group participation facilitated by the lack of nonverbal cues and structure of the virtual workgroups. Cohen and Ellis (2008) also found evidence of more equal participation of women and men postgraduates in virtual learning environments.

Finally, and from a more practical point of view, forging virtual associations allows women who may be juggling domestic responsibilities the opportunity to build networks in their own time (Blisson and Kaur Rana, 2001). Researchers have suggested that women business owners suffer a 'network deficit' because the demands of motherhood and other caring duties deprives them of the time to build social capital (Munch, McPherson et al. 1997). This problem is exacerbated in the case of the non-traditional entrepreneur because business owners operated in male-typed sectors work significantly longer weekly hours than business owners in female-typed sectors (Sappleton, 2009).

In spite of the promise of virtual networking, others have argued that the belief that gender is invisible online is a myth because the language that users employ is itself gendered. Herring (1996) cites "netiquette", or the accepted norms of online conduct as an example demonstrating this. Beliefs about the proper way to communicate online are founded on systems of values and expectations that vary from individual to individual, "yet it is typically the most powerful or dominant group whose values take on a normative status" (Herring, 1996: 115). Thus, the default guidelines for netiquette in mixed-sex online communities encourage male-typed behaviours (candor, debate, freedom from rules and imposition), and discourage and devalue female-typed behaviours (politeness, consideration for others, support and helpful advice). Herring concludes that these partially incompatible systems of values regarding online conduct and expectations of others exaggerates gender differences, provokes male-female conflicts, and has implications for how comfortable women feel in mainstream electronic forums.

Others (Wellman, Salaaf et al. 1996) have argued that online invisibility is undermined by intrinsic cues to social identity (for example usernames, servers, email addresses, signatures and so on) or gendered language styles. Brosnan (2006) suggests that although email messages purport to dissociate the sender from her identity, the content of emails remain gendered. He argues that, because of sex-role socializaton, women respond to the facelessness of email by attempting to reproduce the socio-emotional elements of face-to-face interaction. Thus, women's emails are more polite than men's are; they are more supportive and affect-laden they make greater use of emoticons. On the other hand, men's emails are efficient, short, and concise. In making recommendations to female users in male-dominated organisations, Brosnan suggests that they should refuse to have an email account, and keep emails to male colleagues short and concise ("an average of 5 to 13 words is normal" (Brosnan, 2006: 266) but that female-to-female email communication should remain unaltered.

CONCLUSION

The last decade has seen an unprecedented growth in the number of women-owned firms operating in typically male industry locations. Like all entrepreneurs, these women must include networking as part of their regular business activities if their organizations are to thrive. Indeed, a proliferation of women's business networks has paralleled the growth in nontraditional enterprises. The popularity of all-female social networks at both the organizational and extra-organizational level can be accounted for by the principle of homophily, as well as women's historical exclusion from 'old boys' networks' (Fisher 2004). However, this chapter has contended that while these networks fulfill important expressive roles, the instrumental function of women-only networks may be less successful. Because there are relatively few women in prominent roles in nontraditional industries, face-to-face networks may not include a critical mass of women who can provide much-needed information, resources and gateways to powerful others (Knouse and Webb, 2001). The alternative course of action – forging links with male colleagues and counterparts – might be similarly unproductive because of gender stereotyping. The result may be that women nontraditional business owners are locked into a damned-if-they-do-damned-if-they-don't situation - termed here the *segregation/stereotyping bind*.

This chapter has considered ways in which virtual networking might aid nontraditional women business owners in overcoming this bind to build collaborative working relations with all-important male ties. A review of the literature suggests that online networking gives nontraditional women entrepreneurs more power over self-representation and "gendered language", thereby enabling relationships to develop that might otherwise have been stunted because of gender stereotyping. In particular, the anonymity of computer-based networking offers the greatest potential to empower women who are in a minority in their field. As the markers of gender are weakened, true competences and abilities can be recognized, independent of position or appearance (Sproull and Kiesler, 1991). Additionally, the flexibility and convenience of this media may assist women in establishing connections with prestigious or powerful others in their own time.

SCHOLARLY CONTRIBUTION AND SUGGESTIONS FOR FUTURE RESEARCH

By highlighting the challenges faced by women in male dominated business sectors, this chapter fills an important lacuna in the current literature on women and entrepreneurship. Research on gender in entrepreneurship has been focused on two-group comparisons of sex differences, but has paid little difference to *within-category* differences. The tendency of previous work to treat women as a monolithic, cohesive category has resulted in unclear and conflicting findings. Relationships are formed by interpersonal communication which is determined to a large extent by contextual and structural factors. Therefore, networks cannot be fully understood without knowledge of the communicative processes that take place within them, and the relationship between these and situational characteristics. Highlighting the ways that interactional processes are created by gendered contextual settings is important because as Mirchandani (1999: 225) has pointed out, "while there has been some reflection on the difference which the sex of business owner makes, this reflection has not been contextualized within theoretical understandings of the ways in which entrepreneurial work is situated within gendered processes which form and are formed through relationships between occupation, organizational structure and the sex of the worker".

Gender segregation is an important yet often overlooked feature of business ownership and self-employment. Policymakers and academics have viewed business ownership as an 'alluring escape route" (Heilman and Chen 2003) from labor market discrimination for women, but there is growing evidence that entrepreneurial segregation contributes to gender inequality in similar ways to occupational divisions of labor. For example, women earn less than men in self-employment and business ownership, and segregation makes a significant contribution to earnings disparities (Hundley 2001; Lowrey 2005). The narrow segments of the economy into which women's businesses are crowded makes their firms much smaller than those owned by men, in terms of employment and sales as well as less profitable (Watson 2002; Miller, Besser et al. 2006/7; Verheul, Caree et al. 2009) and less sustainable (Robb 2002; Headd 2003). Finding ways to overcome the barriers to entry into male-dominated sectors should therefore be higher up the research agenda.

Given the recent evidence of the considerable weakening in the gender-based 'digital divide' in Internet access[5], computer mediated networking appears to offer significant potential. Future research should test empirically whether women nontraditional business owners are indeed able to overcome gender stereotyping via virtual networks. Rich information about the content of relations forged and maintained via electronic means would be particularly useful. Comparison of interpersonal interactions between same- and cross-sex dyads would shed light on the level of resources nontraditional women entrepreneurs are able to secure from their ties. Qualitative research could help us to gain insights into communicative processes in virtual networks.

REFERENCES

Ahl, H. (2006). Why Research on Women Entrepreneurs Needs New Directions. *Entrepreneurship: Theory & Development*, *30*(5), 595–622. doi:10.1111/j.1540-6520.2006.00138.x

Aldrich, H. (1989). Networking Among Women Entrepreneurs. In Hagan, O., Rivchun, C., & Sexton, D. (Eds.), *Women-Owned Business* (pp. 103–132). New York: Praeger.

Aldrich, H., & Zimmer, C. (1986). Entrepreneurship Through Social Networks. In Sexton, D. L., & Smilor, R. W. (Eds.), *The Art and Science of Entrepreneurship* (pp. 3–23). Cambridge, MA: Ballinger.

Aldrich, H. E., Carter, C., et al. (2002). *With Very Little Help From Their Friends: Gender and Relational Composition of Nascent Entrepreneurs' Startup Teams*. 22nd Annual Entrepreneurship Research Conference, Babson College, MA.

Aldrich, H. E., & Elam, A. B. (1997). Strong Ties, Weak Ties and Strangers: Do Women Owners Differ from Men in Their Use of Networking to Obtain Assistance? In Birley, S., & Macmillan, I. C. (Eds.), *Entrepreneurship in a Global Context*. London: Routledge.

Almack, J. C. (1922). The Influence of Intelligence on the Selection of Associates. *School and Society, 16*(410), 529–530.

Anderson, A. R., & Miller, C. J. (2003). "Class Matters": Human and Social Capital in the Entrepreneurial Process. *Journal of Socio-Economics, 32*, 17–36. doi:10.1016/S1053-5357(03)00009-X

Bates, T. (2002). Restricted Access to Markets Characterizes Women-Owned Businesses. *Journal of Business Venturing, 17*, 313–324. doi:10.1016/S0883-9026(00)00066-5

Baum, J. A. C., & Silverman, B. S. (2004). Picking Winners or Building Them? Alliance, Intellectual and Human Capital as Selection Criteria in Venture Financing and Performance of Biotechnology Startups. *Journal of Business Venturing, 19*, 411–436. doi:10.1016/S0883-9026(03)00038-7

Belliveau, M. A. (2005). Blind Ambition: The Effects of Social Networks and Institutional Sex Composition on the Job Search Outcomes of Elite Coeducational and Women's College Graduates. *Organization Science, 16*(2), 134–150. doi:10.1287/orsc.1050.0119

Bird, S. R., & Sapp, S. G. (2004). Understanding the Gender Gap in Small Business Success. *Gender & Society, 18*(1), 5–28. doi:10.1177/0891243203259129

Birley, S. (1985). The role of Networks in the Entrepreneurial Process. *Journal of Business Venturing, 1*(1), 107–117. doi:10.1016/0883-9026(85)90010-2

Blake, M. K. (2006). Gendered Lending: Gender, Context and the Rules of Business Lending. *Venture Capital, 8*(2), 183–201. doi:10.1080/13691060500433835

Blisson, D., & Kaur Rana, B. (2001). *The Role of Entrepreneurial Networks: the Influence of Gender and Ethnicity in British SMEs*. The 46th International Conference for Small Business, Taipei, Taiwan.

Borooah, V. K., & Collins, G. (1997). Women and Self-Employment: An Analysis Of Constraints and Opportunities in Northern Ireland. In Deakins, D., Jennings, P., & Mason, C. (Eds.), *Small Firms: Entrepreneurship in the Nineties* (pp. 72–88). London: Paul Chapman Publishing.

Brass, D. J. (1985). Men's and Women's Networks: A Study of Interaction Patterns and Influence in an Organization. *Academy of Management Journal, 28*(2), 327–343. doi:10.2307/256204

Brophy, D. J. (1989). Financial Women-Owned Entrepreneurial Firms. In Hagan, O., & Rivchun, C. S. D. (Eds.), *Women-Owned Businesses* (pp. 55–75). New York: Praeger.

Brosnan, M. J. (2006). Gender and Diffusion of Email: An Organizational Perspective. In Barrett, M., & Davidson, M. J. (Eds.), *Gender and Communication at Work* (pp. 260–269). Aldershot: Ashgate.

Brown, B., & Butler, J. E. (1995). Competitors as allies: a study of entrepreneurial networks in the U.S. wine industry. *Journal of Small Business Management, 33*(3), 57–66.

Bruni, A., Gherardi, S., & Poggio, B. (2004). Doing Gender, Doing Entrepreneurship: An Ethnographic Account of Intertwined Practices. *Gender, Work and Organization, 11*(4), 406–429. doi:10.1111/j.1468-0432.2004.00240.x

Brush, C., & Carter, N. M. (2004). *Clearing the Hurdles: Women Building High-Growth Businesses*. Upper Saddle River, NJ: Financial Times Prentice Hall.

Buhai, S., & van der Leij, M. (2006). *A Social Network Analysis of Occupational Segregation*. Tinbergen Institute Discussion Paper 016/1 Tinbergen Institute.

Cabrera, S. F., & Thomas-Hunt, M. (2007). Street Cred and the executive woman: The effects of gender differences in social networks on career advancement. In Correll, S. J. (Ed.), *The Social Psychology of Gender* (pp. 123–147). Elsevier Science Press. doi:10.1016/S0882-6145(07)24006-8

Carli, L. L. (2001). Gender and Social Influence. *The Journal of Social Issues*, *57*(4), 725–741. doi:10.1111/0022-4537.00238

Carli, L. L. (2006). Gender Issues in Workplace Groups: Effects of Gender and Communication Style on Social Influence. In Barrett, M., & Davidson, M. J. (Eds.), *Gender and Communication at Work* (pp. 69–83). Aldershot: Ashgate.

Cejka, M. A., & Eagly, A. H. (1999). Gender-Stereotypic Images of Occupations Correspond to the Sex Segregation of Employment. *Personality and Social Psychology Bulletin*, *25*, 413–423. doi:10.1177/0146167299025004002

Center for Women's Business Research (2008). *Key Facts about Women-Owned Businesses* (2008 Update).

Cohen, M. S., & Ellis, T. J. (2008). *The Asynchronous Learning Environment (ALN) as a Gender-Neutral Communication Environment. 38th ASEE/IEEE*. NY: Saratoga Springs.

Coviello, N. E., & Munro, H. J. (1995). Growing the Entrepreneurial Firm: Networking for International Market Development. *European Journal of Marketing*, *29*(7), 49–61. doi:10.1108/03090569510095008

Coyle, H. E., & Flannery, D. D. (2005). *Gendered Contexts of Learning Female Entrepreneurs in Male-Dominated Industries within the United States*. Summer Institute of the National Center for Curriculum Transformation Resources on Women. Turkey.

Cross, J. L. M., & Lin, N. (2008). Access to Social Capital and Status Attainment in the United States: Racial/Ethnic and Gender Differences. In Lin, N., & Erickson, B. H. (Eds.), *Social Capital* (pp. 364–379). Oxford: Oxford University Press.

Feld, S. L. (1982). Social Structural Determinants of Similarity among Associates. *American Sociological Review*, *47*(6), 797–801. doi:10.2307/2095216

Fertala, N. (2005). *Do Birds of a Feather Flock Together and Perform Economically Better? A Study of the Homophily Paradox Among Immigrant Entrepreneurs in Germany*. The Twenty-Fifth Annual Research Conference, Babson College, Wellesely, MA.

Fisher, A. (2004). *Why Women Rule The latest numbers show that they are starting more new businesses than men and growing them faster. What's going on?* Fortune Small Business.

Florin, J., & Lubatkin, M. (2003). A Social Capital Model of High-Growth Ventures. *Academy of Management Journal*, *46*(3), 374–384. doi:10.2307/30040630

Glick, P., & Fiske, S. T. (1999). Sexism and Other 'Isms': Interdependence, Status, and the Ambivalent Content of Stereotypes. In Swann, J. W. B., Langlois, J. H., & Gilbert, L. A. (Eds.), *Sexism and Stereotypes in Modern Society* (pp. 193–221). Washington, DC: American Psychological Association. doi:10.1037/10277-008

Granovetter, M. (1973). The Strength of Weak Ties. *American Journal of Sociology*, *78*(6), 1360–1380. doi:10.1086/225469

Granovetter, M. (1985). Economic Action and Social Structure: The Problem of Embeddedness. *American Journal of Sociology*, *91*, 481–510. doi:10.1086/228311

Green, E., & Cohen, L. (1995). 'Women's Business': Are Women Entrepreneurs Breaking New Ground or Simply Balancing the Demands of 'Women's Work' in a New Way? *Journal of Gender Studies*, *4*(3), 297–314. doi:10.1080/09589236.1995.9960615

Hansen, E. L. (1995). Entrepreneurial Networks and New Organization Growth. *Entrepreneurship. Theory into Practice*, ***, 7–19.

Hansen, E. L. (2000). Resource Acquisition as a Startup Process: Initial Stocks of Social Capital and Organizational Foundings. In *Proceedings of the Twentieth Annual Entrepreneurship Research Conference*. Babson College.

Hanson, S., & Blake, M. (2008). Gender and Entrepreneurial Networks. *Regional Studies*, *43*(1), 135–149. doi:10.1080/00343400802251452

Headd, B. (2003). Redefining business success: Distinguishing between closure and failure. *Small Business Economics*, *21*(1), 51–61. doi:10.1023/A:1024433630958

Heilman, M. E., & Block, C. J. (1989). Has Anything Changed? Current Characterizations of Men, Women, and Managers. *The Journal of Applied Psychology*, *74*(6), 935–942. doi:10.1037/0021-9010.74.6.935

Heilman, M. E., & Chen, J. (2003). Entrepreneurship as a Solution: the Allure of Self-Employment for Women and Minorities. *Human Resource Management Review*, *13*, 347–364.

Herrick, J. W. (1999). And Then She Said: Office Stories and What They Tell Us about Gender in the Workplace. *Journal of Business and Technical Communication*, *13*(3), 274–296. doi:10.1177/105065199901300303

Herring, S. (1996). Posting in a Different Voice: Gender and Ethics in Computer-Mediated Communication. In Ess, C. (Ed.), *Philosophical Perspectives on Computer Mediated Communication* (pp. 115–143). Albany, NY: State University of New York Press.

Hundley, G. (2001). Why Women Earn Less than Men in Self-Employment. *Journal of Labor Research*, *22*(4), 818–828. doi:10.1007/s12122-001-1054-3

Ibarra, H. (1992). Homophily and Differential Returns: Sex Differenecs in Network Structure and Access in an Advertising Firm. *Administrative Science Quarterly*, *37*(3), 422–447. doi:10.2307/2393451

Kelly, E. P., & Young, A. O. (1993). Sex Stereotyping in the Workplace: A Manager's Guide. *Business Horizons*, *36*(2), 23–29. doi:10.1016/S0007-6813(05)80034-5

Kim, P. H., & Aldrich, H. E. (2005). *Social Capital and Entrepreneurship*. Hanover, MA: Now.

Knouse, S. B., & Webb, S. C. (2001). Virtual Networking for Women and Minorities. *Career Development International*, *6*(4), 226–228. doi:10.1108/13620430110397541

Liao, J., & Welsch, H. P. (2001). *Social Capital and Growth Intention: The Role of Entrepreneurial Networks in Technology-Based New Ventures*. MA: Wellesely.

Liñán, F., & Santos, F. J. (2007). Does Social Capital Affect Entrepreneurial Intentions. *International Advances in Economic Research*, *13*(4), 443–453. doi:10.1007/s11294-007-9109-8

Lind, M. R. (1999). The Gender Impact of Temporary Virtual Work Groups. *IEEE Transactions on Professional Communication*, *42*(4), 276–285. doi:10.1109/47.807966

Lind, M. R. (2001). An Exploration of Communication Channel Usage by Gender. *Work Study*, *50*(6), 234–240. doi:10.1108/00438020110403338

Lipman-Blumen, J. (1976). Toward a Homosocial Theory of Sex Roles: An Explanation of the Sex Segregation of Social Institutions. In Blaxall, M., & Reagan, B. (Eds.), *Women and the Workplace: The Implications of Occupational Segregation* (pp. 15–32). Chicago: The University of Chicago Press.

Ljunggren, E., & Kolvereid, L. (1996). New Business Formation: Does Gender Make a Difference? *Women in Management Review*, *11*(4), 3–12. doi:10.1108/09649429610122096

Loscocco, K. A., Robinson, J., Hall, R. H., & Allen, J. K. (1991). Gender and Small Business Success: An Inquiry into Women's Relative Disadvantage. *Social Forces*, *70*, 65–87. doi:10.2307/2580062

Lowrey, Y. (2005). *US Sole Propriertorships: A Gender Comparison 1985-2000*. Washington, D.C.: Small Business Administration Office of Advocacy.

Marlow, S., & Carter, S. (2008). Constructing Female Entrepreneurship Policy in the UK: Is the USA a Relevant Benchmark? *Environment and Planning. C, Government & Policy*, *26*(2), 335–351. doi:10.1068/c0732r

Marlow, S., & Strange, A. (1994). Female Entrepreneur - Success By Whose Standards? In Tanton, M. (Ed.), *Women in Management: A Developing Presence* (pp. 172–184). London: Routledge.

Martin, L. M., & Wright, L. T. (2005). No Gender in Cyberspace? Empowering Entrepreneurship and Innovation in Female-run ICT Small Firms. *International Journal of Entrepreneurial Behaviour & Research*, *11*(2), 162–178. doi:10.1108/13552550510590563

McGowan, P., & Hampton, A. (2007). An Exploration of Networking Practices of Female Entrepreneurs. In Carter, N. M., Henry, C., Cinneide, B. O., & Johnston, K. (Eds.), *Female Entrepreneurship: Implications for Education, Training and Policy* (pp. 110–134). London: Routledge.

McKnight, D. H., & Cummings, L. L. (1998). Initial Trust Formation in New Organizational Relationships. *Academy of Management Review*, *23*(3), 473–490.

McPherson, J. M., & Smith-Lovin, L. (2001). Birds of a Feather: Homophily in Social Networks. *Annual Review of Sociology*, *27*, 415–444. doi:10.1146/annurev.soc.27.1.415

McPherson, M., & Smith-Lovin, L. (1987). Homophily in Voluntary Organizations: Status Distance and the Composition of Face-to-Face Groups. *American Sociological Review*, *52*(3), 370–379. doi:10.2307/2095356

Mehra, A., & Kilduff, M. (1998). At the Margins: A Distinctiveness Approach to the Social Identity and Social Networks of Underrepresented Groups. *Academy of Management Journal*, *41*(4), 441–452. doi:10.2307/257083

Miller, N. J., & Besser, T. L. et al. (2006/7). Do Strategic Business Networks Benefit Male- and Female-Owned Small Community Businesses. *Journal of Small Business Strategy*, *17*(2), 53-74.

Mirchandani, K. (1999). Feminist Insight on Gendered Work: New Directions in Research on Women and Entrepreneurship. *Gender, Work and Organization*, *6*(4), 224–235. doi:10.1111/1468-0432.00085

Montgomery, J. D. (1991). Social Networks and Labour-Market Outcomes: Toward an Economic Analysis. *The American Economic Review*, *81*(5), 1408–1418.

Morrison, A. M., & von Glinow, M. A. (1990). Women and Minorities in Management. *The American Psychologist*, *45*(2), 200–208. doi:10.1037/0003-066X.45.2.200

Munch, A., & McPherson, M. A. (1997). Gender, Children, and Social Contact: The Effects of Childrearing for Men and Women. *American Sociological Review*, *62*(4), 509–520. doi:10.2307/2657423

Nohria, N. (1992). Information and Search in the Creation of New Business Ventures: the Case of the 128 Venture Group. In Nitin, N., & Eccles, R. G. (Eds.), *Networks and Organizations* (pp. 240–261). Boston, MA: Harvard Business School Press.

Owen, C. L., & Todor, W. D. (1993). Attitudes Toward Women as Managers: Still the Same. *Business Horizons*, *36*(2), 12–16. doi:10.1016/S0007-6813(05)80032-1

Ozgen, E., & Baron, R. A. (2007). Social sources of information in opportunity recognition: Effects of mentors, industry networks, and professional forums. *Journal of Business Venturing*, *22*(2), 174–192. doi:10.1016/j.jbusvent.2005.12.001

Popielarz, P. A. (1999). (In)Voluntary Association: A Multilevel Analysis of Gender Segregation in Voluntary Organizations. *Gender & Society*, *13*(2), 234–250. doi:10.1177/089124399013002005

Renzulli, L. A., & Aldrich, H. (2000). Family Matters: Gender, Networks and Entrepreneurial Outcomes. *Social Forces*, *79*(2), 523–546. doi:10.2307/2675508

Reskin, B. F. (1993). Sex Segregation in the Workplace. *Annual Review of Sociology*, *19*, 241–270. doi:10.1146/annurev.so.19.080193.001325

Robb, A. M. (2002). Entrepreneurial Pefromance by Women and Minorities: The Case of New Firms. *Journal of Developmental Entrepreneurship*, *7*(4), 384–397.

Roomi, M. A. (2007). *Role of Human and Social Capital in the Growth of Women-owned Enterprises*. Glasgow: Institute for Small Business and Entrepreneurship.

Rudman, L. A. (1998). Self-Promotion as a Risk Factor for Women: The Costs and Benefits of Counterstereotypical Impression Management. *Journal of Personality and Social Psychology*, *74*(3), 629–345. doi:10.1037/0022-3514.74.3.629

Sappleton, N. (2009). Women Non-traditional Entrepreneurs and Social Capital. *International Journal of Gender and Entrepreneurship*, *1*(3), 192–218. doi:10.1108/17566260910990892

Schein, V. E. (1994). Managerial Sex Typing: A Persistent and Pervasive Barrier to Women's Opportunities. In Davidson, M. J., & Burke, R. J. (Eds.), *Women in Management: Current Research Issues* (pp. 41–52). London: Paul Chapman Publishing.

Schein, V. E. (2001). A Global Look at Psychological Barriers to Women's Progress in Management. *The Journal of Social Issues*, *57*(4), 675–688. doi:10.1111/0022-4537.00235

Smeltzer, L. R., & Fann, G. L. (1989). Gender Differences in External Networks of Small Business Owner/Managers. *Journal of Small Business Management*, *27*(2), 25–32.

Smith, P. L., & Smits, S. J. (1992). Female Business Owners in Industries Traditionally Dominated by Males. *Sex Roles*, *26*(11/12), 485–496. doi:10.1007/BF00289870

Sproull, L., & Kiesler, S. (1991). *Connections: New Ways of Working in the Networked Organization*. Cambridge, MA: The MIT Press.

Srinvasan, R., Woo, C. Y., et al. (1994). Performance Determinants for Male and Female Entrepreneurs. In *Proceedings of the Fourteenth Annual Entrepreneurship Research Conference*. Babson College, MA.

Suseno, Y. (2008). Examining the Role of Social Capital in Female Professional's reputation Building and Opportunities Gathering: A Network Appriach. In Aaltio, I., Kyro, P., & Sundin, E. (Eds.), *Women Entrepreneurship and Social Capital: A Dialogue and Construction* (pp. 147–166). Copenhagen.

Thomson, R. (2006a). The Effect of Topic of Dicussion on Gendered Language in Computer-Mediated Discussion. *Journal of Language and Social Psychology, 25*(2), 167–178. doi:10.1177/0261927X06286452

Thomson, R. (2006b). Gender and Electronic Discourse in the Workplace. In M. Barrett & M. J. Davidson (Eds.), *Gender and Communication at Work* (pp. 239-249). Aldershot: Ashgate: 239-249.

Travers, C., & Pemberton, C. (1997). Women's Networking Across Boundaries: Recognizing Different Cultural Agendas. *Women in Management Review, 12*(2), 61–67. doi:10.1108/09649429710162820

(2002). *U.S. Census Bureau.* Washington, D.C.: Survey of Women Business Owners.

Uzzi, B. (1996). The Sources and Consequences of Embeddedness for the Economic Performance of Organizations: The Network Effect. *American Sociological Review, 61*(4), 674–698. doi:10.2307/2096399

Verheul, I., & Caree, M. (2009). Allocation and Productivity of Time in New Ventures of Female and Male Entrepreneurs. *Small Business Economics, 33*(3). doi:10.1007/s11187-009-9174-x

Verheul, I., & Risseeuw, P. (2002). Gender Differences in Strategy and Human Resource Management: The Case of Dutch Real Estate Brokerage. *International Small Business Journal, 20*(4), 443–476. doi:10.1177/0266242602204004

Watson, J. (2002). Comparing the Performance of Male- and Female-Controlled Businesses: Relating Outputs to Inputs. *Entrepreneurship: Theory and Practice, 26*(3), 91–100.

Weiler, S., & Bernasek, A. (2001). Dodging the Glass Ceiling? Networks and the New Wave of Women Entrepreneurs. *The Social Science Journal, 38*(1), 85–103. doi:10.1016/S0362-3319(00)00111-7

Wellman, B., & Salaaf, J. (1996). Computer Networks as Social Networks: Collaborative Work, Telework and Virtual Community. *Annual Review of Sociology, 22*, 213–238. doi:10.1146/annurev.soc.22.1.213

Welter, F., & Trettin, L. (2006). The Spatial Embeddedness of Networks for Women Entrepreneurs. In Fritsch, M., & Schmude, J. (Eds.), *Entrepreneurship in the Region* (pp. 35–59). New York: Springer. doi:10.1007/0-387-28376-5_3

West, C., & Fenstermaker, S. (1993). Power, Inequality and the Accomplishment of Gender: An Ethnomethodological View. In England, P. (Ed.), *Theory on Gender/Feminism on Theory*. New York: Aldine.

Witt, P. (2004). Entrepreneurs' Networks and the Success of Start-Ups. *Entrepreneurship & Regional Development, 16*, 391–412. doi:10.1080/0898562042000188423

Wood, L. M. (1999). The Use of Electronic Networking by Australian Small Business Women. In *Proceedings of the Australian Community Networking Alliance, Balart, ACNA*.

ENDNOTES

[1] The NWBC considers women-owned firms in the following broad industrial classifications to be nontraditional: Agriculture, Mining, Construction, Manufacturing, Trans-

portation/Communications and Wholesale Trade. Traditional industries for women are: Retail Trade, Finance/Insurance/Real Estate and Services.

2   Others have distinguished between inbreeding and baseline homophily (e.g. Montgomery, J. D. (1991). "Social Networks and Labour-Market Outcomes: Toward an Economic Analysis." *The American Economic Review 81*(5), 1408-1418, Buhai, S., & van der Leij, M. (2006). *A Social Network Analysis of Occupational Segregation*. Tinbergen Institute Discussion Paper 016/1 Tinbergen Institute.)

3   Recent research suggests that around 80% of household purchases are made by women.

4   Even feminist historians agree that women's suffrage could not have been achieved without the support of men.

5   Research reported in Cohen and Ellis (2005) indicates that 68% of American men have online access compared to 66% of American women.

A previous version of this chapter was published in the International Journal of E-Politics, Volume 2, Issue 2, edited by Celia Romm, pp. 18-36, copyright 2011 by IGI Publishing (an imprint of IGI Global).

Chapter 11
Would Elizabeth Cady Stanton Blog?
Women Bloggers, Politics, and Political Participation

Antoinette Pole
Montclair State University, USA

ABSTRACT

This study examines the role of women political bloggers and how they use their blogs for purposes related to politics, public policy, and current events. Based on a combined purposive-snowball sample, in-depth interviews were conducted with 20 women political bloggers in October 2006. Findings show respondents blog about a range of topics, not necessarily unique to women. Generally, women use their blogs to inform their readers, check the media, engage in advocacy efforts, and solicit charitable contributions from their readers and more specifically, women ask their readers to vote and contact elected officials. Data show women deal with a range of challenges blogging most notably discrimination. Though a majority of women political bloggers reported they did not face discrimination, interviewees qualified their responses saying they witnessed discrimination and discriminatory attitudes, suggesting the political blogosphere is somewhat inhospitable to women.

DOI: 10.4018/978-1-60960-759-3.ch011

INTRODUCTION

In February 2007, the ascension and decline of two women political bloggers garnered national attention, when just days after being hired to blog for Democratic presidential candidate John Edwards they were forced to resign. Popular liberal bloggers, Amanda Marcotte of *Pandagon* (http://www.pandagon.net/, formerly http://pandagon.blogsome.com/) and Melissa McEwan of *Shakespeare's Sister* (http://shakespearessister.blogspot.com/) were hired by the Edwards campaign as Blogmaster and Netroots Coordinator respectively. In these capacities, the women were tasked with attracting liberal supporters, while building a blog audience on behalf of the campaign. On February 6, 2007, having blogged for only one week, Marcotte and McEwan were accused of writing anti-Catholic posts by the President of the Catholic League, Bill Donohue (Broder, 2007). While the Edwards campaign supported both bloggers, they resigned amidst a maelstrom of negative publicity (Marcotte, 2007). During her short tenure with the Edwards campaign, Marcotte simultaneously continued blogging at Pandagon. An explanation of her actions appeared in *Salon* (http://www.salon.com/) in which Marcotte wrote, "Reasonable people," I thought, "can tell the difference between a personal blog post and those I'll write for the campaign"(Marcotte, 2007). Yet this was not the case.

The above narrative illustrates the public fate of two popular feminist political bloggers. It underlines the scrutiny and challenges they faced while working for the Edwards campaign. Marcotte explains,

What I also failed to understand was how much McEwan and I would stick out. I was aware that I didn't exactly fit the image people have of bloggers who join campaigns—the stereotype being 30-something nerdy young white men who wear khakis and obsess over crafting their Act Blue lists. I wasn't aware that not fitting the image would attract so much negative attention. In fact, I mostly saw this all as a baby step in the direction of diversity, since McEwan and I differed from the stereotype mostly by being female and by being outspoken feminists (Marcotte, 2007).

Whether or not the challenges these women faced are attributable to gender or their ideological position is debatable. This incident drew substantial attention in the mainstream media and the blogosphere; it only partially describes women political bloggers' experiences, however.

To date few studies examine the role of women bloggers within the realm of politics, public policy and current events. Filling the gaps in the literature, this research asks: what is unique about women political bloggers and how do women political bloggers use their blogs for purposes related to politics and participation? Based on in-depth interviews conducted in October 2006 with 20 women, this research attempts to identify the demographics of these bloggers, about which topics these women blog, and how they use their blogs in the context of politics. It also seeks to identify what challenges, if any, women political bloggers face, and whether they experience exclusion and discrimination in the blogosphere.

LITERATURE

Political Blogs

A growing body of work examines the rise of political bloggers, detailing how bloggers use this medium to participate in politics both online and/or in-person, and how bloggers use their blogs to mobilize readers. While a broad literature frames the research on blogging more generally, focusing on politics and participation this study draws upon descriptive and exploratory studies, and to a lesser degree theory building research.

Earlier works conducted by McKenna and Pole (2004, 2008) investigate the activities of A-list—

the most popular bloggers—and average political bloggers. Findings from these studies indicate that bloggers engage in a variety of activities including informing readers, reporting errors and omissions in the media, engaging in advocacy efforts, and soliciting charitable contributions. Specifically, bloggers encourage their readers to engage in a variety of political activities including asking readers to vote (70%), contact an elected official (64%), and sign petitions (46%) (McKenna & Pole, 2008). Complementing these works, Wallsten (2007) develops a system of classifying political blogs. He argues that blogs are used as "transmission belts, soapboxes and mobilizers" (Wallsten, 2007). Political blogs link to other sites, provide a forum for discussion and encourage readers to mobilize. This study of women political bloggers builds on the findings from these studies.

Emphasizing well-trafficked blogs, a study conducted by Perlmutter (2008) examines the impact of blogs on the American political system. He asserts that blogs are not especially powerful in the traditional sense of politics since they lack financial, moral and social leverage to induce readers to participate. Despite this, he suggests that blogs "improve democracy and enrich political culture" (Perlmutter, 2008). In contrast, Pole's (2010) study of political blogging focuses not on well-trafficked blogs, but on the political actors and average citizens who blog about politics. Data show that political bloggers are in fact using their blogs for purposes related to politics, and in particular blacks and Lesbian-Gay-Bisexual-Transgender bloggers use their blogs to encourage readers to vote, contact elected officials, and sign petitions (Pole, 2010).

Women and Political Blogs

Another body of literature emphasizing women bloggers also informs this research (Herring, 2003; Herring et al., 2004; Herring & Paolillo, 2006; Kennedy et al., 2005; Pedersen & Macafee, 2007). While studies explore the role of gender more broadly, few accounts document the role of women political bloggers. Moreover, these accounts tend, more often than not, to explore the role of blogs in the context of language and feminist theory rather than investigating specifically whether and how blogs are used for politics.

The role of women bloggers in the context of political discourse and the public sphere highlights two prominent debates. Keenly noted by Mitchell (2007), studies of women bloggers typically ask, "where are the women bloggers," progressing to "which women, and why." Addressing this not infrequent refrain, "where are the women," Osell (2007) suggests that historically women writers have been invisible. Using feminist consciousness Osell (2007) surveys bloggers asking whether they blog anonymously and why. Results of bloggers[1] show 105 out of 141 respondents blogged anonymously. Reasons for remaining anonymous included the desire to blog about work and personal experiences, while maintaining privacy. According to the author, "the association of men with the public world and women with the private world is reflected, to some extent, in bloggers' choice of topics, and that the latter—rather than diffidence about the public nature of blogging—drives some women's choices to use aliases" (Osell, 2007). A connection between blogging and attracting readers based on one's appearance is discussed by Ratliff (2007), who studies political bloggers. She notes that many of the prominent male political bloggers link to each other, thereby producing a "fraternity-like atmosphere." Accordingly, Ratliff (2007) argues that women who blog about their sexuality, might feel compelled to do so in order to maintain readers. She further asserts that when women's physical attractiveness is discussed by male bloggers, this serves as a "subtle exclusionary tactic, and women end up being discouraged from participating in discussions in the comments on these blogs" (Ratliff, 2007). These studies exemplify the role of women in the public and private spheres, providing context for research on women political bloggers.

A pointed discussion of politics and women political bloggers is undertaken by Nolan (2007). She examines the role of women in politics depicting a range of women from Hillary Rodham Clinton to Condoleezza Rice. Drawing on politics, Nolan (2007) shows that women's voices are louder in the political blogosphere because the Internet acts as an equalizing force. This and the aforementioned research provide context for this study of women political bloggers. Exploring how women political bloggers use their blogs to participate and encourage their readers to do so, illuminates the potential of this medium across different modes of participation, in a variety of political settings.

IMPORTANCE OF POLITICAL BLOGS

Political blogs have transformed the political landscape in the United States. Early on bloggers were credited with unseating Senate Majority Leader Trent Lott (Shactman, 2002), and more recently, 2008 presidential candidates used their blogs to advertise and mobilize volunteers (Pole, 2010). Efforts to mobilize individuals, raise awareness, encourage readers to contact elected officials, vote, and register to vote are just some of the ways blogs have been used as vehicles for participation. The empirical implications of these efforts are noteworthy. Elections, for example, can be won or loss. Though ultimately loosing the general election, Ned Lamont's successful 2006 primary victory highlights the power of bloggers, who were credited with Lamont's primary victory (Cohen, 2007). Additionally, nearly all major media outlets have created blogs alongside their more traditional methods of disseminating the news, underlining their saliency. Given their growth, coupled with their import, we must consider whether and how this impacts women political bloggers.

Figures documenting the number of blogs and readers vary considerably. In March 2008, one report indicated 184 million blogs were created and 346 million people read them globally (Solis, 2009). Other sources offer more conservative estimates. *Technorati* (http://technorati.com), a well-known blog aggregator, reports that since 2002, 133 million blogs exist. Given the pervasiveness of blogs, their significance in the political blogosphere should not be underestimated. While the number of blogs that focus on politics, public policy and current events remains unknown, the previously mentioned examples highlight the consequences of this growing medium.

Although the differences between men and women political bloggers is beyond the scope of this study—instead this study focuses on what is unique about women political bloggers—it is important to note the significance of gender disparities within the political blogosphere, given the implications this has on civic engagement and politics. A survey of 147 political bloggers shows just 30 percent of respondents are women (McKenna & Pole, 2008), mirroring broader blogosphere findings from the 2008 and 2009 "State of the Blogosphere" (Technorati State of the Blogosphere 2008, 2009). If few women blog about politics, what affect does this have on political discourse, participation and the political landscape? Marcotte's explanation, "I didn't exactly fit the image people have of bloggers who join campaigns—the stereotype being 30-something nerdy young white men" underlines the politics of gender, which arise even online. Whether and how women political bloggers participate in politics, provides a snapshot toward furthering our understanding of women's use of information technology in the context of politics and participation. This study provides a baseline to which subsequent studies can be compared.

Challenges to participating in politics are not uncommon. Women in the US likely face more obstacles than men, because they are primary care givers to children, combined with their presence in the workforce. Among traditional forms of political participation, men are slightly more active than women with men averaging 2.3 acts and women two acts (Verba et al., 2005). Pronounced disparities also exist between men and women's online activities (Fallows, 2005). This study presents an opportunity to discover what challenges, if any, women face in using blogs for purposes related to politics. Time and money are two of the most commonly mentioned obstacles to participating in politics. Blogging's flexibility enhances opportunities to participate. Bloggers are not bounded by time, geography, or monetary constraints, again common obstacles to offline forms of participation. Absent these obstacles, bloggers can have, and have had, a real impact on US politics, underlining the importance of this medium. For example, women can encourage readers to contact elected officials or they can become engaged in shaping public policy through their blogs without having to leave home.

Finally, recent studies (Herring, 2003; Herring et al., 2004; Herring & Paolillo, 2006; Kennedy et al., 2005) detail how women blog for reasons unrelated or only partially related to politics. Analyses that focus on women political bloggers emphasize the context of language and feminist theory. These studies narrowly illustrate the importance of women political bloggers in particular circumstances—such as overtly sexualized blogging—but they do not account for broader political acts like voting and contacting elected officials, or working on a political campaign like Marcotte and McEwan. This research contributes to the literature by providing a better understanding about what issues women political bloggers blog and how women political bloggers use their blogs to undertake a range of political activities.

METHODS

This paper examines the role of women political bloggers, the unit of analysis. It is primarily exploratory and descriptive. The overarching research questions ask: what is unique to women political bloggers; how do women political bloggers use their blogs to participate in politics; and how do they encourage their readers to participate? To a lesser degree this research also is explanatory.

Data for this project are based on in-depth interviews (see Appendix A: Interview Instrument for Women Bloggers) with 20 women political bloggers conducted in October 2006.[2] Methodologically, in-depth interviews were most appropriate for several reasons. Asking "how" and "why" questions can easily be accomplished with in-depth interviews. These types of questions are common components of exploratory research (Yin, 2008). Other methods do not lend themselves to asking these types of questions. Interviews provide respondents with an opportunity to offer more textured, nuanced responses than do surveys. Due to the absence of an identifiable sampling frame, survey research was not possible. To evaluate how women blogged about politics, content analysis was not deemed viable since this approach assesses blog content rather than investigating the blogger, which is the unit of analysis.

Because the universe of women political bloggers is unknown a sample of women political bloggers was generated using a combined purposive-snowball sampling approach (Patton, 2002). Snowball sampling is appropriate for generating a study sample with known characteristics "women political bloggers." This is a widely accepted method of sampling in studies of blogging. According to Hindman et al., "any site more than three clicks away from any of the top 200 Google or Yahoo results on a given topic is definitely off the beaten track, and not likely to have any substantial impact" (Hindman et al., 2003). Blogs identified by search engines and blog aggregators are most visible, with visibility decreasing exponentially as one moves further from the top blogs. Scholars need not catalog thousands of blogs, but instead can credibly focus on the most heavily linked ones since they are the most widely read and influential. Based on the rankings of bloggers from *Technorati* (http://technorati.com) and *The Truth Laid Bear* (http://

truthlaidbear.com), a group of women political bloggers were identified from among the top 100 bloggers. Of these blogs, more women political bloggers were identified through their blogrolls. Finally, during interviews respondents were asked to identify other women political bloggers who might be interested in participating in the study.

Criteria for inclusion in the study included, women who blogged about politics, public policy, and current events; residents of the US who blog in English; and evidence of posting within a week of being interviewed. Since this study focuses on politics and political participation, only bloggers with an emphasis on these areas were selected for inclusion in the sample and ultimately the study. Political bloggers are operationalized as individuals whose blogs focus on politics, public policy and/or current events. An invitation to participate was sent via e-mail. The aforementioned goals of the study were outlined and respondents engaged in self-selected participation. To ensure that respondents were in fact political bloggers, a cursory review of their blog was undertaken and two filter questions were asked (Appendix A, questions 1 and 2).

To solicit participants, an e-mail was sent to approximately 45 women political bloggers. Interested bloggers responded to the e-mail and 20 interviews were conducted via telephone in October 2006. On average, interviews lasted approximately one hour. Interviewees were asked 25 questions and all but three questions were open-ended. The interview questions were designed to assess about what issues women political bloggers blog; in what types of activities do they ask their readers to engage; whether or not they feel blogging is a form of political participation; what challenges they face, if any; and whether or not they face exclusion and discrimination. During the interview the responses to all questions were typed almost verbatim as the interview was being conducted. Upon receiving recommendations of other women to interview, potential interviewees were contacted via e-mail with a note indicating the name of the blogger who referred them, as well as the original e-mail detailing the parameters of the study. Of the 20 women political bloggers interviewed in 2006, 16 out of 20 still were blogging in January 2010.

Appropriate for qualitative studies that are exploratory, descriptive or empirical in nature, this study relies on grounded theory which provides predictions, explanations, interpretations and applications (Glaser & Strauss, 1967). While deductive methods of research seek to confirm or reject existing theories of social research, grounded theory is an inductive process that begins with data and concludes with the development of theory. The first step in grounded research is sample selection. According to Glaser and Strauss (1967), theoretical sampling differs from statistical sampling in that the former relies upon the saturation of categories to discover theory, whereas the latter relies upon distributions of people or categories that require verification—regardless of saturation. Sample sizes differ dramatically, with the former not requiring large sample sizes. Data collection in grounded theory is not predicated on generalizability, but rather on the generation of theory. Interviews ceased once categories reach "theoretical saturation," (Glaser & Strauss, 1967) which occurred with 20 interviews.

By systematically examining the content, categories emerge as do the properties of these categories (Glaser & Strauss, 1967). Content analysis was performed using the transcripts from the interviews by indentifying major themes and quantified themes for each interview question. A variety of themes emerged around topics typically covered by the mainstream media ranging from the war in Iraq to Katrina. Other themes also emerged including: abortion, child rearing, gender, family, feminist issues, law and gender, media and gender, reproductive rights, and sexual double standard.

FINDINGS

Snapshot of Women Political Bloggers and Their Blogs

Overall, the 20 women interviewed for this study are well-educated, middle-aged and occupationally varied. Illustrated in Figure 1, the data show that 35 percent of women (7 out of 20) possess a bachelor's degree, and 60 percent of interviewees (12 out of 20) hold a degree equivalent to, or higher than, a master's degree. One-third of the sample earned a Ph.D., JD, or MD. Respondents range in age from 23 to 55, with an average age of 42. The occupations of women political bloggers vary from a secretary to a psychiatrist. A large number of respondents, for example, are professors or writers.

The women in this study were selected because they blog about politics, public policy and current events. On average, respondents indicated that 75 percent of their blog content pertained to politics, public policy and current events with a range of 45 to 100 percent of content focusing on the aforementioned (Figure 2). A relatively balanced representation of partisan preferences appears in the sample, with 13 of the 20 women describing their orientation as left, liberal or Democrat, another three reporting their orientation as right or Republican, and the balance indicating their orientation as Independent or other (Figure 3). Commenting on her political affiliation a blogger lamented "I think being a conservative blogger is more difficult than being a female blogger on the left," thus illustrating the partisan tensions in this segment of the blogosphere. Further evidence of this was echoed by another responded declaring, "I feel that other women are out to destroy me because I'm not a proper woman—a Democrat—and I'm not supporting the politics of the left. I'm viewed as a traitor and I don't toe the line. I'm conspicuous."

An important component of blogs is traffic to the blog and readership. Traffic corresponds to a blog's popularity levels. Not unlike the popular bloggers Marcotte and McEwan, a majority of the women in this sample are "A and B-list" bloggers with highly trafficked blogs (Pole, 2010).[3] Nearly 60 percent of the women interviewed received 1,000 or more unique page visits per day.

Figure 1. n = 20 Note: Source Women Political Blogger (2006)

Figure 2. Note: Source Women Political Blogger (2006)

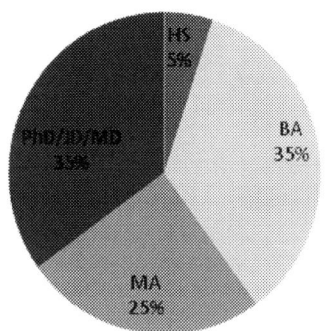

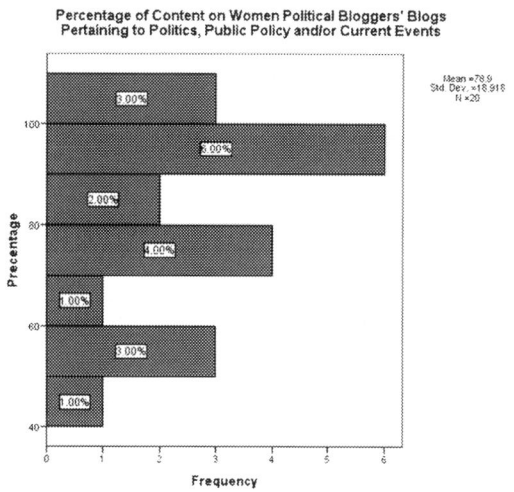

Figure 3. n = 20 Note: Source Women Political Blogger (2006)

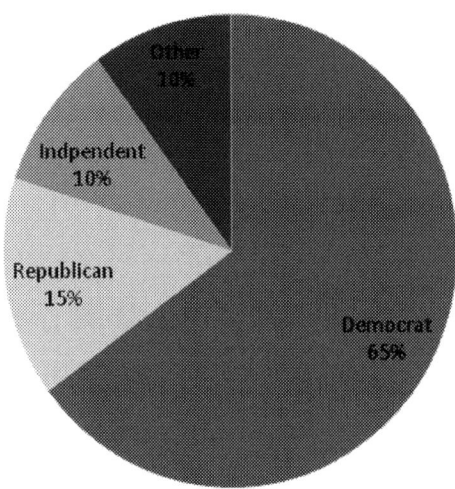

Political Orientation of Women Political Bloggers

Four of the 20 bloggers are considered A-list bloggers with 13,500 unique visits per month.

Politics: Topics About Which Women Blog and Political Activities

Women blogged about a variety of topics ranging from the war in Iraq to reproductive rights (Table 1). When asked whether the topics about which they blog are unique to women, respondents said that the topics are not necessarily unique to women. One respondent said,

No, the issues are not unique because they are connected to women. Women are more interested in the issues that I'm covering. Women care more about abortion than do men; not that men don't care, but the direct impact is on women and generally, women are more vested.

Table 1. Topics About Which Women Political Bloggers Blog

Topics Unique to Women Political Bloggers
abortion
child rearing
gender
family
feminist issues
law and gender
media and gender
reproductive rights
sexual double standard
Other Topics Women Political Bloggers Blog About
Abu Garib
campaigns/elections
consumerism
congressional races
Duke rape case
economics
education
environment
health care
immigration
economic justice
euthanasia
Guantanamo Bay
Katrina
LGBT issues
military
poverty
racism
religion and politics
state and local politics
Supreme Court decisions
torture bill
United Nations
voting
war in Iraq

Similarly, another woman said that while the issues were not unique, "I think a lot of issues get more attention from women. For example, the conflict between home and work is always framed as a women's issue. Education affects everyone, but traditionally women pay more attention to this." Several bloggers, however, said that they blog from a woman's perspective or focus on issues that some might consider "women's issues" though they arguably affect both men and women. During an interview one woman reported, "I do [blog about issues unique to women]. I find that I blog more about feminist issues, abortion, women's rights and the violation of women's rights abroad." Other respondents echoed this sentiment stating, "I talk about feminism and cultural feminism," while another blogger asserted, "Obviously it [my blog] is things that affect me and I am a woman. When addressing women's issues, we probably look at things a little differently." Many bloggers elaborated on how particular issues might impact women.

The data show that all women political bloggers use their blogs to inform their readers and to report errors or omission in the media. Not all bloggers, however, use their blogs to engage in advocacy efforts or to solicit charitable contributions. Still, 90 percent of women (18 out of 20) engage in advocacy efforts and 80 percent of women (16 out of 20) solicit charitable contributions. To understand in greater detail how women use their blogs for purposes related to politics, interviewees were asked whether or not they asked their readers to engage in a variety of political activities (Table 2). Respondents mentioned encouraging their readers to contact an elected official, to vote and sign a petition most frequently. During the Terri Schiavo controversy one blogger recalled asking her readers to contact their member of congress. In addition to providing a link to members of congress, she also requested readers contact the president and their state representatives. Still another respondent provided great detail recounting, "The torture bill was something that attracted a lot of attention. I

Table 2. Percentage of Activities Bloggers Asked Readers to Undertake

Activity	Women Bloggers (n = 20)
Contact an elected official	80 (16)
Vote	75 (15)
Sign a petition	60 (12)
Register to vote	45 (9)
Attend a rally	45 (9)
Attend a political party meeting	40 (8)
Attend a political fundraiser	--- ---

Note: Source Women Political Blogger (2006)

encouraged people to vote no on this bill including calling Harry Reid and Hillary Clinton. While this didn't work out, we will work to repeal the bill." Underlining the measures some bloggers take to engage in advocacy efforts, another woman retold of efforts to protest against human rights.

We had a blogathon to raise money for Amnesty International. My co-author and I posted every half hour for 24 hours. We pulled some strings and had some people do some guest posts, but it was still pretty taxing. I didn't write anything in advance.....Part of this was a bit of a protest because we saw the conservative punditry [calling for] relaxing human rights. This was the impetus for doing this.

Challenges, Exclusion and Discrimination in the Blogosphere

Women overwhelmingly expressed challenges to blogging. A majority of respondents asserted that sexism in the blogosphere was rampant. Interviewees repeatedly remarked that nasty comments were commonplace. Invectives such as "shut-up you dumb-bitch" and "cunt" were echoed throughout the interviews. In the comment section of one woman's blog the following appeared: "I bet you're a fat ugly dike." Women faced not only defamatory comments, but attempts at intimidation. Despite this however, one respondent remarked that she used attempts at intimidation to her advantage. She continued to respond to comments, contending that if she did not, the opponents would continue to bully her.

Other challenges also are evident. Many respondents reported not being linked to by other bloggers—largely white males—or having robust traffic levels. Others noted contending with trolls. Individuals frequenting blogs with which they disagree, who await opportunities to write disparaging comments, are commonly known as trolls. In conjunction with these challenges—arguably problems bloggers face regardless of gender—are other obstacles including one's political affiliation, education, familial obligations, and anonymity. As noted previously, a woman political blogger who identified with the conservative party, she said received flack from other women political bloggers because she did not ascribe to the politics of the left. Another challenge women in this sample mentioned, was balancing blogging and their familial obligations.

When asked whether or not they face exclusion in the blogosphere, a majority of women (13 out of 20) indicated that they did not. Underlining the welcoming nature of the blogosphere, one respondent said that the blogosphere had been a great community. Among those however, who reported feeling excluded, one respondent said, "I'm openly excluded by Daily Kos. This isn't only me, but my co-bloggers too. We criticized and nitpicked him. He pulled us off his blogroll after we negatively reviewed his book." Similarly another interviewee asserted that one's gender led to exclusion stating, "I think there is exclusion if you give signs of being a woman, but not by all men. You have to get past this and you have to give your credentials." Still others who mentioned feeling excluded, attributed this to not being linked to by other bloggers, or suggesting there were partisan differences that led bloggers from the left to exclude women blogging from a conservative perspective. Exclusion in these instances had little or nothing to do with gender. Illustrating this, a respondent suggested that exclusion was, "probably self–inflicted. I don't have a blogroll of other bloggers. The informal arrangement in the blogosphere is linking. I don't put people on a blogroll so I don't get linked as much."

Half of the women political bloggers in this study reported that they did not face discrimination. Many respondents qualified this statement, stating that they personally did not face discrimination, however they experienced discriminatory attitudes. One woman remarked that she did not recall feeling discrimination by other bloggers, but instead by readers. Keenly noted by one blogger, "I don't feel that I face gender discrimination in my day-to-day life. There is however systematic discrimination in the world. I think that the blogosphere is a bit more egalitarian—more so than other discourses, but it's not a perfect meritocracy." Discrimination for respondents was ideological or sexist in nature. Underlining the latter, an interviewee recounted,

People tend to think they can talk about my physical appearance. I've had people post my photo on conservative sites, and other sites in which people have indicated they would like to f- -k me and sexually assault me. Most of the women I know have faced this. I would say this hasn't happened to men. I think it happens to both old and young women. Women are mocked for their physical appearance or age....I think as women we are delegitimized. We can't really win no matter what we look like. The Ann Althouse controversy—a conservative law professor—posted a picture of Jessica Valenti with President Clinton and accused her of sticking her breasts out. It was completely ridiculous. This is an example of giving female bloggers a taller hurdle to get through.

DISCUSSION

Not unlike other political bloggers (McKenna & Pole, 2008), the women in this study are well-educated and concentrated in occupations that emphasize writing, namely academia and journalism. More than half of the respondents interviewed in this study earned a master's or higher degree, compared to less than half of average political bloggers (McKenna & Pole, 2008) who hold a master's degree or higher. Because this study relies on a snowball sampling, these data are not generalizable to all women political bloggers. However, it is not surprising that women political bloggers are well-educated given that blogging requires a facility for writing. Though the women I interviewed were, on average, slightly older than other political bloggers, this again might be attributed to the snowball sample. Studies detailing online activities across gender, suggest that women still lag behind men on two measures "having read a blog" and "creating a blog" (Fallows, 2005; Smith, 2008), which support the findings from this study.

Respondents estimate that three-quarters of the content on their blogs pertains to politics, public policy and current events. More specifically, the topics about which women blog are extensive. Because a majority of the women I interviewed reported that the issues they blogged about are not necessarily unique to women, it implies that the political discourse on these blogs is extensive, potentially appealing to a diverse audience. Blogging provides a space for discussing a myriad of topics from a woman's vantage point, free of time, editorial and monetary constraints which often inhibit participation. Yet, the failure of women to be linked to male bloggers still poses a formidable obstacle if the blogosphere is to be used a means of facilitating political discussions and an exchange of ideas. Despite this, blogs, according to the women I interviewed, are an excellent way for women to become involved in politics and to have a voice. Blogging is a flexible medium allowing women to read and discuss a wide range of issues or a subset of women's issues at their convenience. So too, it creates a sense of community with shared interests. Several women indicated that they were reassured by knowing that other women had similar experiences.

Women political bloggers use their blogs to inform their readers, check the mainstream media and engage in advocacy efforts in ways which were similar to average political bloggers (McKenna & Pole, 2008), suggesting that women political bloggers behave similar to other political bloggers, regardless of gender. Encouraging readers to make charitable contributions is undertaken more frequently by women political bloggers than other political bloggers. This is not unexpected given that the literature on women and philanthropy indicates that women tend to be more engaged in philanthropy than men.

These data reveal that women political bloggers do indeed encourage their readers to participate in politics with more than half of all bloggers reporting that they use their blogs to encourage readers to contact elected officials, to vote, and to sign petitions. Average rates of participation on these activities for average Americans (Verba et al., 1995) is not nearly as high as political bloggers, suggesting that bloggers themselves may be more civically engaged than other individuals. All but one interviewee indicated using her blog to mobilize readers. While previous studies of participation, mostly notably Putnam (2000), report low rates of civic engagement, these findings show that perhaps people are not as disaffected and unconnected as we think. While traditional methods of participating offline still occur, online forms of participating likely augment more traditional ways of participating in politics. Blogging serves both a direct and indirect conduit for participating. Directly, mobilization efforts have been mounted and political discourse occurs on a daily basis. Indirectly, bloggers ask their readers to undertake a variety of activities. Respondents frequently provided detailed accounts of instances in which they asked their readers to vote, contact elected officials, and to sign petitions. One blogger said, "I've promoted candidates and legislative reforms. I've encouraged people to become active in their communities. I've encouraged parties to be more open to certain things. I've advocated social and political positions...." This quotation illustrates the potential for using blogs creatively to participate in politics online.

The above findings support Wallsten's (2007) characterization of political blogs as "transmission belts, soapboxes and mobilizers." In addition to providing links to other sites, women political bloggers best illustrate the notion of soapboxes and mobilizers. Women write about a wide range of topics on their blogs that are not necessarily unique to women, and they receive comments, illustrating an exchange of ideas. Though the topics are not necessarily unique, many topics that traditionally receive less mainstream media coverage—and which are not necessarily of interest to the larger public—can be discussed at length. One woman asserted her ability to shape the policy agenda because of her blog, exemplifies blogs as soapboxes. The data also highlight respondents as mobilizers who use their blogs to promote candidates and organize grassroots efforts. Similarly, these findings illustrate how respondents use their blogs to mobilize their readers to vote and contact elected officials.

Obstacles to blogging occur regardless of one's gender, as do instance of exclusion. Though subtle, discrimination appears not infrequently. A majority of respondents reported that they personally did not face discrimination. Several respondents however, qualified their answers saying that there were discriminatory attitudes in the blogosphere or other bloggers faced discrimination. Gender discrimination in the blogosphere, while not rampant does exist, despite what theoretically should be an environment absent traditional constraints. When asked what challenges they faced, more than half of the bloggers indicated that they faced sexism or sexists remarks. Respondents also asserted that male bloggers tended not to link to them. The lack of linking might be intentional or unintentional. Even if men unintentionally exclude women this is arguably a form of discrimination. Discrimination and exclusion may be overt—in the form of invectives or intentionally refusing to link to a blogger. And, it may also be far more subtle, such as the failure to link to women political bloggers. Still the effect is arguable the same—a fragmented blogosphere, absent a cross-fertilization of ideas.

LIMITATIONS

This study contains several limitations. First, the sample size is limited to 20 respondents. As a result, the findings cannot be generalized to all women political bloggers. A larger study of political women should be conducted. Compiling a sample of political bloggers remains difficult due to the absence of a sampling frame. Still, this exploratory study provides a framework for future research. Second, the results from this study might be biased toward better-educated and more popular women political bloggers. It might best depict A-list and some B-list women political bloggers. As noted previously, Hindman et al. (2003) indicate that selection based on the most popular bloggers is likely to have a greater impact. This is certainly the case in politics, in which many of the most popular bloggers are followed not only by a devoted cadre of readers, but by the mainstream media. Third, while data collection occurred in October 2006, there are few, if any, studies detailing how women political bloggers in the US use their blogs for purposes related to politics, so this study fills a valuable gap. Of note, as of January 2010, 13 of the 20 interviewed were still blogging about politics. Lastly, this study focuses on women political blogs who write in English and reside in the US. Future research should examine other countries.

CONCLUSION

This study of women political bloggers mirrors the findings from McKenna and Pole's (2008) study of average political bloggers with regard to how bloggers use their blogs to inform readers, report errors in the media, engage in advocacy efforts and solicit charitable contributions. Further, women political bloggers asked their readers to engage in a variety of activities ranging from contacting elected officials to voting with similar frequencies.

Despite these similarities, these data offer two noteworthy departures from other studies of political blogging. First, a greater percentage of women political bloggers reported asking readers to make a charitable contribution than average political bloggers (McKenna & Pole, 2008), suggesting that women's blogs might a useful conduit for raising money. While beyond the purview of this study, using blogs to raise money for not only charitable activities, but for political activities merits future examination. Second, discrimination appears to be prevalent among women political bloggers, with half of respondents noting that they faced discrimination or witnessed it. This is cause for concern. Blogging about politics, public policy and current events might pose greater obstacles for women political bloggers than for others. Prescriptions for reducing and eliminating discrimination in the political blogosphere ought to be explored. Meanwhile, blogging offers a virtual space to discuss politics and an opportunity to encourage civic engagement, both online and off, despite incidents of discrimination.

The empirical implications of this work abound. Coverage of women's issues that might otherwise receive little to no media attention occurs not infrequently through this medium. Absent ongoing or even intermittent coverage of women's issues, blogs provide an opportunity to engage in ongoing virtual discussion. Blogs also provide a virtual forum for shaping public debate and the public agenda, which in turn can influence public policy. McEwan and Marcotte's participation in the Edward's campaign provides anecdotal evidence of the challenges to participating on and offline, which are further supported by these data. Blogging offers a flexible, inexpensive alternative to shaping political discourse, while encouraging reader participation. Yet, despite these benefits, it appears that women are underrepresented in the political blogosphere. Though definitive data detailing the composition of the political blogosphere are unavailable, underrepresentation may adversely affect political discourse, discussion and the exchange of ideas. Ultimately online and offline forms of participation might suffer as well if women feel intimidated, excluded or discriminated. Of course, future research should endeavor to address these questions more fully.

Going beyond feminist constructs, this work fills a gap in the literature on women who blog about politics. This piece highlights online participatory practices that often become translated into offline forms of political activities by cataloguing specific activities building on existing studies of blogging. This research contributes to our understanding of the specific activities undertaken by women political bloggers and the activities they encourage their readers to undertake.

Finally, future studies of political blogging should examine how women political bloggers fare compared to other political bloggers, such as men and other underrepresented groups. Research should expand upon the sample size, and it should seek to make connections between blogging and more traditional forms of political activity undertaken by political bloggers. Political blogging remains a rapidly maturing medium that continues to alter politics and political discourse in the United States.

ACKNOWLEDGMENT

I want to extend a special note of thanks to Matthew Moore for his help in suggesting the revised title.

REFERENCES

Broder, J. (2007, February 7). *Edwards bloggers cross the line.* Retrieved from http://www.nytimes.com/2007/02/07/us/politics/07edwards.html?scp=2&sq=Amanda+Marcotte&st=nyt

Cohen, D. (2007, August 30). *Building social capital through online communities: The strategy of Ned Lamont's 2006 Senate Campaign.* Paper presented at the Annual Meeting of the American Political Science Association, Chicago, IL.

Fallows, D. (2005). *How men and women use the internet.* Retrieved from http://www.pewinternet.org/Reports/2005/How-Women-and-Men-Use-the-Internet/06-Activities-and-Trends/02-Men-and-women-are-equally-likely-to-do-many-online-activities.aspx?r=1

Glaser, B. G., & Strauss, A. L. (1967). *The discovery of grounded theory.* New York, NY: Aldine Transaction.

Herring, S. C. (2003). Gender and power in online communication. In Holmes, J., & Meyerhoff, M. (Eds.), *The handbook of language and gender* (pp. 202–228). Oxford, UK: Blackwell. doi:10.1002/9780470756942.ch9

Herring, S. C., & Paolillo, J. C. (2006). Gender and genre variation in weblogs. *Journal of Sociolinguistics.*

Herring, S. C., Scheidt, L. A., Bonus, S., & Wright, E. (2004). Bridging the gap: A genre analysis of weblogs. In *Proceedings of the Thirty-Seventh Hawaii International Conference on System Sciences* (p. 11). Washington, DC: IEEE Computer Society.

Hindman, M., Tsioutsiouliklis, K., & Johnson, J. A. (2003, March 31). *"Googlearchy": How a few heavily-linked sites dominate politics on the web.* Paper presented at the Annual Meeting of the Midwest Political Science Association.

Kennedy, T. L. M., Robinson, J. S., & Trammell, K. (2005, October 5-9). *Does gender matter? Examining conversations in the blogosphere.* Paper presented at Internet Research 6.0: Internet Generations, Chicago, IL.

Marcotte, A. (2007). *Why I had to quit the John Edwards campaign.* Retrieved from http://www.salon.com/news/feature/2007/02/16/marcotte/index.html

McKenna, L., & Pole, A. (2004). *Do blogs matter? Weblogs in American politics.* Paper presented at the Annual Meeting of the American Political Science Association, Chicago, IL.

McKenna, L., & Pole, A. (2008). What do bloggers do: An average day on an average political blog. *Public Choice, 134*(1), 97–108. doi:10.1007/s11127-007-9203-8

Mitchell, S. (2007). *Access to technology: Race, gender, class bias.* Retrieved from http://bloggingfeminism.blogspot.com/

Nolan, C. (2007). *Women's voices are louder online.* Retrieved from http://bloggingfeminism.blogspot.com/

Olsen, D., Berlin, E., Olsen, E., McLean, J., & Sussman, M. (2009). *State of the blogosphere.* Retrieved from http://technorati.com/blogging/feature/state-of-the-blogosphere-2009/

Osell, T. (2007). *Where are the women? Pseudonymity and the public sphere, then and now.* Retrieved from http://www.barnard.edu/sfonline/blogs/osell_01.htm

Patton, M. Q. (2002). *Qualitative research & evaluation methods.* Thousand Oaks, CA: Sage.

Pedersen, S., & Macafee, C. (2007). Gender differences in British blogging. *Journal of Computer-Mediated Communication, 12*(4). doi:10.1111/j.1083-6101.2007.00382.x

Perlmutter, D. (2008). *Blog wars*. New York, NY: Oxford University Press.

Pole, A. (2009, April). *Clinton and Obama: Blogging the 2008 presidential primaries*. Paper presented at the Midwestern Political Science Association. Chicago, IL.

Pole, A. (2010). *Blogging the political: Politics and participation in a networked society*. New York, NY: Routledge.

Putnam, R. (2000). *Bowling alone: The collapse and revival of American community*. New York, NY: Simon and Schuster.

Ratliff, C. (2007). *Attracting readers: Sex and audience in the blogosphere*. Retrieved from http://www.barnard.edu/sfonline/blogs/ratliff_01.htm

Shactman, N. (2002). *Blogs make the headlines*. Retrieved from http://www.wired.com/culture/lifestyle/news/2002/12/56978

Smith, A. (2008). *New numbers for blogging and blog readership*. Retrieved from http://www.pewinternet.org/Commentary/2008/July/New-numbers-for-blogging-and-blog-readership.aspx

Solis, B. (2009). *Are blogs losing their authority to the statusphere*. Retrieved from http://www.techcrunch.com/2009/03/10/are-blogs-losing-their-authority-to-the-statusphere/

Verba, S., Lehman Schlozman, K., & Brady, H. E. (1995). *Voice and equality: Civic voluntarism in American politics*. Boston, MA: Harvard University Press.

Wallsten, K. (2007). Political blogs: Transmission belts, soapboxes, mobilizers, or conversation starters? *Journal of Information Technology & Politics*, 4(3), 19–40. doi:10.1080/19331680801915033

White, D., & Winn, P. (2008). *State of the blogosphere*. Retrieved from http://technorati.com/blogging/feature/state-of-the-blogosphere-2008/

Yin, R. K. (2008). *Case study research: Design and methods*. Thousand Oaks, CA: Sage.

ENDNOTES

[1] Bloggers were composed of 27 men and 114 women.

[2] While the timeframe for this study was selected randomly, it precedes the 2006 mid-term elections. Because of the timing, this may have yielded a greater proportion of posts on issues salient to election cycle.

[3] A-list bloggers receive thousands of unique visits a day, have high visibility and are classified as the top 100 bloggers based on traffic and links to their blog, whereas B-list bloggers receive less traffic and fewer links than A-list bloggers (Pole, 2010).

APPENDIX: INTERVIEW INSTRUMENT

Women Political Bloggers

1. Do you discuss politics, public policy or current events on your blog, and if so, what percentage of your blog is devoted to politics, public policy or current events?
2. What issues do you blog about? [Prompt: race/ethnicity, gender, economics, local politics, international politics?]
3. Are the issues you blog about unique to women?
4. Do have a target audience that your blogging to? Do you know who your readers are? [Prompt: Are your readers other women?]
5. How much influence or agency do you feel you have because of your blog? On a scale from 1 to 10, 1 being little to no influence and 10 being a lot of influence, please tell me how much influence you feel that you have as result of writing a blog?
6. Do you feel you face any challenges as a women blogger?
7. When did you start your blog, and why did you start a blog?
8. What do you hope to gain from blogging?
9. Have you asked your readers to engage in the following activities:
   ___ To vote
   ___ To register to vote
   ___ To contact an elected official
   ___ To sign a petition
   ___ To attend a rally, protest or march
   ___ To attend a political fundraiser
   ___ To attend a political party or local community meeting
10. Do you feel that writing a blog is form of participating in politics? Why?
11. Have you used your blog to engage in any of the following activities:
    ___ To inform readers
    ___ To point out errors or omissions in the media
    ___ To engage in political advocacy
    ___ To raise money charitable cause
12. If you have engaged in political advocacy, can you give me a few examples of this? How did you do this? What was the cause/reason? What were the results?
13. How can blogs best be used for women? [Can blogs be used to raise awareness in communities for women's issues?]
14. Do you engage in community organizing or any other types of organizing on your blog? If yes, how so?
15. Are you in contact with other bloggers? If so, who are you in contact with? Are these bloggers women bloggers?
16. What do see as the benefits of blogging?
17. What impact will blogging have on your readers?
18. What are your traffic levels?
19. What is your gender?
20. What is your age?

21. What is your occupation?
22. How would you describe your political orientation?
23. What is the highest degree you have completed?
24. Why do you think there aren't more women bloggers? [Is it because of the technology? Is difficult for women/minorities to get into these networks?] Why is there this big boys club?
25. Do you feel excluded by other bloggers? Do you feel that you have faced discrimination in the blogosphere?
26. Is there anything else that you would like to add that I haven't asked you?

This work was previously published in International Journal of E-Politics (IJEP), Volume 2, Issue 2, edited by Celia Romm Livermore, pp. 37-54, copyright 2011 by IGI Publishing (an imprint of IGI Global).

Chapter 12
Gender Differences in Social Networking Presence Effects on Web-Based Impression Formation

Leslie Jordan Albert
San Jose State University, USA

Timothy R. Hill
San Jose State University, USA

Shailaja Venkatsubramanyan
San Jose State University, USA

INTRODUCTION

With growth in the use of computers and the Internet, we now live in a world where there are two spheres of existence–a physical sphere and a digital sphere. Many now rely on the Web as a reflection of reality for finding facts. For example, many turn to the Web to get the address of a store rather than consult traditional yellow pages in book form. This existential dichotomy between our physical and digital spheres of existence gives rise to a number of issues. One interesting issue is that people are able to perceive others and form opinions about them based solely on the information available about those people on the Web.

While it may be tempting to characterize the Web as essentially equivalent to traditional print media as a basis for forming impressions, a critical look reveals fundamental differences. For example, relative to traditional media, Web-based sources are easier to access, far more searchable, much more amenable to aggregation, and have a longer lifespan because of the refresh-ability afforded by their digital format. Further, these sources are growing exponentially with Web 2.0 technologies that allow anyone to create and post

DOI: 10.4018/978-1-60960-759-3.ch012

their own content. Kalyanaraman and Sundar (2008) suggest that today's Web-based sources and tools provide options for self-expression and self-presentation that are unprecedented. As the information paradigm changes, it follows that the processes, and thus outcomes, are being affected and it thus behooves us to explore them toward a better understanding and use.

One important aspect of the paradigm shift is that, ironically, though the self-publishing phenomenon enabled by the Web lacks the well-established integrity controls of traditional print media, it has become an important source of information for decision making in both personal and professional contexts. For example, a recent study sponsored by Microsoft surveyed 275 US hiring managers and human resource professionals about their hiring practices and found that 85% of these recruiters work for firms that have hiring policies requiring them to investigate potential candidates online and 70% admitted to turning down potential employees based on the information they found on the Web (Cross-tab, 2010). Given the importance of decisions being made based on impressions created solely from Web searches, it is imperative that we understand the characteristics of both the searcher and the searched that may impact formation of these impressions.

We examine this issue in greater detail using the concepts of ePersona and ePerception (Venkatsubramanyan & Hill, 2007). ePersona refers to searchable digital information about a particular person from a variety of sources – personal home pages, social and professional networking sites, organizational Web pages, news articles, blogs and others. ePerception is a term coined by Vazire and Gosling (2004) but was defined and further developed to refer to the perception formed by people about others depending primarily, if not exclusively, on the ePersona (Venkatsubramanyan and Hill, 2007; 2009a, 2009b). Building upon their model of impression formation, we explore how the gender of the perceiver, and the presence or absence of social networking activity for an ePersona, affect the formation of an ePerception. Additionally we examine how these effects differ by the gender of the searched individual, the "target", as compared with that of the perceiver. The results are both interesting and valuable, particularly given the integral role gender issues play in the social power and political dynamic of the workplace as well as broader social contexts, because better understanding the inherent biases enables awareness that can empower the affected individuals.

The chapter is organized as follows. First we describe the theoretical background, present Venkatsubramanyan and Hill's (2009b) model of Web-based impression formation, develop our research question and put forward the propositions used to guide our research. Next we describe the experimental methodology for an empirical study designed to address the research question, followed by a discussion of results and their interpretation. We conclude with a summary and a discussion of the study's limitations, implications and directions for future research.

THEORETICAL DEVELOPMENT

In existing models of impression formation, the perceiver is the person forming an impression, while the target is the person about whom the impression is formed. People form impressions about others based on primary (or direct) and secondary (or indirect) sources of information. Primary sources of information include personal interactions (face-to-face or otherwise) including both verbal and behavioral cues. Secondary sources of information include sources such as hearsay (opinions expressed by others), photographs, voice recordings, official records, news articles, biographies, and others, now including Web-based information.

Traditional models are grouped into two main categories: trait-based and stereotype-based. In the trait-based models, such as Asch's Configural

Figure 1. Web-based perception model: ePersona and ePerception

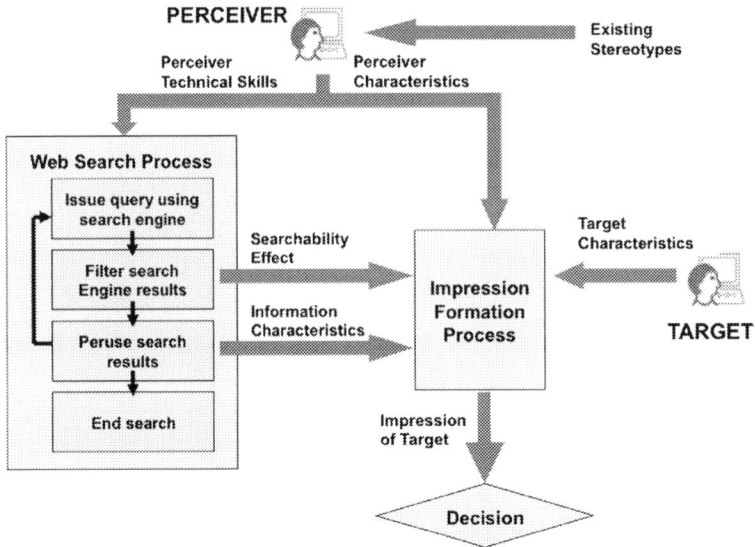

Model and Anderson's weighted-average model, various traits of the target come together in the perceiver's mind to form a unified impression (Brewer, 1988). According to Asch, there are two types of traits–central traits (traits that have a strong effect on interpretation of other traits) and peripheral traits (traits that do not significantly affect subjects' impressions of the perceived personality). Other researchers have found a primacy effect where traits that appear first have more impact in final impression (Widmeyer & Loy, 1988). On the other hand, stereotype-based models theorize that people rely on social categories, or stereotypes, to form impressions since stereotyping reduces the amount of information to which perceivers must attend. Using a stereotype, a perceiver may infer the person's personality attributes without having to attend carefully to that person's behavior (Sherman, Lee, Bessenoff & Frost, 1998).

These existing perception models provide the foundation for a model of Web-based perception that recognizes the Web as a new medium that "changes the game." The uniqueness of the Web as a source of information and medium of communication can be seen in emerging studies of impression formation in the digital age (Hancock, 2001; Jacobson, 1999; Markey & Wells, 2002; Walther, 1997; McKenna & Bargh, 2000). Venkatsubramanyan and Hill (2007) describes this model of ePerception (Figure 1) that extends traditional models of perception to account for differences effectuated by the digital information domain, specifically, by the way Web-based search impacts our perceptions of others.

As shown in the figure, there is a perceiver and a target as in traditional perception models (Brewer, 1988). The characteristics of both the perceiver and the target feed into a traditional impression formation process. In the digital domain, perceiver characteristics go beyond traditional notions of personality, emotional state, and social characteristics to include factors such as level of information literacy, online experience and comfort level, and search skill expertise. For instance, a study conducted by Ford, Miller and Moss (2005) concluded that cognitive styles, levels of prior Internet experience and perceptions, study approaches, age and gender affect retrieval effectiveness. As in face-to-face contexts, existing stereotypes may also influence perceivers though some research suggests that this influence differs in the digital sphere (Lee, 2004).

The target and the ePersona are shown as separate entities since the ePersona may also be affected by factors beyond the characteristics, behavior and control of the target such as the perceived currency and aesthetic of information sources, links between pages, and production quality of the information (e.g. picture clarity of images or videos).

The searchability effect refers to the impact of the search process itself (apart from the results) on the impression being formed (and subsequently the decision being made). Filtering search results then plays into searchability as cognitive effort is theorized to impact the impression formation process and outcome. Perusing the search results themselves then impacts the process through the target information, influenced by the characteristics of the information sources themselves. The perceiver considers all these factors to finally form an impression of the target, which then flows into the decision making process (such as hiring the target for a prospective employment position).

The theoretical model (Figure 1) provides a framework for empirical investigation of the impact of the Web on impression formation. The central research questions raised by this model are – (a) How do characteristics of Web-based search results impact impression formation? (b) How does the search process, by itself, affect impression formation? (c) How do the characteristics and skills of perceivers influence their perceptions of ePersonas? and (d) What characteristics of targets are most influential in the creation of an ePersona and perceivers' impressions of that ePersona? Our study of the literature revealed that the first question is currently being studied in some quarters (e.g. Vazire & Gosling, 2004). Studies related to the remaining questions, however, are lacking in current literature. In this study, we investigate the roles of perceiver and target characteristics, specifically gender and social networking presence, in the formation of ePerceptions.

Our research question is thus: How do differences in perceiver gender and target gender affect the impressions formed based upon an ePersona's social networking activity?

Perceiver and Target Characteristics: Gender

As seen in Figure 1, the proposed model of Web-based perception suggests that the individual characteristics of both the perceiver and the target play an important role in the impression formation process. Although there are many characteristics of interest, we chose to focus this study on the effects of perceiver and target gender for two main reasons. First, studies such as Gefen and Straub (1997) suggest that men and women may perceive and use communication technologies, such as email, differently. One explanation of these differences is that women may perceive a higher degree of social presence in online contexts, as Richardson and Swan (2003) found in their study of online courses. Females may also perceive online information sources differently than males. For example, Huffaker and Calvert (2005) analyzed gender identity and language issues based on entries in the Web 2.0 online journal phenomenon "blogs" (Web logs) and found this new media format surfaced gender-based differences that broke from stereotypical expectations. Venkatsubramanyan and Hill (2009b) found that women's decision making processes may be more influenced by a potential teammate's social networking activity than are men's suggesting that perceiver gender may contribute to differences in ePerceptions.

Second, gender stereotypes may also influence impression formation. Computer mediated communication (CMC) has been shown to reduce the available amount of "individuating information," cues that allows us to differentiate between group members (Lee, 2004). Although one might expect gender to play less of a role when social cues are limited, research suggests differently (Lea & Spears, 1991). The Social Identity model of Deindividuation Effects (SIDE) suggests that group members with insufficient individuating information for forming perceptions will fill the void by assigning stereotypical traits to other members of their group. SIDE also suggests that a lack of individuating information about others leads to a greater affinity with one's own group (i.e. those of the same gender, race, nationality, etc.) and a greater likelihood that one will exhibit the stereotypical behaviors attributed to that group (Postmes & Spears, 2002). This is particularly relevant to the formation of ePerceptions as Web-based searches do not allow targets to provide supplemental individuating information and thus gender stereotypes and gender affinity may play a greater role in the formation of ePerceptions than they would in impressions formed face-to-face. Further, research also suggests that gender stereotype effects are so powerful that even minimal gender cues can encourage the assignment of gender stereotypes to others. Lee (2004) found that gendered cartoon avatars alone were sufficient for individuals to assign stereotypical behaviors and attributes to group mates in CMC contexts even when those individuals were told that the gender of the avatar may or may not match the actual gender of the group mate. Similarly, Nass, Moon and Green (1997) found that subjects assigned gender stereotypes to computers that exhibited male and female voices and responded accordingly. These studies all suggest the importance of both perceiver and target gender in the formation of ePerceptions.

ePersona Characteristics: Social Networking Activity

As with individual characteristics of the perceiver and target, we suggest that characteristics of a target's ePersona also impact perceivers' impressions of the target. The ePersona includes all digital information that may be gathered about an individual via a search engine. This information may be generated by the target, such as personal Web pages and blog entries, or by third parties, as in the case of news articles, public records and others' Web posting about the target. In this study, the ePersona aspect on which we focus is social networking, due to its popularity among participants (Facebook alone claims to have over 200 million users as of 2009) and among organizations looking to vet potential employees (CareerBuilder, 2008) as well as literature, cited above, indicating that gender and social networking effects may be interrelated.

Beyond their wide acceptance, social networking sites also offer unique information about individual targets. Unlike personal Web pages, which consist primarily of target-controlled information, social networking sites combine target and third party-generated information about target individuals. According to Warranting Theory (Walther & Parks, 2002), perceivers value others' opinions and views of an individual over that individual's self-evaluation. Applying Warranting Theory to ePerception suggests that the key role that third party-generated content plays in social networking, in the form of friends' comments, for example, would render those sites more salient to the impression formation process than target-controlled sites and indeed Walther, Van Der Heide, Hamel, and Shulman (2009) found this to be true. Futhermore, Back, Stopfer, Vazire, Gaddis, Schmukle, Egloff, and Gosling (2010) suggests that it is difficult for individuals to misrepresent themselves on social networking sites as any attempts to do so may be viewed and corrected by others and thus social networking sites are perceived as more accurately representing target individuals, a fact that may further increase the impact of social networking activity on Web-based impression formation. Lastly, their own gender may be expected to affect the way perceivers view social networking participation as suggested by Walther, Van Der Heide, Kim, Westerman, and Tong (2008) in a study that analyzed reactions to both the comments and attractiveness of friends in Facebook and found males and females differently impressed.

Combining these prior studies with the theoretical model presented in Figure 1, we begin our investigation of the roles of gender and social networking activity in ePerception with a set of propositions to be assessed experimentally. Propositions, rather than hypotheses, are presented here as overarching conceptual statements, each representing a series of testable statements too numerous to list individually.

First we draw from Warranting Theory to propose that social networking engagement, in and of itself, implies a willingness to be perceived through more credible third-party sources of information and this extends naturally to a range of positive associations, so perceivers of both genders will tend to view ePersonas with social networking more favorably than those without:

P1a: *Males will perceive a target ePersona with social networking presence more positively than a target ePersona without social networking presence.*

P1b: *Females will perceive a target ePersona with social networking presence more positively than a target ePersona without social networking presence.*

We then build upon the literature on perception differences between genders to propose that both males and females will assign gender stereotypes to gendered ePersonas and express greater affinity with their own gender group in low-cueing CMC contexts such as search engine results.

P2a: *Females will perceive a female target ePersona without social networking presence more positively than a male target ePersona without social networking presence.*

P2b: *Females will perceive a female target ePersona with social networking presence more positively than a male target ePersona with social networking presence.*

P2c: *Males will perceive a male target ePersona without social networking presence more positively than a female target ePersona without social networking presence.*

P2d: *Males will perceive a male target ePersona with social networking presence more positively than a female target ePersona with social networking presence.*

Based upon prior studies that suggest gender differences in interactions with, and perceptions of, CMC and upon recognized gender-stereotypical responses, we further propose an interaction effect between social networking activity and target gender that will differ for female and male perceivers.

P3a: *Females will perceive the value of social networking differently for male targets than for female targets.*

P3b: *Males will perceive the value of social networking differently for female targets than for male targets.*

EXPERIMENTAL METHODOLOGY

In this study we use personality dimensions as developed in Venkatsubramanyan and Hill (2007) to assess the perceiver's impression of a target ePersona as a potential project teammate. Venkatsubramanyan and Hill (2007) began with the traditional five-factor model of personality traits (Watson, 1989). The five-factor model comprises a hierarchical organization of five basic personality dimensions: Extraversion, Agreeableness, Conscientiousness, Neuroticism, and Openness to Experience. These five basic factors break down further into 106 personality dimensions. Through pilot testing the authors reduced the original personality dimension list down to a set of nine dimensions deemed by subjects as the most applicable for assessing potential teammates based solely on Web-based perceptions. These include (a) Commitment to Excellence, (b) ability to work as an Effective Team Member, (c) ability to Manage Multiple Tasks, (d) ability to Handle Conflict, (e) having a strong interest in Working with People, (f) Managing Anger, (g) ability to Take Direction, (h) Curiosity, and (i) ability to Adapt to New Situations. The current study employs these same personality dimensions to assess perceivers' impressions of ePersonas.

Our experimental design used university students to evaluate ePersonas for the purpose of selecting potential team members for a class project. Subjects were recruited from several upper division undergraduate business courses and offered course credit for their participation. 210 participated in the study including 104 males and 106 females and a wide range of business specialties (Finance, Management, etc.). The task of selecting and evaluating a potential teammate was chosen for two main reasons. First, most college students have experience working in teams on course projects, and in many cases had to select those teammates themselves, thus the task was an appropriate one for our subjects. Second, we suggest that the process of vetting potential teammates parallels the decision making processes required to assess and select one or more individuals from a pool of candidates and that these processes are largely independent of context. Examples of such decisions include selecting members for work-based teams, creating a short list of candidates for job interviews and choosing among professional service providers. The specific task of selecting a potential teammate is merely the scenario we used to develop a better understanding of how we form impressions of others based upon Web searches and we suggest that this scenario does not greatly limit the generalizability of our findings.

The experiment was conducted online with each subject randomly assigned to one of four manipulations representing the different potential teammates: John Doe 1, John Doe 2, Jane Doe 1 and Jane Doe 2. Each subject was provided a list of Google-style search results pertaining to his or her assigned potential teammate ePersona and was informed that the individual's name had been changed to preserve anonymity. The search result links were disabled so that subjects' information regarding the potential teammate was limited to the search results lists alone. If any link was clicked upon by the subject, a pop up message saying "DNS server is down. Please try again later." was generated. Search results for John Doe 1 and Jane Doe 1 each contained 10 results, 8 of which were links to social networking sites. The results lists for John Doe 2 and Jane Doe 2 also contained 10 links though none of the results for these two ePersonas were to social networking sites. Please see Appendices A and B for screen captures of the search results for Jane Doe 1 and Jane Doe 2.

After looking at the search results page, subjects were asked to report the number of minutes spent reviewing the results to get an estimate of the amount of cognitive effort exerted in assessing the ePersona. Subjects were then asked to score the target as a potential team member on the nine factors described above. A five point Likert (Likert, 1932) scale ranging from "-2" (very unfavorable impression) to "+2" (very favorable impression) was used for each of these factors. Additionally, subjects were asked to rate the desirability of the target as a potential team member and their confidence in these desirability ratings. Also included in the rating was the question:

"If you are forced to make a decision at this point with no further information, would you select this person to be on your team – yes or no?"

Answers to this yes or no question were coded in the form of 1 or 0 for analysis purposes. On a scale of 1 to 5, subjects were asked to indicate how confident they felt about their decision. Subjects were asked to provide qualitative feedback as answers to these two questions:

1. What kind of teammate would the student be?
2. What do you think influenced your decision?

Demographic information was also collected about the subjects including factors such as age, gender, major, number of team projects performed in the past, number of computer related courses, number of years of computer experience, number of years of experience searching the Internet, frequency of Web search, and a self-rating of his or her own level of Web search skills.

ANALYSIS AND DISCUSSION

To begin the data analysis, the qualitative, open-ended responses were reviewed to confirm that the subjects were sufficiently engaged in the scenario and to identify any notable trends. Several observations surfaced:

1. Nearly all subjects responded with written feedback and in every case the message was sensible and earnest, showing an investment in the scenario that lends veracity and credibility to the results, e.g.

"Due to John Doe's Web presence, I feel that he would be a good team member since he seems to have the desire to be connected with friends and associates. Having a LinkedIn, account shows that he has contacts that have worked or are working in business environment. Plus, it seems that he has also created a Website for himself, which shows that he is technically savvy."

2. A small but noticeable minority of respondents expressed frustration that there was not enough information for them to feel comfortable with the decision (mostly for the ePersonas with no social networking presence) but this reinforces that they were invested in the scenario and reminds us that we are studying first impressions that don't always lead to outcomes in and of themselves, e.g.

"There is nothing to assure that she is a good candidate for the team; therefore it would not be wise to base the decision on the information shown in the Google search results."

3. Overwhelmingly, respondents did attend to social networking presence strongly and extrapolated that to various trait implications, including those associated with the teammate decision, e.g.

"The search results show that he has strong social network connections. Therefore, he should be an active, sociable and energetic person. He seems to be teammate-material."

4. A number of respondents derived surprisingly broad and strong negative character traits from the ePersonas that were essentially neutral, merely lacking social networking presence, e.g.

"Not outgoing, unsocial, shy, quiet, and keeps to himself."

"The fact that there's almost nothing out there about her... She seems like she probably doesn't have much ambition or desire to do anything or accomplish anything."

"Not social from the fact he doesn't use any SNS which is very strange these days."

This suggests that at least some social networking presence may now be considered the norm with deviations leading to negative associations.

5. It was clear that in addition to the negative associations of too little social networking presence, there is a recognition that too much

may be bad, at least in the project/teammate scenario, e.g.

"The fact that she is on most of the social networks. I would want her on the team because of her people skills that she could possibly have; however, it makes me worry that she would spend too much time online rather than working on the project."

"Jane doe would be a slacker. All of the pages described for this potential team mate were social sites, she would probably get distracted easily and not be focused on getting work done."

Based on the qualitative analysis then, it is clear that the data is reliable and reflects the intended manipulations. Further, the apparent strength of impact underscores the value of the research. In addition, the expectation of social networking presence and the sensitivity to the level or degree to which it appears in an ePersona raise unforeseen issues informed by this study but needing further research.

The quantitative data were analyzed with respect to the research model and propositions, yielding additional insights from both expected and unexpected results. Standard statistical significance tests were applied for comparison of group mean pairs corresponding to the propositional statements.

The results are summarized in tables that list the experimental measures as rows and the various treatment group comparisons in columns. The equal sign ("=") appears in cells where the means of the corresponding treatment groups (column) showed no significant difference (at the 0.05 confidence level) for that measure (row). For cells where significant differences were found, the group with the higher (always "better") mean is indicated with an abbreviation followed by the "^" symbol (e.g. "M^" indicates a higher/better mean for males.) Some cells are shaded to highlight matches across columns for a given measure. The experimental measures (rows) include teammate desirability, the nine personality dimensions relevant to the task scenario as previously identified by Venkatsubramanyan and Hill (2007) and two additional measures: 1) Rating Confidence which captures the degree of confidence subjects felt in their decision and 2) the binary Yes/No decision to include the potential teammate represented by the ePersona, in the fictitious project team.

Table 1 relates to Proposition 1 which posited that both males (1a) and females (1b) would perceive target ePersonas including social networking presence more positively than those without. The proposition was based on relevant literature and the research model and indeed, the results support this expectation. Both female and male perceivers showed a preference for ePersonas with social networking presence across several dimensions.

Though there was some overlap between females and males in the specific dimensions affected, there were even more dimensions where the two genders differed. Both male and female perceivers gave advantage to social networking personas in judging the Work with People and Curious dimensions. But females treated ePersonas similarly across four (4) additional dimensions (Desirability, Manage Multiple Tasks, Handle Conflict and Adapt to New Situations) while males did so for only one (1), Manage Anger, which was apparently not salient for females. And so males registered significant advantage for social networkers across only three (3) dimensions, as opposed to six (6) for females, and yet, only the males showed significance in the Yes/No decision outcome measure, one that ostensibly reflects a comprehensive summation of desirability and the nine dimensions. This might be interpreted plausibly as males feeling more freedom to be decisive, even without delving as deeply and finding as much convincing evidence as the females, due to males' confidence that they will have the power to control the outcome and diffuse any consequences that would possibly derive.

In any case, the overarching implication of Table 1 is that social networking presence does

Table 1. Comparison of social networking influence for female vs. male perceivers

Impression Dimensions	Female Perceivers (results group both F & M targets) SN vs. no SN	Male Perceivers (results group both F & M targets) SN vs. no SN
Desirability	SN^	=
Effective Team Member	=	=
Excellence	=	=
Manage Multiple Tasks	SN^	=
Work with People	SN^	SN^
Handle Conflict	SN^	=
Manage Anger	=	SN^
Curious	SN^	SN^
Adapt to New Situations	SN^	=
Ability to Take Direction	=	=
Additional Measures		
Rating Confidence	=	=
Yes/No Decision	=	SN^

(where "=" indicates no significant difference at .05 level; SN^ = Social Networking higher at .05)
(Note: n = 104 Males + 106 Females = 210 perceivers total; F & M = Female & Male targets)

make a difference, for perceivers of both genders, but in subtly different ways. Looking deeper, the next set of propositions explores how the target gender plays a role within the context of social networking presence or lack thereof for female perceivers (P2a/b).

Table 2, providing significance findings for *female perceivers only*, suggests a number of interesting observations with respect to Propositions 2a and 2b, that females will perceive females more positively than males for targets without or with social networking presence. The leftmost two columns of results correspond to P2a and b, respectively, and show no support for either. No significant differences were found in comparing female to male targets without social networking presence (P2a). For ePersonas with social networking presence (P2b), significance was found only for a single dimension, Effective Team Member, and male targets were rated higher, not lower, as predicted in the proposition. This may owe to the overriding influence of established stereotypes when other cueing is indiscriminate as it was in this treatment with all targets showing social networking presence equally (target gender was the only variable).

But the result is intriguing in that it surfaced only in the social networking presence treatment (P2b) and not in the treatment that was identical except for the absence of social networking presence (P2a). The logical implication is an interaction effect–these female perceivers viewed male targets differently from female targets in the context of social networking presence only, at least on this single dimension–target gender and social networking presence effect differently when both are present than in the case of either alone. This is consistent with the observation from Table 1 that male and female perceivers view the value of social networking presence differently, but it adds a nuance–that those differences may further vary, depending on target gender, as suggested by Propositions P3a and P3b. To delve further into the interaction implications, the analysis explored how females viewed the added value of social networking for female as opposed to male targets (P3a), as seen in the rightmost two columns above.

Table 2. Significance for female perceivers

	Target Characteristics			
	Gender advantage effect within No SN and SN target groups		SN advantage effect within F and M target groups	
Impression Dimensions	No SN: F vs. M	SN: F vs. M	F: SN vs. no SN	M: SN vs. no SN
Desirability	=	=	=	SN^
Effective Team Member	=	M^	=	SN^
Excellence	=	=	=	SN^
Manage Multiple Tasks	=	=	=	SN^
Work with People	=	=	SN^	SN^
Handle Conflict	=	=	=	=
Manage Anger	=	=	=	=
Curious	=	=	SN^	SN^
Adapt to New Situations	=	=	SN^	SN^
Ability to Take Direction	=	=	=	=
Additional Measures				
Rating Confidence	=	=	=	=
Yes/No Decision	=	=	=	=

(where "=" indicates no significant difference at .05 level; SN^ = Social Networking higher; M^ = Males higher)
(Note: n = 106 Females; SN = Social Networking; F = Female; M = Male)

The most notable point that surfaces is that female perceivers' impressions were affected by social networking across more dimensions for male targets than for female targets. Significant effects from social networking were found only for three (3) of the ten dimensions for female targets (Work with People, Curious, and Adapt to New Situations). And the same three (3) dimensions showed significant effects for male targets but four (4) additional dimensions also show significant impact from social networking for the males, for a total of seven (7) affected among the ten (10) measured.

For male targets, these female perceivers' impressions of the Desirability, Effective Team Member, Excellence and Manage Multiple Task dimensions showed significant positive effect from social networking that was not found in their perceptions of female targets *where the sole difference was gender*. This implies an interaction effect–the presence of social networking affects females' perception differently, and more comprehensively for the males they are evaluating than the females. In other words, female perceivers give a greater social networking premium to males than females and/or penalize males more than females for the lack of social networking presence. Indeed closer inspection of the data suggests that it is a combination of both as the specific means for at least five (5) of the dimensions show a pattern of wider spread for the male targets–those with social networking outscore the social networking females while those without score lower than their female counterparts.

Thus, Table 2 provides convincing support for P3a and the existence of an interaction effect, for female perceivers, between target gender and social networking presence. In this case, the female perceivers are clearly more affected by social networking presence in male ePersonas than in female ePersonas. Specifically the findings suggest that, for female perceivers, the stereotypical

Table 3. Significance for male perceivers

	Target Characteristics			
	Gender advantage effect within No SN and SN target groups		SN advantage effect within F and M target groups	
Impression Dimensions	No SN: F vs. M	SN: F vs. M	F: SN vs. no SN	M: SN vs. no SN
Desirability	=	=	=	=
Effective Team Member	=	=	=	=
Excellence	=	=	=	=
Manage Multiple Tasks	=	=	SN^	=
Work with People	=	=	SN^	SN^
Handle Conflict	=	=	=	=
Manage Anger	=	=	=	SN^
Curious	=	=	SN^	=
Adapt to New Situations	=	=	=	=
Ability to Take Direction	=	=	=	=
Additional Measures				
Rating Confidence	=	=	=	=
Yes/No Decision	=	=	=	SN^

(where "=" indicates no significant difference at .05 level; SN^ = Social Networking higher)
(Note: n = 104 Males; SN = Social Networking; F = Female; M = Male)

view of males as less socially adroit than females is strongly reinforced when male targets lack social networking presence and this effect is excessively negative with respect to the non-social networking female targets. Interestingly, the positive effect of social networking presence on female perceivers' impressions of male targets is so exaggerated that this same stereotype is strongly negated for male targets engaged in social networking, perhaps because the elevated social investment is unexpected for males and defies the stereotypical view held by the females, even when compared to social networking females.

The obvious next question is whether the same phenomenon holds for male perceivers as examined in Table 3.

Table 3 replicates Table 2 but for male perceivers and is interesting for similar reasons and by contrast with Table 2's results for female perceivers.

First, as with female perceivers, the leftmost two columns of Table 3 show that there is no support for the propositions that males would show a target gender preference when ePersonas lack or include social networking presence (P2c and P2d, respectively). In contrast to the results for female perceivers however, there is no stereotypical effect that gives the contrasting gender, female, an advantage in any dimension whether social networking presence is apparent or not.

In further similarity to female perceivers, male perceivers show some interaction effect between target gender and social networking presence so P3b is somewhat supported but this is far less pronounced than in the case of female perceivers. For most of the ten (10) dimensions, there is no significant difference in their perceptions of either males or females, based on social networking presence. For one (1) of the ten dimensions, Work with People, significance is found both for male *and* female targets, so male perceivers are

Table 4. Significance comparison between male & female perceivers

	Target Characteristics			
	Gender advantage effect within No SN and SN target groups		SN advantage effect within F and M target groups	
Impression Dimensions	No SN: F vs. M	SN: F vs. M	F: SN vs. no SN	M: SN vs. no SN
Desirability	=	=	=	SN^ by F only
Effective Team Member	=	M^ by F only	=	SN^ by F only
Excellence	=	=	=	SN^ by F only
Manage Multiple Tasks	=	=	SN^ for F by M (but not by F) and SN^ for M by F (but not by M)	
Work with People	=	=	SN^ by both M&F	SN^ by both M&F
Handle Conflict	=	=	=	=
Manage Anger	=	=	=	SN^ by M only
Curious	=	=	SN^ by both M&F	SN^ by F only
Adapt to New Situations	=	=	=	=
Ability to Take Direction	=	=	=	=
Additional Measures				
Rating Confidence	=	=	=	=
Yes/No Decision	=	=	=	SN^ by M only

(where "=" indicates no significant difference at .05 level; SN^ = Social Networking higher; M^ = Males higher)
(Note: n = 104 Males/106 females; SN = Social Networking; F = Female; M = Male)

consistent across target gender on this dimension (as are females according to Table 2).

For these male perceivers, then, the only target gender differences due to social networking presence are found in Manage Multiple Tasks (advantage only for female targets), Manage Anger (advantage only for male targets), and Curious (advantage only for female targets). A difference is also found however, with advantage for males only, in the additional experimental binary measure, the Yes/No decision, referring to the scenario task of choosing a teammate. Again, as with the similar result from Table 1, the implication is that male perceivers are more comfortable than female perceivers, making the choice decision based on less evidence, possibly due to greater confidence in their power to control the eventual consequences.

In summary, Table 3 suggests that for males, as for females, there is some interaction between target gender and social networking presence. The effect is different however, from that for female perceivers in that it is not as consistent or as comprehensive across the range of impression dimensions. And yet, for male perceivers, it is seen in the Yes/No measure which intuitively would seem to encapsulate the range of dimensions in being the action outcome.

Table 4, comparing the significance findings from Tables 2 and 3, helps summarize and also highlights interesting contrasts.

Only for the Work with People dimension do male and female perceivers give social networking advantage to both male and female targets alike, though they both are consistent across target genders with respect to showing no significant social networking effect for three (3) other dimensions: Handle Conflict, Adapt to New Situations and Ability to Take Direction. Both also give social networking advantage to females for Curious, but only female perceivers do so for male targets. Thus, we see again that there is some overlap and some difference in how male and female perceivers are impressed by social networking, depending target gender. Target gender changes the way they both view social networking in ePersonas but in different ways and to different degrees. Perhaps the most interesting inconsistency is that, for the Manage Multiple Tasks dimension, male perceivers give a social networking advantage only to female targets while the reverse is true–female perceivers give it only to male targets. This is quite extraordinary and when combined with other gender differences identified above suggests some intriguing ideas about underlying phenomena.

We speculate that many of the observed differences in the way males and females assign value to social networking presence by target gender are attributable to differences in the stereotypes upon which they draw and the central role of sociality in those stereotypes. Unlike impressions formed through rich engagements, such as face-to-face interaction, impressions formed from search results are based upon a relative paucity of informative cues. Yet, it appears that most subjects assigned significant influence to the available information and felt comfortable basing their selection decisions upon it. While one might wonder if subjects' comfort with their decisions was due, in part, to the fictional nature of the decision context, subjects' comments support the legitimacy of both the scenario and their decision-making processes. Indeed most subjects appeared to be invested and engaged in the task *and* comfortable making their decision based on search results. This we believe implies a quality unique to the Web-based form of impression formation–that perceivers over-attend to the information they find on the Web and feel unwarranted confidence in resultant decisions, probably due to an unfounded faith in the veracity of Web-based information and the misperceived sufficiency of search-based representations of that information. At the same time, the SIDE model, discussed above, suggests that these perceivers are likely falling back, however unconsciously, on gender stereotypes to supplement the sketchy information provided.

It follows then that stereotypes may play an exaggerated role in Web-based impression formation and this study can be interpreted accordingly. For example, female perceivers gave benefit for social networking presence only to male targets on several of the measures and we suggest that this stems from the contradiction they saw between the evidence of social networking participation and their stereotypical view of males as less socially adept than females. This view, exaggerated by the lack of other individuating information, may have led females to over-reward males who were perceived to defy the stereotype based upon their social networking activity. Additionally, females penalized males more than females for lacking social networking presence across a number of factors, further supporting this interpretation–the lack of social networking presence reinforced the stereotypical female view of males as socially deficient and led to the disproportionate denigration of the "socially lacking" males.

The fact that this phenomenon was not observed for male perceivers makes sense in light of the specific social networking context of this study and the way it relates to gender-based stereotypes. A lack of social aptitude is a major component of females' stereotypical view of males and thus the selection decisions of female subjects were more influenced by the social networking presence, or the lack thereof, for males than for females. But if males indeed have lower social awareness, then sociality may play a less influential role in males' stereotyping of others and thus may leave their selection processes less affected by social networking presence. This suggestion is consistent with the findings reported Table 3 above. If one were to replicate this experiment, focusing on a more influential dimension of males' stereotypical views of females than sociality, then we might expect to see the same phenomenon–a stronger effect of their stereotypes coming through in their ratings of females than for males.

So it may be that Web-based impression formation is more susceptible to gender bias than are more traditional modes of impression formation and the effect may be stronger. "Googling" someone may give a false sense of confidence, leading perceivers to believe they know they know "enough" to make a selection decision as their limited knowledge of a target is likely supplemented, however unconsciously, by stereotypes. If this is indeed the case, this is a hazard of life in the digital domain best addressed through awareness and understanding.

CONCLUSION

In summary, the data lend support to some of the propositions and suggest some interesting nuances in the similarities and differences between male and female perceivers and the way their ePerception is affected by target gender, social networking presence, and interactions between the two. Previous research identifying the effects of social networking presence and perceiver gender were further substantiated and extended to incorporate ePersona gender and the analysis surfaced intriguing phenomena.

Specifically this study found that both males and females tend to perceive a potential teammate with social networking activity more favorably than one without. However, results also indicate that males and females perceive and assign that social networking benefit differently. Female perceivers in our study viewed those engaged in social networking sites as more curious and desirable and more capable of multi-tasking, handling conflict, adapting to new situations and interacting well with others than those with no involvement in social networks. Meanwhile, male perceivers viewed social networking activity as an indication that a potential teammate is more curious, better able to manage anger and more likely to work well with others. These findings offer support for Proposition 1b and some support for Proposition 1a.

Our findings also suggest that social networking activity is a more influential factor than is target gender as neither male nor female perceivers significantly preferred one gender over the other when controlling for targets' involvement in social networks. Therefore we find no support for Propositions 2a-2d.

There does appear however, to be an interaction effect between target gender and social networking presence. Female perceivers showed a preference for male targets with social networking presence over males without social networking presence on seven of the ten assessed characteristics and a preference for social networking females on three of the target characteristics. Of particular interest is that female perceivers appear to assign a greater value to the social networking activities of male targets than that of female targets and assign a greater penalty to non-social networking males than they do to non-social networking females. This finding supports Proposition 3a.

Male perceivers also had more favorable impressions of those with social networking activity, scoring social networking males more favorably than non-social networking males on three characteristics and scoring social networking females higher than non-social networking females on three characteristics. Thus there does appear to be some support for Proposition 3b however the interaction effect of target gender and social networking activity is not as strong for male perceivers as it is for female perceivers.

These findings hold implications for practice, pedagogy and further research. For managers, the observed interaction for perceivers of both genders (especially females) between social networking presence and target gender (same as perceiver or different) amounts to a bias that may distort impressions of job applicants, for example, and could lead to sub-optimal hiring decision outcomes in the workplace. In particular, hiring supervisors should recognize the benefits and penalties a female interviewer may place upon male applicants based solely upon their social networking activities. Although the student-based subject pool admittedly limits the degree of generalizability to professional environments, one can argue that these individuals are no more than a year or two away from entering the workforce and they may be expected to carry their biases with them into positions where they will evaluate potential job applicants for their work teams. It thus behooves them, and their future managers, to develop an awareness of their biases with the aim of minimizing them and/or compensating for them in their decisions. And it behooves job applicants to be aware of them, particularly the strong negative reactions to the lack of social networking presence, but also the sensitivity to excessively high levels of such activity, as evidenced in the qualitative responses, and to consider tuning their ePersonas accordingly, to the extent possible, when job seeking.

Findings concerning individuals' use and valuation of social networking activity in impression formation hold implications for instructors as well. This study suggests that Web-based information may well play a role in students' selection of teammates for class-based team projects be they face-to-face teams in traditional classrooms or virtual teams in online courses. The availability of such influential information on the Internet may be particularly problematic for instructors looking to minimize students' preconceived opinions of potential teammates based upon factors such as social networking activity, which may play little or no role in the ability of a student to contribute meaningfully to a group. Although our sample was limited to business students, our findings may apply to other student groups as well given that most Millennials (a term often used to label those born roughly between 1980 and 2000) are technically savvy and it is arguable that no one group within this generation is any more or less likely to turn to the Internet for information when making such selection decisions.

While this and similar studies have shed some light upon the impacts of social networking activity on our perceptions of others, many questions regarding the role of Web-based information in impression formation remain unanswered. Focusing specifically on social networking, some extensions of this study would entail exploring the question of how much social networking presence is beneficial (or detrimental), the consequences of possessing a common name or sharing a name with a celebrity (making it hard to disambiguate the right identity), and the role of perceiver characteristics beyond gender on the impressions formed. Future research could also investigate the same research questions posed by our study to determine if the findings indeed apply to other populations such as professionals, non-business majors, and non-Milliennials.

Additionally, there remains the larger question of how Web-based searches fundamentally "change the game" in impression formation, as speculated in the discussion above. Based on the results of this study, we speculate that ePerceivers may tend to read more into search results than they should, relying on gender stereotypes (and possibly others) and thereby introducing greater bias into their impressions than they would in face-to-face contexts. Of course, this phenomenon is likely still evolving as the Web generation matures and users become savvier about the reliability, quality and quantity of Web-based information and as society's stereotypical views of males and females shift over time. And indeed some of the behaviors and perceptions that underlie those stereotypes may themselves be expected to be fundamentally transformed by Web-based phenomena such the social networking revolution. The evolution of these phenomena will create moving targets that provide challenging but important opportunities for future research that further investigates the impact and formation of Web-based impressions.

ACKNOWLEDGMENT

We thank the reviewers and editors for their insights and assistance with this manuscript. This research was supported by the Behavioral Research Group of the College of Business at San Jose State University.

REFERENCES

Back, M. D., Stopfer, J. M., Vazire, S., Gaddis, S., Schmukle, S. C., Egloff, B. & Gosling, S. D. (2010). Facebook profiles reflect actual personality not self-idealization. *Psychological Science*. Advance online publication. doi:10.1177/0956797609360756

Brewer, M. B. (1988). A dual process model of impression formation. In Scull, J. R. K., & Wyer, R. S. (Eds.), *Advances in social cognition* (Vol. 1, pp. 1–36). Hillsdale, NJ: Lawrence Erlbaum Associates, Inc.

CareerBuilder. (2008). One-in-five employers use social networking sites to screen job candidates. Retrieved from http://www.careerbuilder.com/share/aboutus/pressreleasesdetail.aspx?id=pr459&sd=9%2F10%2F2008&ed=12%2F31%2F2008&siteid=cbpr&sc_cmp1=cb_pr459_&cbRecursionCnt=1&cbsid=894900f6873d4775af1c4681201d0758-275320933-R5-4

Cross-tab Marketing Services. (2010). Online reputation in a connected world. 1-23. Retrieved from http://www.microsoft.com/privacy/dpd/research.aspx

Ford, N., Miller, D., & Moss, N. (2005). Web search strategies and human individual differences: A combined analysis. *Journal of the American Society for Information Science and Technology*, 56(7), 757–764. .doi:10.1002/asi.20173

Gefen, D., & Straub, D. W. (1997). Gender differences in the perception and use of e-mail: An extension to the technology acceptance model. *Management Information Systems Quarterly*, 21(4), 389–400. .doi:10.2307/249720

Hancock, J. T., & Dunham, P. J. (2001). Impression formation in computer-mediated communication revisited. *Communication Research*, 28(3), 325–347. .doi:10.1177/009365001028003004

Huffaker, D. A., & Calvert, S. L. (2005). Gender, identity, and language use in teenage blogs. *Journal of Computer-Mediated Communication*, 10(2), article 1. Retrieved from http://jcmc.indiana.edu/vol10/issue2/huffaker.html

Jacobson, D. (1999). Impression formation in cyberspace: Online expectations and offline experiences in text-based virtual communities. *Journal of Computer-Mediated Communication*, 5(1), 461–479. Retrieved from http://jcmc.indiana.edu/vol5/issue1/jacobson.html.

Kalyanaraman, S., & Sundar, S. S. (2008). Impression-formation effects in online mediated communication. In Konijn, E. A., Tanis, M., Utz, S., & Linden, A. (Eds.), *Mediated interpersonal communication* (pp. 217–233). Hillsdale, NJ: Lawrence Erlbaum Associates.

Lea, M., & Spears, R. (1991). Computer-mediated communication, de-individuation and group decision-making. Special issue: Computer-supported cooperative work and groupware. *International Journal of Man-Machine Studies*, 34(2), 283–301. .doi:10.1016/0020-7373(91)90045-9

Lee, E. J. (2004). Effects of gendered character representation on person perception and informational social influence in computer-mediated communication. *Computers in Human Behavior*, 20(6), 779–799. .doi:10.1016/j.chb.2003.11.005

Likert, R. (1932). A technique for the measurement of attitudes. *Archives de Psychologie*, 140, 1–55.

Markey, P. M., & Wells, S. M. (2002). Interpersonal perception in internet chat rooms. *Journal of Research in Personality, 36*, 134–146. .doi:10.1006/jrpe.2002.2340

McKenna, K. Y. A., & Bargh, J. (2000). Plan 9 from cyberspace: The implications of the internet for personality and social psychology. *Personality and Social Psychology Review, 4*(1), 57–75. .doi:10.1207/S15327957PSPR0401_6

Nass, C., Moon, Y., & Green, N. (1997). Are machines gender neutral? Gender-stereotypic responses to computers with voices. *Journal of Applied Social Psychology, 27*(10), 864–876. .doi:10.1111/j.1559-1816.1997.tb00275.x

Postmes, T., & Spears, R. (2002). Behavior online: Does anonymous computer communication reduce gender inequality? *Personality and Social Psychology Bulletin, 28*(8), 1073–1083. .doi:10.1177/01461672022811006

Richardson, J., & Swan, K. (2003). Examining social presence in online courses in relation to students' perceived learning and satisfaction. *Journal of Asynchronous Learning Networks, 7*(1), 68–88. http://www.sloan-c.org/publications/jaln_main.

Sherman, J. W., Lee, A. Y., Bessenoff, G. R., & Frost, L. A. (1998). Stereotype efficiency reconsidered: Encoding flexibility under cognitive load. *Journal of Personality and Social Psychology, 75*(3), 589–606. http://www.apa.org/pubs/journals/psp/. doi:10.1037/0022-3514.75.3.589

Vazire, S., & Gosling, S. D. (2004). e-Perceptions: Personality impressions based on personal websites. *Journal of Personality and Social Psychology, 87*, 123–132. http://www.apa.org/pubs/journals/psp/. doi:10.1037/0022-3514.87.1.123

Venkatsubramanyan, S. & Hill, T. (2009a). An empirical investigation into the effects of web search characteristics on decisions associated with impression formation. *Information Systems Frontiers*. Advance online publication. doi: 10.1007/s10796-009-9177-9

Venkatsubramanyan, S., & Hill, T. (2009b). Gender differences in social networking presence effects on web-based impression formation. *AMCIS 2009 Proceedings*. Paper 364, http://aisel.aisnet.org/amcis2009/364.

Venkatsubramanyan, S., & Hill, T. R. (2007). Evaluating potential team members using search results: An experimental study. *International Symposium on Information Systems (ISIS), Hyderabad, India, December 2007*.

Walther, J. B. (1997). Group and interpersonal effects in international computer-mediated collaboration. *Communication Research, 23*, 342–369. .doi:10.1111/j.1468-2958.1997.tb00400.x

Walther, J. B., & Parks, M. R. (2002). Cues filtered out, cues filtered in: Computer-mediated communication and relationships . In Knapp, M. L., & Daly, J. A. (Eds.), *Handbook of Interpersonal Communication* (3rd ed., pp. 529–563). Thousand Oaks, CA: Sage.

Walther, J. B., Van Der Heide, B., Hamel, L. M., & Shulman, H. C. (2009). Self-generated versus other-generated statements and impressions in computer-mediated communication: A test of warranting theory using Facebook. *Communication Research, 36*(2), 229–253. .doi:10.1177/0093650208330251

Walther, J. B., Van Der Heide, B., Kim, S.-Y., Westerman, D., & Tong, S. T. (2008). The role of friends' appearance and behavior on evaluations of individuals on Facebook: Are we known by the company we keep? *Human Communication Research, 34*(1), 28–49. .doi:10.1111/j.1468-2958.2007.00312.x

Widmeyer, W. N., & Loy, J. W. (1988). When you're hot, you're hot! Warm-cold effects in first impressions of persons and teaching effectiveness. *Journal of Educational Psychology, 80*, 118–121. .doi:10.1037/0022-0663.80.1.118

KEY TERMS AND DEFINITIONS

ePersona: The searchable digital information about a particular person from a variety of sources – personal home pages, social and professional networking sites, organizational Web pages, news articles, blogs and others.

ePerception: The perception formed by people about others depending primarily, if not exclusively, on the ePersona.

Perceiver: In the context of impression formation, the perceiver is the person forming an impression of another individual.

Target: In the context of impression formation, the target is the person about whom an impression is formed.

The Social Identity model of Deindividuation Effects (SIDE): A theory that suggests perceivers will use stereotypes in forming impressions of others for which they have limited differentiating information.

Warranting Theory: A theory that suggests perceivers, when forming impressions of an individual target, will place a greater value on the opinion of a third party over that individual's self-evaluation.

APPENDIX A: SEARCH RESULTS FOR JANE DOE 1

Google
Web Images Video News Maps more »
Jane Doe 1 [Search] Advanced Search / Preferences

Web — Results 1 - 10 of about **3,340,000** for Jane Doe 1 [definition]. (0.19 seconds)

Jane Doe 1 - San Jose, CA | Facebook
Jane Doe 1 is on Facebook. She....Facebook gives people the power to share and makes the world more open and connected.
www.facebook.com/people/Jane Doe 1/1031931835 Cached - Similar pages

Personal Page
Jane Doe 1. This is my personal home page...
http://www.students.sjsu.edu/~Jane Doe 1 Cached - Similar pages

MySpace.com - Jane Doe 1 - www.myspace.com/195342149
MySpace profile for Jane Doe 1 with pictures, videos, personal blog, interests, information about her and more.
profile.myspace.com/index.cfm?fuseaction=user.viewProfile&friendID=195342149
Cached - Similar pages

Friendster - Jane Doe 1
Friendster: ; location: San Jose, CA
profiles.friendster.com/48608703 Cached - Similar pages

Jane Doe 1 | pcdr01 | Reunion.com Member Profile
Find information on Jane Doe 1,San Jose,CA.
www.reunion.com/pcdr01 Cached - Similar pages

Jane Doe 1 - LinkedIn
View Jane Doe 1's professional profile on LinkedIn. LinkedIn is the world ...
www.linkedin.com/pub/dir/Jane Doe 1 Cached - Similar pages

US Census....Bay area....Published, December 2000
.....Jane Doe 1 b.1984, San Jose, CA, USA...Student5 b. 1956, San Leandro, CA, USA
https://www.uscensus.gov/... Cached - Similar pages

Jane Doe 1 |San Jose, CA | Classmates.com
Jane Doe 1 2008 graduate of Local High School in San Jose, CA is on ...
www.washingtonpost.com/wp-dyn/content/discussion/2008/02/27/DI2008022701866.html
Cached - Similar pages

Student55 | Facebook
... with Student 6, Jane Doe 1, Student7....
www.facebook.com/index.cfm?fuseaction=user.viewprofile&friendid=421235
Cached - Similar pages

Jane Doe 1 | Facebook
Jane Doe 1 is on Facebook. Join Facebook to connect with Jane Doe 1 and others you may know. Facebook gives people the power to share and makes the ...
www.facebook.com/people/Jane Doe 1/1300044488 - Cached Cached - Similar pages

APPENDIX B: SEARCH RESULTS FOR JANE DOE 2

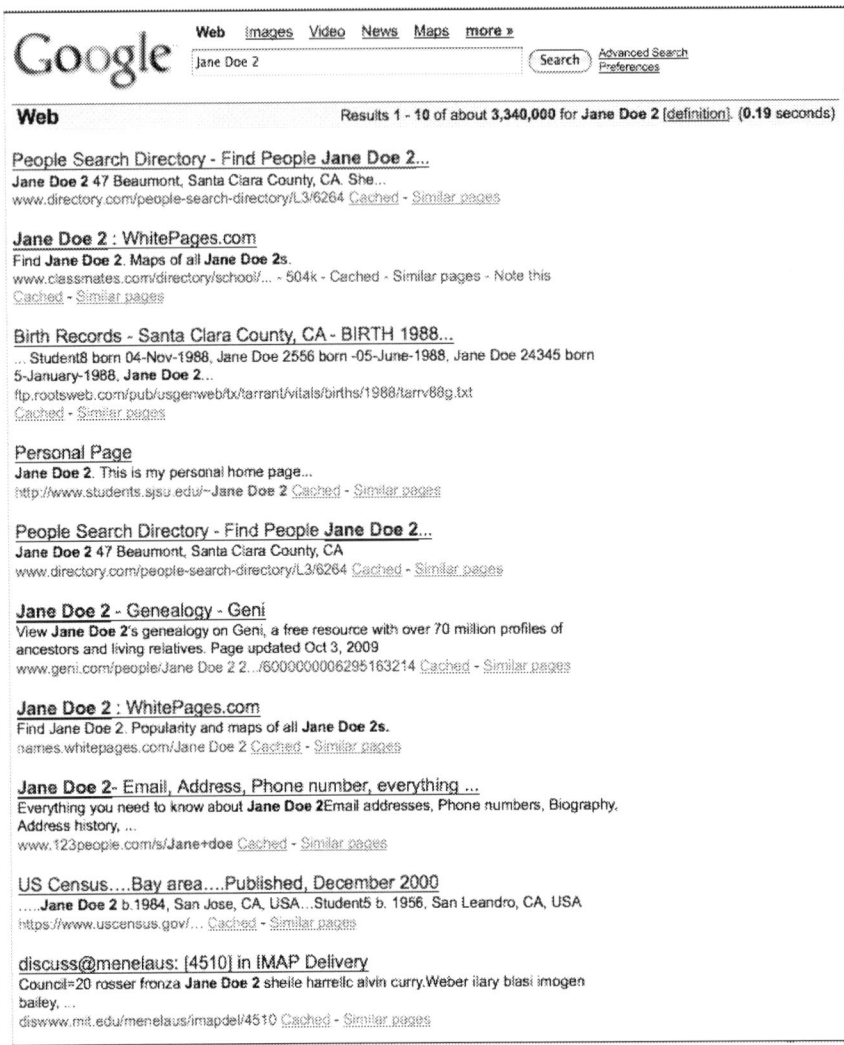

A previous version of this chapter was published in the International Journal of E-Politics, Volume 2, Issue 2, edited by Celia Romm, pp. 55-73, copyright 2011 by IGI Publishing (an imprint of IGI Global).

Section 3
Gender and eDating

Chapter 13
E-Dating:
The Five Phases on Online Dating

Monica T. Whitty
Nottingham Trent University, UK

ABSTRACT

Online dating continues to grow in popularity as a way for individuals to locate a potential romantic partner. Researchers have examined how people present themselves on these sites, which presentations are more likely to lead to success, the effectiveness of the matchmaking tools that some companies employ, the stigma attached to using these sites and the types of people who are drawn to online dating. However, there is an absence of scholarly work on how these relationships progress compared to traditional models of courtship. This chapter sets out a model for the phases of online dating and compares this model with Givens' (1979) work on a traditional model of courtship. It is argued here the phases of online dating are very different to other courtship models. These differences pose new challenges and create new benefits to those who elect to find a partner via one of these sites.

WHERE CAN ONE FIND A ROMANTIC PARTNER ON THE NET?

Before moving on to consider online dating sites, it is important to understand that these sites emerged because it became obvious that there was a need and a market for such sites. As is well-known, the Internet was not originally set up as a social space,

DOI: 10.4018/978-1-60960-759-3.ch013

but rather as a space to transfer data. However, not long after (even in its most primitive textual form), friendships and romances began to blossom. People were meeting each other in all sorts of places, MUDs and MOOs (multi-users dungeons or domains, which are essentially role-playing sites), bulletin boards, chat rooms, newsgroups and gaming sites.

Bulletin board systems (BBs) were possibly the first place where romances initiated on the

Internet. DeVoss (2007) has succinctly described how these relationships initiated in these sites and quite rightly points out that one could not escape gender roles in these spaces (even when people swapped gender). There are many anecdotal stories about how people played with self-narratives in these sites, one of which DeVoss describes is the story of a woman who called herself Princess:

Her registry claimed that she as 5'1", a 'curvaceous' 102 pounds, had deep brown eyes, and waist-length black hair. She flirted online, and male users would virtually flirt to enter her private chat room. As online time passed, the men... wanted to meet her...Over and over again, she resisted their efforts to entice her off the BBS and into physical public space. She wrapped herself in an elaborate story—that she was hiding from a violent ex-husband who was involved in organised crime...she wasn't allowed to give out her real name, phone number, address...

Somehow, however, one of the more aggressive male users...found out where she lived...only to find her the exact opposite of her online identity... They reported her to be phenomenally hefty, incredibly ugly, partially toothless, and surrounded by a gaggle of children. (p. 21-22)

Of course, just as with many face-to-face encounters, not everyone online is looking for a traditional heterosexual relationship. BBs were also a meeting place for people with more deviant sexual tastes. Wysocki (1998) reported her examination of "how and why individuals participate in sexually explicit computer boards; and to see if sex on-line is a way of *replacing* face-to-face relationships or a way of *enhancing* them" (p. 426). She found that participants enjoyed the anonymity the Internet affords. Moreover, given their lack of time in their personal life it allowed them to find individuals with similar sexual interests and to share sexual fantasies within the confines of their own homes. Many of the people she surveyed were in fact happy to keep these sexual relationships online. More recently, Wysocki and Thalken (2007) examined adult Web sites to find that, in particular, older individuals (a mean age of 40 years), were more likely to use S & M adult Web sites. Most people using the sites included photographs; however, what took these researchers by surprise was that many of the photos included face shots (indicating no attempt to ensure complete anonymity). The mix of people using these sites did so to either engage in online sexual fantasies or to find partners who shared their fantasies and were prepared to enact them off-line.

Romances and friendships have also been known to develop in MUDS and MOOs, chat rooms and newsgroups. For example, in the mid-90s, Parks and Floyd (1996) found that two-thirds (60.7%) of their newsgroup sample had formed a personal relationship with someone they had met for the first time online. Of these, 7.9% stated that this was a romantic relationship. Utz (2000) found that 76.7% of the MUD users she surveyed reported forming a relationship online that developed off-line, of which, 24.5% stated this was a romantic relationship. Whitty and Gavin (2001) found that individuals form real friendships in chat rooms and that some of these participants preferred that these relationships remain online.

THE CONSTRUCTION OF ONLINE DATING SITES

Given the numbers of people seeking others online for love and sex, it is little wonder that companies have tried to formalize this process as well as to make money from people who are prepared to seek out romance on the net. Online dating sites continue to abound online and increase in popularity. Yahoo.com claims almost 380 million visitors per month to their online dating site (Pasha, 2005), and FriendFinder.com say they have over 3.5 million active members (Dating Sites Reviews.com, 2006).

During the early days of the Internet, given the restricted technology capabilities and bandwidth, online dating sites looked more like newspaper personal ads. Individuals would read a profile and contact the person on the site to learn more about them and to gauge whether the other was also interested. Men were much more likely to subscribe to these sites than women and companies allowed women onto these sites for free to ensure men had an adequate selection.

These days, the sites are still typically set up to have their users construct a personal ad for themselves. The amount of information and detail people can add is obviously less restrictive due to increased bandwidth. Clients can, and generally do, show at least one photograph of themselves, and can also add video and voice to their profiles. Online daters can present information about themselves in a number of ways. They can rate themselves or check boxes indicating attributes such as their age, gender, location, job and physique (e.g., a choice ranging from slim to overweight). Some of these questions, such as age and gender are often a compulsory requirement. In addition, clients are usually given an opportunity to add to and expand upon this information. For example, they may elaborate on their hobbies and musical interests, or the type of person they are attempting to attract.

Not all online dating sites expect their clients to do all the matching work themselves. Some sites do the matching for the client. For example, some online dating sites will ask the client to fill out descriptive details and sometimes a personality scale. These sites claim to be able to 'scientifically' match individuals. The assumption is that there is a formula to matching appropriate people. This view, however, is somewhat contentious. Houran, Lange, Rentfrow, and Brukner (2004), for instance, have argued that such compatibility tests have provided little psychometric support. Nevertheless, it is argued here that even if these test lack scientific rigor they still assume some of the laborious work from the site users. In contrast, other sites provide a more flexible approach, whereby clients can opt to fill out such tests and be presented with profiles of clients deduced to be suitable matches or instead the client can wade through the sea of possibilities and select for themselves.

In addition to the general online dating sites such as, eHarmony, True.com, Match.com and so forth, there are also more specialized online dating sites which gather like-minded individuals together. For example, there are sites designed specifically for Christians, Jews, vegans, Goths or spiritual people. Such sites are similar to social groups which one might join in the hope of finding others that share the same values or interests. Moreover, it potentially cuts out some of the work associated with the search for the perfect other. These sites are discussed in more detail later in the chapter. For now, this chapter turns to examine how relationship development on an online dating site compares with traditional dating. It does so by firstly examining the traditional off-line courting process.

THE FIVE PHASES OF OFFLINE DATING

Givens' (1979) developed a five-stage model to explain the traditional courting process. In his model the first phase is the 'attention phase'. During this phase an individual (typically a woman) will try to attract a person of the opposite sex by displaying non-verbal signs of attraction, such as primping, object caressing and using quick glances towards and away from the target. The second phase is the 'recognition phase', where flirting behavior consists of head cocking, pouting, primping, eyebrow flashes and smiles. Givens suggests that 'interaction' does not occur until the third phase. This is when conversation is initiated. During this stage, participants appear highly animated, displaying laughing or giggling. Interestingly, Givens notes that men are generally hesitant to approach women without some initial indication of interest from the woman. The fourth phase in his model is 'sexual arousal', which is finally followed by 'resolution'. As Whitty (2003) has argued, Givens' work not only demonstrates that women make the first move, but also highlights the importance of non-verbal cues in the signaling of sexual attraction. These signals are crucial in the game of flirting. Given that flirtatious behavior features heavily when one is trying to initiate contact with a potential romantic mate off-line, this chapter now turns to examine flirtation in a little more detail.

Flirtation

Flirtatious behavior is evident throughout the early phases that Givens has delineated. Success in finding a potential partner off-line might be in some ways attributed to how skilled an individual is at flirting. Moore (1985) observed individuals in bars and identified 52 facial expressions and gestures that women display when they flirt. Although these actions might be subtle, they can be enough to indicate attraction. Moreover, it is important to note that one gesture in isolation is typically not enough to indicate attraction. *Sometimes a smile is just a smile.* While flirting is not necessarily a prelude to sexual interaction, researchers, such as Moore (1985; 1998), have found that women who display them are more likely to be approached by a man who shares the same sentiment.

In order for flirtation to work, it needs to be ambiguous. This is why it typically involves a number of carefully orchestrated body gestures rather than a frank statement, such as 'I really fancy you'. As will be examined in more detail later on in this chapter, this poses an obvious problem online where the body is typically not physically present (note: Webcams, videos, and photos can make the body more present). Moreover, the 'dance' can be so subtle that it is not always conscious to the flirtee. First, however,

I move to develop a model that delineates the phases involved in online dating.

THE FIVE PHASES OF ONLINE DATING

Theorists have devised models to explain formal matchmaking services, such as personal ads, video dating and computer matchmaking. Ahuvia and Adelman (1992), for example, have devised the SMI model (searching, matching, and interacting model). They parallel matchmaking services with basic market functions. For instance, in the market place initially 'searching' is required, that is, gaining information essential for exchange (in regards to matchmaking this means searching for information about a potential other). Second, 'matching' is required to bring together compatible exchange partners (in regards to matchmaking this would mean bringing together two singles that seem well matched). Ahuvia and Adelman swap the term 'transacting' for 'interacting'. They do so because transacting requires a negation of implementing an exchange and to do so with matchmaking an interaction needs to take place.

Internet dating does seem to include the three phases highlighted by Abuvia and Adelman. However, these phases do not necessarily neatly fit into the sequential order they propose. Matching, for instance can happen at two points. First, the site might suggest matches from a specific formula devised to 'scientifically' match individuals and then the client might search through the site's choices to decide who they believe is an appropriate match. Alternatively, the client might begin by searching through the sea of profiles until they find profiles that are suitable matches. Next, contact is made on the site initially to indicate interest in another and the other has to reciprocate mutual interest. From there the two potentials begin to interact and decide whether they wish to progress the relationship further.

There is some utility in drawing from Abuvia and Adelman's model to explain the process of online dating; however, I argue here for a more sophisticated model that also draws from Givens model of the traditional off-line courting process. As outlined below, these phases include 'the attention phase', 'the recognition phase', 'the interaction phase', 'the face-to-face meeting', and finally 'resolution'. These phases have been derived from previous empirical work I have conducted on online dating (see Whitty, in press; 2007; Whitty & Carr, 2006). As is explained below, the phases do involve many of the processes Givens refers to; however, the phases are not necessarily ordered in the same way and each phase arguably requires additional skills.

Phase 1: The Attention Phase

During the 'attention phase', off-line a person (usually the woman) will try to attract a person of the opposite sex by displaying subtle non-verbal signals. With online dating, one does not have an immediate target to display subtle interest in. Instead, the person displays signs of attraction by selecting an attractive photograph to represent themselves. In the main, profiles that do not show a photograph are overlooked (Whitty & Carr, 2006). Moreover, women typically take more time and care selecting an attractive photograph of themselves. Overkill with a photograph, that is, a photograph professionally taken or looking too sexual, is treated skeptically. This might parallel with a woman showing too much interest in the first phase off-line—such a woman might be deemed a little too desperate.

During this attention phase, another unique way online daters might go about attracting others to their profile is via the name they use to represent themselves (what is commonly referred to as a screen name). Some use very flirtatious names (e.g., Imcute or Bubbly), while others select names that reflect their personal identity (e.g., mountainclimber). Others might select a non-flirtatious name (e.g., Jt28 or Smith48). Interestingly, in a recent study by Whitty and Buchanan (2007), it was found that certain screen names are deemed more attractive than others, and that men and women are more motivated to make contact with individuals with different types of screen names. It was found that men more than women are attracted to and motivated to contact screen names which indicate physical attractiveness (e.g., 'Hottie' and 'Greatbody') and that women more than men are attracted to screen names which demonstrate intelligence ('Wellread' or 'Welleducated'). Overall, less flirtatious names were perceived by the majority of people as less attractive names.

The photograph and the person's screen name is usually the first bit of information others encounter on the site. If they like what they see, they might move to the next phase or they might (and typically do) seek out more information about the person. Therefore, continuing to hold another's attention prior to any interaction can be quite an arduous task for an online dater, and arguably they need to be quite skillful at crafting a profile (see Whitty, 2007).

As research has found, online daters tend to give much thought to their profiles. Sometimes a dater will trial a certain profile to see what types of people they attract or if indeed they attract anyone. Whitty (2007, p. 64) presents an extract from an interview where Lynne (named changed for confidentiality) explained what a painstaking task this was for her:

I: *O.K., looking at your own profile, how did you decide what information to include about yourself?*

L: *After many hours and trial and error, I changed it a few times ... I did want to describe myself in a way that would give a bit of a cross section of me and depending on what sort of results that I got ... with each profile I would then sort of go and revise*

> that. Having said that, I have only revised it I think three times.
>
> **I:** But you feel like you got different responses accordingly to different profiles?
>
> **L:** Yeah a little bit.

As Whitty (in press) reports, online daters often make great efforts to construct an attractive profile. In addition to a photo, the types of information daters said were important included their interests and hobbies (53%) and a description of their personality (35%). Furthermore, a number of the people she interviewed stated that it was important to inject humor into their profile (17%), state their occupation (13%), how intelligent they are (12%) and make their profile appear unique and different (12%).

It has been argued that a successful profile is not just a profile that stands out in a crowd by one which also appears 'real' or genuine and trustworthy (Whitty, 2007). Whitty (2007) suggests that online daters should ascribe to the BAR (balance between and attractive and real self) approach. Presenting an ideal presentation of the self does not work on online dating sites—as we see in the fourth phase, if people do not live up to their profiles they typically do not earn a second date. Presenting a real profile also means avoiding writing anything that appears too clichéd. For example, describing one's most romantic date as going for a stroll on the beach or sipping wine by log fires are seen by many as either naïve, untrustworthy or unoriginal. Ironically, by sounding too romantic, an online dater appears less real or authentic and hence a less appealing candidate. One participant in Whitty's (2007, pp. 64-65) research aptly described why he/she would avoid such profiles:

> **T:** *I tend to stay away from those people with sort of cliché stuff. I think it appears in a lot of profiles ...*
>
> **I:** *What would be some of the clichés that you would be turned off by?*

> **T:** *With some, on some profiles it has a very sexual overtone, which puts me off totally. Sometimes it is like a passage of clichés, walks on the beach, romantic evenings, romantic getaways, a bottle of wine, and nice crackling fire. It just doesn't ring true, it just sounds like a, it doesn't seem very real.*

Phase 2: The Recognition Phase

As with off-line courting, the second phase of online dating requires more flirtation and some recognition. Online dating sites have attempted to mimic this step in the construction of their Web sites. Rather than immediately e-mailing a member of the site that the client finds attractive, instead, online daters are often given the option to virtually flirt. That is, many sites give the option to send a 'form' note via the site, often referred to as a 'wink' or a 'kiss'. Akin to flirting off-line, this can appear less intrusive and more subtle than a more detailed e-mail introducing oneself and asking for the person to, in turn, self-disclose to them. Although this is probably an important and necessary step, it is argued later in this chapter that online dating sites' attempts to incorporate flirtation is still fairly clunky and this could well be a problem that some companies might want to address.

Phase 3: The Interaction Phase

Similar to off-line, the next phase in the online dating courting process involves interaction. This of course takes place online. It might initially take place through: an exchange of e-mails via the site; a site's instant messaging program; personal e-mail accounts; or an instant messaging program found off the site. It might then move on to phone (usually mobile phone) or SMS texting. As with courting off-line, this stage can also be very flirtatious. Although the traditional physical off-line cues are not present, substitutes can be found for these non-verbal cues (Whitty, 2003; 2004). One can describe in more detail how they look or the types of physical actions they would like to enact. Individuals also flirt by using emoticons, such as smiley faces and winks, and acronyms (e.g., LNK which represents love and kisses or QT which represents cutie).

As Whitty (2003; 2004) has found, individuals can indeed flirt online, despite the absence of traditional cues. Moreover, sometimes they can feel freer to flirt and express their sexuality than they might in face-to-face encounters. Similarly, Alapack, Blichfeldt, and Elden (2005) state that online the "living flesh is being transformed into a 'legible' body" (p. 52). They also state that a special online flirting language has evolved. In line with previous work (e.g., Whitty, 2003; Ben-Ze've, 2004) they also state that cyber-flirting can trigger physical and sexual reactions.

Empirical research has found that not only do individuals flirt online, but they do so in gender-defined ways (Whitty, 2004). Her survey of 5,697 participants found that women were more likely than men to flirt by using online substitutes for non-verbal cues. For example, they admitted to using more emoticons and acronyms to flirt online and described themselves as physically attractive. Men, in contrast, were more likely to initiate contact.

Typically this third phase is relatively short. While, some flirtation and self-disclosure takes place for online daters during the interaction, this phase is more about verifying information and setting up the face-to-face meeting (or one or both of the pair decide to drop out of the process). Given that profiles already provide a curriculum vitae about the individual, additional information about the person does not need to be disclosed. Hence, during this phase if the pair feels comfortable enough they move to setting up a face-to-face date. Conducting a face-to-face date rather than spending more time conversing and getting to know one another is preferable, as most online daters would agree that one cannot test out real 'chemistry' in an online environment. Spending too much time conversing without agreeing to a date is frowned upon by many—as such a person is often deemed a tease or insincere.

Phase 4: The Face-to-Face Meeting

The fourth phase is roughly equivalent to Givens' fourth stage titled the 'sexual arousal' stage. For online daters, the first date very much determines if there will be any further dates. This meeting is different to a 'traditional' first date where sexual attraction has already been established and the couple plan to spend a romantic evening together. Instead it is a meeting where the pair test out physical chemistry and see how well the person matches up with the profile. For safety reasons the meeting usually takes place in a public space (e.g., cafe shop, bar) and is most likely scheduled for a restricted amount of time (so that individuals can make a quick escape if the interaction does not run smoothly). Some devise contingency plans that if the meeting works out well it will continue onto a proper date—(e.g., dinner). Given the fears that individuals will not live out to their profiles, some online daters are savvy enough to check out their date from a vantage point and if they do not turn out looking exactly how they had hoped (e.g., if the photo is a few years out of date or they look somewhat larger than described), the person might not go through with the date (sometimes politely calling on the other's mobile with an excuse as to why they could not make it).

It has been found that 68% of participants believe that the first face-to-face meeting is a screening out process (Whitty, in press). If there is much discrepancy between the person and the profile, the person is usually judged as untrustworthy and not worthwhile pursuing. Common lies told on these sites tend to be about physical appearance. For example, it has been found that men are more likely to lie about their height and status, while women are more likely to lie about their weight (Whitty, in press; Whitty, & Carr, 2006; Hancock, Toma, Ellison, 2007).

A person might have written a profile that they can live up to in the flesh. However, during this first meeting, if one of both parties does not believe there is any physical chemistry or 'sexual arousal' (as Givens refers to), then the two might remain friends or discontinue any further contact.

Phase 5: Resolution

As with Givens' five stages of courting, resolution is the final stage of the online dating courting process. After the first meeting, individuals know whether they are sexually attracted to and whether they want to learn more about their date. If they are still uncertain, they might set up a couple more dates; however, they will generally do so while checking out other options on the site. If they are confident they are interested in moving forward they will typically take themselves off the site. Problems arise between couples if expectations are not meet and one is discovered to be still actively using the site when the other has taken themselves off.

STRENGTHS AND WEAKNESSES OF ONLINE DATING

This chapter now turns to the pros and cons of finding a romantic partner via an online dating site. It does so by paying particular attention to the five phases outlined above.

Flirting on the Net

As explained earlier, off-line flirting consists of a gamut of non-verbal behaviors. Moving through Givens' phases partly requires an ability to carefully orchestrate these gestures (e.g., amount and timing). Despite the absence of the physical body in cyberspace, individuals are able to cyberflirt (Whitty, 2003). As demonstrated, flirtation is evident in a number of the online dating phases.

In the first phase, people need to be quite strategic in their self-presentation. The advantage here (especially for the socially awkward and shy) is that this is achieved asynchronously—giving individuals ample time to consider how they might best present themselves (e.g., photos developed in photoshop, choice of screen name, and details about themselves). The disadvantage is that different skills are required to the one's traditionally used off-line. Some may be oblivious as to which presentation of self will lead to the best result. Cyberspace arguably provides a safer and more playful space for self-presentations (Whitty, 2003). While this has its advantages in other places online, if people are too creative with their online dating profiles (e.g., making themselves appear more physically attractive, younger, wealthier, more interesting), then although this might attract others in the first instance, it is far less likely that they will sustain interest. Having to deal with numerous rejections in the end can have a negative effort on the individuals' self-esteem.

Flirtation continues in the second phase where individuals can send a flirtatious 'form note' through the site, often referred to as 'winks' or 'kisses'. As described earlier, when people flirt face-to-face, they usually display an array of non-verbal behaviors that may or may not be interpreted as attraction. Figuring out whether the other person is mutually attracted and trying to capture another's attention is part of the fun. Moreover, the end goal of flirting is not necessarily an attempt to form a relationship or even a sexual encounter. Online dating sites have obviously added the 'virtual flirt form note' into their design as a way to mimic off-line flirtation. Sending a virtual kiss or a wink appears less intrusive than an e-mail. However, it is not as ambiguous as a display of non-verbal gestures in an off-line setting. Moreover, if people contact others on a dating site they do so in order that they can meet others to form a relationship with. The intentions are far less ambiguous compared to flirting in other places. Sending a virtual note is a clear message that the person is interested in another and holds some hope that a relationship might develop. Therefore, it is argued here that online dating sites' attempts to mimic off-line flirting is still a little clunky and theoretically is not as effective as flirting in other situations (online or off-line).

One way around this problem might be to incorporate flirtatious applications similar to those used by social networking sites. For instance, many of Facebook's applications both enable and encourage flirtation—in a way that does not have to be construed as such. One can send friends virtual gifts (even virtual drinks). Gifts sent online do not have to signal attraction; however, if you were to give someone a physical gift off-line, such as flowers, a teddy bear or handcuffs, the intent would become more obvious. One can dedicate a song to someone online. A more obvious, but still playful application is comparing friends online. For example, this application asks which of two friends you would prefer to date—while this remains equivocal, it does present a flirtatious cue to another. An individual can also send a friend a nomination for a virtual prize. Sending them a nomination for 'the most likely to worship spongebob' would not perhaps be perceived as flirtatious as 'best looking male', or 'most likely to appear in playboy'.

The First Date: Safer but More Clinical

As explained earlier, for online daters the first date is usually set up in a public space for a restricted amount of time. People using these sites do so as they are aware that they are meeting strangers and they are cautious of the type of people they might be meeting. They are also unlikely to give away too much information that might reveal where they live or work. This is potentially safer than meeting potential romantic partners in a bar after a few drinks. However, the disadvantage of the first meeting is that it is much less a date and more like a job interview. Having to live up to the other person's expectations can be a challenge. Moreover, in normal courting situations each person does not typically know as much about the person whom they are considering dating.

Self-Disclosures: Too Much Too Soon?

One of the most popular theories to explain relationship development is *social penetration theory* (Altman & Taylor, 1973). Social penetration theory is an incremental theory which argues that relationships move to greater levels of intimacy over time. According to this theory, how greater intimacy is achieved is typically through depth and breadth of self-disclosure. Breadth of self-disclosure refers to discussing a range of topics, such as information about one's family, career, and so forth. Depth refers to the more central core of one's personality; that is, the more unique aspects of one's self. The timing of how much one self-discloses is crucial to determining whether a relationship will continue to proceed. Rushing self-disclosure in the early stages of a relationship can seem unnatural and desperate and can lead to an abrupt end. According to social penetration theory, in the early phases of relationship development, one moves with caution, discussing less intimate topics and checking in the conversations for signs of reciprocity. Gradually, one feels safer to reveal aspects of themselves.

As can be seen in the five stages of online dating, there is far less opportunity for relationships to develop in the way proposed by the social penetration theory. On an online dating site the profiles are set up in such as way to reveal both depth and breadth. For instance, within the profiles, individuals typically have to provide information about surface levels aspects of themselves, such as eye color, drinking and smoking habits, relationship status, number and types of pets and occupation. In addition, they are given space to write more in depth about themselves, where they are asked to describe their personality, interests (what they read, music they listen to and so forth), their ideal date, and their political persuasion. They are encouraged on these sites to open up about all aspects of themselves. The sites argue that by doing so that they will attract the most appropriate person. Given the amount of information individuals are presented with it is no surprise the conversations that take place via e-mail, telephone and so forth prior to the first meeting are more to clarify information about the person as well as to arrange the meeting. Therefore, as Whitty (in press) suggests, online dating is arguably even more removed from what people are accustomed to when it comes to developing a relationship. There is less opportunity to test the waters gradually and check for reciprocity, instead, reciprocity is determined prior to communication with the individual. The advantage this gives online daters is that they are granted more control over their self-presentations. However, in turn, because the profile compiles all the information about the person in one chunk, it is easier to check. Hence, others are less forgiving when there is a mismatch between the person they meet face-to-face and the person they were presented with in the profile.

Sexual Attraction: Takes a While to Get There

As pointed out in the online dating model, determining 'physical chemistry' or 'sexual attraction' does not usually take place until the couple meet face-to-face. Off-line flirting usually takes places because one or both individuals find the other physically attractive. Individuals probably do not make it to the interaction phase if there is not at least some physical chemistry there. Of course knowing for sure if there is a desire to take the attraction further does not take place usually until phase 4 in Givens' model.

In comparison to face-to-face, online daters can only guess and hope that they find the person they have meet online to be someone they are physically attracted to and in turn if that person reciprocates that attraction. Judging from a couple of photographs (which may well be out of date or developed in photoshop) is a tricky process and individuals can only hope that they will find the person physically attractive when they meet in the flesh. Moreover, research has also found that judgments are often skewed because people increase their expectations about the type of person who might be attracted to them (Whitty & Carr, 2006). So unlike off-line, where individuals initially find the other attractive and then move to get to know the person, for online daters it is more about knowing the person and then determining if there is any physical attraction. Spending the time interacting and then meeting can seem like a waste of time for individuals if they find there is not any physical chemistry when the pair meet face-to-face. Some online dating sites are encouraging users to also place videos of themselves on the site— this might go some way towards reducing the problem of delayed determination of physical attraction.

Spoiled for Choice

A unique aspect of online dating is that the site presents the client with a sea of possibilities—far more apparent choices than they are ever going to be presented with in most other situations. Every other person is there for the same reason: to find a romantic and/or sexual partner (typically a long-term romantic partner). The goal is not ambiguous, which as stated above has certain advantages and disadvantages. Being presented with so many options has been compared to being in a sweet shop (Whitty & Carr, 2006). However, the problem with this set up is that not all persons on the site are real potentials. The same type of person with the same type of physical attributes that one normally attracts off-line is realistically the only type of person who is going to reciprocate attraction. What online daters tend to do is to raise their expectations as to what they can realistically attract (Whitty & Carr, 2006).

The other problem with being presented with so many choices is that the online dater feels they can be hyper-critical of a person when they meet them face-to-face. Given they have a record of that person's claims if their date does not match up with their profile they can easily be discarded given that 'there are plenty more fish in the sea'.

Stigmatized Activity

A final problem worthy of note is that online dating has been found to be a stigmatized activity (Wildermuth, 2004). Peris et al. (2002), for instance, have argued that "it is generally assumed that people who enter cyberspace to form interpersonal relationships generally show greater difficulties in social face-to-face situations" (p. 44). Donn and Sherman (2002) found that undergraduate students were more likely than postgraduate students to believe that people who form relationships online were desperate. Wildermuth (cited in Wildermuth, 2004) found that even friends and family members of online daters expressed strong disapproval of their online activities. Not surprisingly, their views negatively impacted online daters' perceptions of the self. In the interviews carried out by Whitty (see Whitty & Carr, 2006; Whitty, 2007; in press), this stigmatization was also reported by online daters. To deal with this stigma, the online daters made friends on the dating site and created what they called a community. This community helped normalize their online dating activities.

This stigmatization also seems to be evident across some cultures. Malchow-Møller (2003), for instance, found that the French men she interviewed in a focus group stated that online dating is akin to personal ads and so like people who use personal ads, online daters must be desperate. She argues that Danes, in contrast, are more likely to perceive online daters as ordinary people. The problem with this study, however, is that the researcher only included two focus groups in her study and it would not be valid to assume that one focus group could represent Danes and the other French, so further research is essential to test these claims. What does seem evident is that in the main, participants talked about online daters as a separate social group (even if some perceive them to be 'normal' people).

A stigmatized person's social identity is devalued in a particular social context (Crocker, Major, & Steele, 1998). Goffman (1963) contended that a stigma discredits a person and reduces them "from a whole and usual person to a tainted, discounted one." Online dating is obviously not the typical way individuals find a mate, but then not all atypical behaviors are stigmatized. So what researchers need to consider is why has online dating been perceived as a stigma? Why are these individuals, akin to those who write personal ads (Ahuvia & Adelman, 1992), perceived by many as desperate? I suggest here that it is not because online dating is a fairly new activity (especially given its similarity with personal ads which have been around for decades) but rather because of the process—as demonstrated in the online dating model proposed in this chapter the courting process follows different steps to the model of dating that individuals are more accustomed to.

CONCLUSION

To conclude, by drawing upon Givens' model of the traditional courtship process, this chapter was able to develop a model to explain the online dating process. This model argued for five phases including: 'the attention phase', 'the recognition phase', 'the interaction phase', 'the face-to-face meeting' and finally 'resolution'. Here it was noted that flirting is not as natural or ambiguous as it can be in other types of courtship (both online and off-line). Moreover, unlike face-to-face attraction, physical chemistry is not determined until much later down the line. The differences in the courtship process provide users of these sites with certain advantages and disadvantages. It is suggested here that online dating companies might want to re-dress some of the problems with the design and usage of their sites in order to enhance their clients' online dating experience.

REFERENCES

Ahuvia, A., & Adelman, M. (1992). Formal intermediaries in the marriage market: A typology and review. *Journal of Marriage and the Family*, *54*(2), 452–463. doi:10.2307/353076

Alapack, R., Blichfeldt, M., & Elden, A. (2005). Flirting on the Internet and the hickey: A hermeneutic. *Cyberpsychology & Behavior*, *8*(1), 52–61. doi:10.1089/cpb.2005.8.52

Bargh, J., McKenna, K., & Fitzsimons, G. (2002). Can you see the real me? Activation and expression of the "true self" on the Internet. *The Journal of Social Issues*, *58*, 33–48. doi:10.1111/1540-4560.00247

Ben-Ze'ev, A. (2004). *Love online: Emotions and the Internet*. Cambridge, UK: Cambridge University.

Cooper, A., & Sportolari, L. (1997). Romance in cyberspace: Understanding online attraction. *Journal of Sex Education and Therapy*, *22*(1), 7–14.

Crocker, J., Major, B., & Steele, C. (1998). Social stigma. In: S. Fiske, D. Gilbert, & G. Lindzey (Eds.), *Handbook of social psychology, 2*, 504-553. Boston, MA: McGraw-Hill.

DeVoss, D. (2007). From the BBS to the Web: Tracing the spaces of online romance. In: M. Whitty, A. Baker, & J. Inman (Eds.), *Online matchmaking* (pp. 17-30). Houndmills: Palgrave Macmillan.

Donn, J., & Sherman, R. (2002). Attitudes and practices regarding the formation of romantic relationships on the Internet. *Cyberpsychology & Behavior*, *5*(2), 107–123. doi:10.1089/109493102753770499

Ellis, B., & Symons, D. (1990). Sex differences in sexual fantasy. *Journal of Sex Research*, *27*(4), 527–555.

Givens, D. (1978). The non-verbal basis of attraction: Flirtation, courtship, and seduction. *Psychiatry*, *41*, 346–359.

Goffman, E. (1963). *Stigma: Notes on the management of spoiled identity*. New York, NY: Prentice Hall.

Gwinnell, E. (1998). *Online seductions: Falling in love with strangers on the Internet*. New York, NY: Kodansha International.

Hancock, J., Toma, C., & Ellison, N. (2007). The truth about lying in online dating profiles. *Proceedings of the ACM Conference on Human Factors in Computing Systems (CHI 2007)*, (pp. 449-452).

Higgins, E. (1987). Self-discrepancy theory. *Psychological Review*, *94*, 1120–1134. doi:10.1037/0033-295X.94.3.319

Houran, J., Lange, R., Rentfrow, P., & Bruckner, K. (2004). Do online matchmaking tests work? An Assessment of preliminary evidence for a publicized 'predictive model of marital success'. *North American Journal of Psychology*, *6*, 507–526.

Kenrick, D., Sadalla, E., Groth, G., & Trost, M. (1990). Evolution, traits, and the stages of human courtship: Qualifying the parental investment model. *Journal of Personality*, *58*, 97–116. doi:10.1111/j.1467-6494.1990.tb00909.x

Lenhart, A., & Madden, M. (2007). Social networking Web sites and teens: An overview. *Pew Internet & American Life Project*. Retrieved August 28, 2007, from http://www.pewInternet.org/pdfs/PIP_SNS_Data_Memo_Jan_2007.pdf.

Malchow-Møller, A. (2003). Internet dating: A focus group investigation of young Danes' and Frenchmen's attitudes towards the phenomenon. *Kontur*, *7*, 11–20.

McKenna, K., Green, A., & Gleason, M. (2002). Relationship formation on the Internet: What's the big attraction? *The Journal of Social Issues*, *58*, 9–31. doi:10.1111/1540-4560.00246

Moore, M. (1985). Non-verbal courtship patterns in women: Context and consequences. *Ethology and Sociobiology*, *6*, 237–247. doi:10.1016/0162-3095(85)90016-0

Moore, M. (1998). Non-verbal courtship patterns in women: Rejection signalling: An empirical investigation. *Semiotica*, *118*, 201–214.

Parks, M., & Floyd, K. (1996). Making friends in cyberspace. *The Journal of Communication*, *46*, 80–97. doi:10.1111/j.1460-2466.1996.tb01462.x

Pasha, S. (2005, August 18). Online dating feeling less attractive. *CNN/Money*. Retrieved April 13, 2006, from http://money.cnn.com/ 2005/ 08/ 18/ technology/ online_dating/ index.htm.

Peris, R., Gimeno, M., Pinazo, D., Ortet, G., Carrero, V., Sanchiz, M., & Ibanez, I. (2002). Online chat rooms: Virtual spaces of interaction for socially-oriented people. *Cyberpsychology & Behavior*, *5*(1), 43–51. doi:10.1089/109493102753685872

Reviews, D. S. com. (2006). Retrieved November, 21, 2007, from http://www.datingsitesreviews.com/ staticpages/ index.php? page=2010000100 -FriendFinder

Rogers, C. (1951). *Client-centered therapy*. Boston, MA: Houghton-Mifflin.

Scharlott, B., & Christ, W. (1995). Overcoming relationship-initiation barriers: The impact of a computer-dating system on sex role, shyness, and appearance inhibitions. *Computers in Human Behavior*, *11*, 191–204. doi:10.1016/0747-5632(94)00028-G

Townsend, J. (1993). Sexuality and partner selection: Sex differences among college students. *Ethology and Sociobiology*, *14*, 305–330. doi:10.1016/0162-3095(93)90002-Y

Townsend, J., & Wasserman, T. (1997). The perception of sexual attractiveness: Sex differences in variability. *Archives of Sexual Behavior*, *26*, 243–268. doi:10.1023/A:1024570814293

Walther, J. (1996). Computer-mediated communication: Impersonal, interpersonal and hyperpersonal interaction. *Communication Research*, *23*, 3–43. doi:10.1177/009365096023001001

Walther, J., Slovacek, C., & Tidwell, L. (2001). Is a picture worth a thousand words? Photographic images in long-term and short-term computer-mediated communication. *Communication Research*, *28*, 105–134. doi:10.1177/009365001028001004

Whitty, M. (2002). Liar, Liar! An examination of how open, supportive and honest people are in chat rooms. *Computers in Human Behavior*, *18*(4), 343–352. doi:10.1016/S0747-5632(01)00059-0

Whitty, M. (2003). Cyber-flirting: Playing at love on the Internet. *Theory & Psychology*, *13*(3), 339–357. doi:10.1177/0959354303013003003

Whitty, M. (2004). Cyber-flirting: An examination of men's and women's flirting behaviour both off-line and on the Internet. *Behaviour Change*, *21*(2), 115–126. doi:10.1375/bech.21.2.115.55423

Whitty, M. (2007). The art of selling one's self on an online dating site: The BAR Approach. In: M. Whitty, A. Baker, & J. Inman (Eds.), *Online matchmaking* (pp. 57-69). Houndmills: Palgrave Macmillan.

Whitty, M. (in press). Revealing the 'real' me, searching for the 'actual' you: Presentations of self on an Internet dating site. *Computers in Human Behavior*.

Whitty, M., & Buchanan, T. (2007, manuscript under preparation). What's in a 'screen' name? The types of screen names online daters find attractive.

Whitty, M., & Carr, A. (2006). *Cyberspace romance: The psychology of online relationships*. Basingstoke: Palgrave Macmillan.

Whitty, M., & Gavin, J. (2001). Age/sex/location: Uncovering the social cues in the development of online relationships. *CyberPsychology & Behaviour, 4*, 623–630. doi:10.1089/109493101753235223

Wildermuth, S. (2004). The effects of stigmatizing discourse on the quality of online relationships. *Cyberpsychology & Behavior, 7*(1), 73–84. doi:10.1089/109493104322820147

Wysocki, D. (1998). Let your fingers to do the talking: Sex on an adult chat line. *Sexualities, 1*, 425–452. doi:10.1177/136346098001004003

Wysocki, D., & Thalken, J. (2007). Whips and chains? Fact or fiction? Content analysis of sadomasochism in Internet personal advertisements. In: M. Whitty, A. Baker, & J. Inman (Eds.), *Online matchmaking* (pp. 178-196). Houndmills: Palgrave Macmillan.

This work was previously published in Social Networking Communities and E-Dating Services: Concepts and Implications, edited by Celia Romm Livermore & Kristina Setzekorn, pp. 278-291, copyright 2009 by ISR Publishing (an imprint of IGI Global).

Chapter 14
How E-Daters Behave Online:
Theory and Empirical Observations

Celia Romm Livermore
Wayne State University, USA

Toni Somers
Wayne State University, USA

Kristina Setzekorn
Smith Barney, Inc., USA

Ashley Lynn-Grace King
Wayne State University, USA

ABSTRACT

Following a review of the literature on e-dating, this chapter introduces the e-dating development model and discusses a number of hypotheses that can be derived from it. Also presented in the chapter are some findings from a preliminary empirical research that explored the hypotheses. The findings supported all the hypotheses, indicating that: (1) male and female e-daters follow different stages in their e-dating evolvement; (2) the behaviors that males and females exhibit as e-daters are different; and (3) the feedback that male and female e-daters receive from the environment is different too. The chapter is concluded with a discussion of the implications from this research to e-dating theory development and empirical research.

INTRODUCTION

We define e-dating as a process that takes place online and that results in the establishment of a personal relationship between two individuals. E-dating can be enabled by an e-dating service, for example, an online company that matches individuals to each other (such as Match.com, YahooPersonals, eHarmony, etc.) or it can take place in other online environments, such as chatrooms, newsgroups, and so forth.

As for the goal of the relationship, we prefer not to use terms such as "romance," "flirting," or "courtship" because we acknowledge that the range of relationships that can be categorized as

DOI: 10.4018/978-1-60960-759-3.ch014

e-dating is wide and can encompass anything between cybersex and marriage.

The focus of this chapter is on the "e-dating career" which we define as the sum-total of all the behaviors and experiences that e-daters undergo while being e-daters. Just like a vocational career, the e-dating career may involve not just the behaviors initiated by the e-dater but also the responses that the e-dater receives from the environment. These may include positive, negative or no responses at all.

We prefer to use the term "career" to describe the e-dating process because this term denotes a long-term experience that can last months and even years. It also implies a series of discrete experiences (just like "jobs") that the individual can consider in retrospect as parts of the whole experience. The term "career" also implies overcoming hurdles, adjusting goals, and modifying strategies to fit new realities, for example, it implies, change, growth, and the development of new perceptions about the self—all of which we consider to be part and parcel of the e-dating experience.

Even though our emphasis in this chapter is on the subjective aspects of e-dating, it is important to consider some of the objective realities of this burgeoning new industry.

Online dating has been around for over two decades. It began in the early 1980s and has exploded into an extremely lucrative form of consumer-to-consumer sector of e-commerce. According to a study conducted by the Online Publishers Association (OPA) and comScore Networks, "U.S. residents spent $469.5 million on online dating and personals in 2004, the largest segment of 'paid content' on the web" (Consumer Search, 2005).

The e-dating industry is dominated by several large companies including Match.com, Yahoo Personals, and American Singles. They are followed by a multitude of other sites offering services for all aspects and preferences for dating, for example, JDate.com (which advertises itself as "the larges Jewish singles network"), ChristianSingles.com, BlackSinglesConnection.com, Gay.com., and so forth. According to TrueDating.com (TrueDating, 2006), Match.com is the world's leading online dating Web site, while Yahoo Personals is a close second, both boasting over 9 million members. The main factors contributing to the success of this industry include the relatively low start-up costs associated with running an online dating service and people's willingness to buy love online.

Even though e-dating is a dominant sector in e-commerce, very limited empirical research is available about it in the scientific literature. Our goal in this chapter is to propose a theory of e-dating development and to present the results of an empirical investigation that explored our theory.

The underlying assumption of our e-dating theory is that e-dating is a stage process in which individuals move through a number of steps or stages. Each stage is characterized by a different set of behaviors, with individuals moving from one stage to another as their e-dating "career" unfolds.

While the steps in the e-dating theory are assumed to be essentially the same for males and females, the two genders are assumed to follow a different sequence of these stages. Thus, the model assumes that because of cultural and environmental reasons, males tend to initiate contact earlier in the e-dating process than females do. Also, the theory assumes that relative to females, males are more active in behaviors that are associated with establishment and maintenance of contact.

In the following sections, we discuss the major theories and empirical research that relate to e-dating and how our theory fits in with the existing research. We proceed to present the e-dating theory, the hypotheses that we developed, and the empirical research that we conducted to explore the hypotheses. We conclude with a discussion of the implications from this research for future exploration of the theory and practice of e-dating.

LITERATURE ON DATING

Dating, or "adult romantic attachment" as this phenomenon is often termed in the scientific literature, is an interdisciplinary area of research. Some of the major contributions to the development of the body of theory in this area came from biology, sociology, economics, communication, and psychology.

In the following sections, we present the major questions that each of these disciplines raised in relation to the meaning of adult romantic attachment and some of the answers that it provided to these questions.

The major contribution of biology is in addressing the question of *why* people are attracted to one another and the reasons for the *strength* of this attraction. Thus, the biological literature on monogamy (Gubernick, 1994) suggests that people (and other animal species) are attracted to monogamous relationships because they solve the problem of paternity certainty. Since ovulation is concealed in women, men reduce the risk that their offspring may not be theirs by establishing and maintaining monogamous (romantic) relationships with women. Another explanation that follows this line of reasoning is that monogamous relationships offer protection for immature offspring, particularly given the relatively long time that it takes humans to mature sexually relative to other primates.

Applied to our research, the major contribution of the biological literature is in confirming the importance of adult romantic attraction because of its relationship to our survival as a species.

The major contribution of sociology is in addressing the question of *who* people tend to be romantically attached to. As indicated by Rosenfeld (2005), the literature on mate selection in sociology since the 1940s has been characterized by an interesting paradox. On one hand, the major theorists Merton (1941) and Davis (1941) promoted the "status-caste exchange" arguments, which perceived marriage as an exchange between unequals (men with money marry women of beauty), but on the other hand, there was little empirical evidence to promote these arguments.

Thus, contrary to this theory, the bulk of the research in the sociological literature in the past 70 years has actually demonstrated that people find mates who are similar to themselves in status, class, and education (Mare, 1991; Kalmijn, 1998); religion (Kennedy, 1952; Johnson, 1980; Kalmijn, 1991); and race (Heer, 1974; Lieberson & Waters, 1988; Kalmijn, 1993; Qian, 1997). In other words, married partners tend to be the same on every dimension except gender.

Applied to our research, the major contribution of the sociological literature is in establishing the principle that romantic attachment is rational and that even when men and women are convinced that they simply "fell in love" their actions seem to follow very predicable and rational patterns that suggest a process of rational decision making rather than pure emotion.

The major contribution of economics is in addressing the question of *what* considerations affect people's choice of specific partners. The central concept in the economic treatment of this issue is the concept of "marriage market," where men and women (agents) are assumed to behave strategically to maximize their outcomes from a relationship. The seminal work in this area by Becker (1974) focused on analyzing the manner in which men and women "sort" each other along characteristics such as income, education, and so forth. The underlying assumption of Becker's work was that each partner's "value" is based on the sum of his or her attributes. In order for an equilibrium or a "match" to take place, the two parties have to be convinced that the sum of their attributes or their "value" is the same (or similar) as that of their partner.

Applied to our research, the major contribution of the economic literature is in reiterating the principle that romantic attachment is a rational process of weighting alternatives and considering choices. Obviously, if this is indeed the case, then it stands to reason that this process would entail strategy, growth and change, which are central to our e-dating model.

The major contribution of psychology is in addressing the question of *how* people establish romantic attachments. One of the major frameworks for the study of romantic relationships in psychology is the adult attachment theory that was originally proposed by Hazan and Shaver in the 1980s (Hazan & Shaver, 1987). The theory is based on an earlier theory designed to explain the emotional bond between infants and their caregivers (Bowlby, 1980). According to Fraley and Shaver, (2000, pp. 134-135) the Hazan and Shaver adult romantic attachment theory is based on the following four propositions:

1. The emotional and behavioral dynamics of infant caregiver relationships and adult romantic relationships are governed by the same biological system. In both cases the motivation for attachment is to promote safety, comfort, and survival. Thus, adults typically feel safer when their partner is nearby, accessible and responsive.
2. The kind of individual differences observed in infant-caregiver relationships are similar to the ones observed in romantic relationships. Following this principle, Hazan and Shaver described three styles of adult romantic attachment: (a) secure, (b) anxious/ambivalent, and (c) avoidant.
3. Individual differences in adult attachment behavior are reflections of the expectations and beliefs people have formed about themselves and their close relationships on the basis of their attachment histories; these "working models" are relatively stable, and, as such, may be reflections of care giving experiences.
4. Romantic love, involves the interplay of attachment, care giving and sex. These three separate systems serve different functions but together they reflect the same early experiences and attachment relationships that the individual had as an infant.

The Hazan and Shaver (1987) adult romantic attachment theory generated a stream of empirical studies, including research on the impact of working models on people's perceptions of their partners' intentions (Collins, 1996), the effect of working models on partner choice (Pietromonaco & Carnelley, 1994), relationship stability (Kirkpatrick & Davis, 1994), and relationship dissolution (Pistole, 1995).

Over the years, a number of criticisms of the theory led to a re-thinking and modification of some of its assumptions, including (1) the suggestion that not all romantic or couple relationships are attachment relationships and that some pair bonding relationships are actually not romantic in nature; (2) the original *three* "working models" that people can apply to relationships, can be better conceptualized as *four* "working models" (secure, preoccupied, fearful-avoidant, and dismissing-avoidant); and (3) the stability of the "working models" may not always apply, as some people seem to "overwrite" their early attachment models with new ones that are based on later attachment experiences (Farley & Shaver, 2000).

The Hazan and Shaver theory (1987) is the major basis on which our e-dating development model is based. Specifically, we accept the principle that "working models" (possibly ones originating from early childhood) affect the manner in which adults establish romantic attachments. Also, given that these "working models" are likely to be different for males and females because of their different socialization, it stands to reason that the behavior of adult male and female e-daters will be different too. Secondly, we accept the principle, proposed by the later versions of the theory (Farley & Shaver, 2000) that "working models" are subject to learning and change and that adult males and females are likely to overwrite them in response to changes in their external environment. Indeed, our conceptualization of e-dating as a change process in which coping mechanisms are adopted to deal with new conditions and circumstances can be seen as an example of the "re-writing" of "working models," as it denotes the constant evolvement of coping mechanisms throughout the e-dating career.

Building on the early literature on attachment and dating, a number of authors have attempted to explain the e-dating process. In the following sections, we consider some of the emerging research in this area and its implications to our e-dating model.

Prescriptive Literature on E-Dating

Much of the early literature on e-dating (Rose, 1999; Edgar, Jr., & Edgar, II, 2003; Greenwald, 2003; Silverstein & Lasky, 2004; Orr, 2004; Berry, 2005; Culbreth, 2005, and so forth) can be described as "prescriptive" because of its focus on instructing readers on how to conduct their e-dating careers effectively.

Still, given that not much formal research is available in the area of e-dating, these works can be credited for a number of important contributions, including: describing the various tasks that need to be undertaken by e-daters (Rose, 1999), offering advice on the technological features of the e-dating services (Silverstein & Lasky, 2004), highlighting the potential dangers of e-dating and explaining how one can protect one-self from the dangers that e-dating entails (Orr, 2004; Berry, 2005). Furthermore, the prescriptive literature can be credited with explaining the strategies that work best for certain "specialized" segments of the e-dating population, such as men (Edgar, Jr., & Edgar, II, 2003), women (Rose, 1999; Greenwald, 2003), or baby boomers (Culbreth, 2005).

Even though these publications have not followed formal research methodology, the insights that can be derived from them, particularly on the behaviors that e-daters typically adopt, are important and have influenced our initial formulation of the stages in our e-dating theory.

Thus, without exception the prescriptive publications agree that e-daters (irrespective of gender) start their e-dating career by creating an online profile. They also agree that this is followed by "search" activities that involve consideration of other e-daters' profiles, and that eventually, e-daters initiate or respond to contacts from other e-daters. The above sources are also in agreement that while all e-dating contacts start in the online environment, eventually some e-dating relationships evolve to telephone and/or face-to-face contact. All of these assumptions are central to our e-dating theory.

Descriptive and Predictive Literature on E-Dating

More recent publications have attempted to go beyond the prescriptive by offering a more scientific methodology for data collection and analysis of e-dating research, with some attempts to offer theoretical models that describe, explain, interpret, and predict e-dating behaviors. The following discussion presents three of the most important approaches to the scientific study of e-dating.

Psychological/Philosophical

One of the most interesting attempts to interpret the nature of e-dating has been proposed by Aaron Ben-Ze'ev (2004). Based on interviews with cyberspace daters, this pioneering study explored the differences between "virtual" and "real" relationships. Some of the features that are listed by Ben-Ze'ev as distinguishing the virtual from the real spaces are:

1. The virtual space has a "seductive" nature to it that makes virtual relationships sometimes more intense and emotionally satisfying than real relationships.
2. The virtual space is more egalitarian in that it allows people to interact with each other irrespective of their demographic, physical

or emotional attributes. Indeed, the virtual space is unique in that it offers its inhabitants the opportunity of "inventing" themselves.
3. Because of its limitless possibilities for self-presentation, the virtual space enables more deceit. Indeed, a very high percentage of Internet "lovers," according to this research, mis-represent (lie) about their attributes, feelings and behavior.
4. The virtual space is addictive and because of this (and other reasons) can be dangerous. Online daters may develop strong emotions toward partners in this environment that can eventually affect their real relationships.
5. Online relationships are incomplete and for this reason, to be complete, they need to migrate from the online to the face-to-face environment or from the virtual to the real space.

The Ben-Ze'ev research focuses on cyber-sex rather than on cyber-love. As such, the applicability of its findings to our e-dating theory is limited. Still, its emphasis on the differences between the two spaces and the need to migrate from one to the other in order for a relationship to become real are important assumptions on which our theory of e-dating development is based.

Economics

One of the very first investigations of e-dating from an economic perspective is the article "What makes you click: An empirical analysis of online dating (Hitsch, Hortacsu, & Ariely, 2005). The data set for this research recorded the activities of 23,000 users in the Boston and San Diego area during a 3 ½-month period in 2003.

The data set included self-reported information about users' age, income, education level, ethnicity, political inclinations, marital status, and so forth. The users also posted pictures of themselves on the site, which the researchers rated and ranked (using objective judges) for "attractiveness." The attractiveness rankings together with self-reported information about users' height, weight, and other physical characteristics enabled the researchers to create a measure of physical attractiveness that went beyond just the picture.

The most important aspect of this research was that the researchers were provided with information about the activities of the e-daters. Thus, as the authors note (Hitsch, Hortacsu, & Ariely, 2005), "At each moment in time, we know which profile they browse, whether they view a specific photograph, and whether they send or reply to a letter from another user. We also have some limited information on the contents of the e-mails exchanged: in particular, we know whether the users exchanged phone numbers or e-mail addresses" (p. 3).

The researchers based their model on exchange theory, with the objective being to assess the impact of some variables (notably, physical attractiveness, but also level of education, income, etc.) on e-daters "success," as measured by the likelihood of their e-mails being responded to and/or by other e-daters agreeing to exchange e-mail address and telephone number with them.

The findings from this research confirmed a number of assumptions that are central to our e-dating model:

1. The percentage of males in the e-dating site (55.5%) was higher than in the general population in both locations (49%) and also higher than the percentage of males among Internet users (also 49%). Other than this demographic difference, all other characteristics of the male and female population of daters were similar to those of the general population and of Internet users in the two cities. Compared to the general population in the two cities, the authors reported that the e-dating users were somewhat younger (25-35), more educated and of higher income than the general population.

2. As a rule, the stated preferences of both males and females seemed to confirm the findings from the sociological literature reported on in the previous sections that people prefer to date others who are similar to them on all demographic variables. The authors report that this was true for education, income and race, particularly for women.
3. The percentages of people who state that their appearance is "above average" (very high) and those reporting that their appearance is below average (very low) suggests that users "inflate" their reports. This same phenomenon is also evident in regards to weight (much lower than the average for women in the general population) and height (much taller than the average for men in the general population), suggesting that men tend to inflate their height and women tend to deflate their weight in their e-dating profiles.
4. Males and females behaved differently online. Men were more likely to browse women's profiles than women were (searching behavior). Men were also more likely to send e-mail messages to women than women were, particularly when the men stated that they were interested in a serious relationship. Indeed, this category of men sent more e-mail messages than any other category of e-daters.
5. Males and females were different in the extent to which their actions produced results. Thus, the likelihood of a male to receive a response to an e-mail message was 40%, while the likelihood of a female (irrespective of looks or any other personal attributes) to receive a response to her e-mail message was 70%. The researchers found that attractiveness makes a difference here, with the least attractive women being 2-4 times more likely to send a first contact e-mail message to a man than the most attractive women. The same difference in selectiveness was also evident in the response rate.
6. The final outcomes for daters were that overall, women were browsed more often, received more first contact e-mails and e-mail containing a phone number and/or e-mail address than men. Thus, while men received on average 2.6 first contact e-mails, women received on average 12.6 e-mails; 54% of all men in the sample NEVER received a first contact e-mail at all, whereas only 19% of all women were never approached by e-mail.
7. As for the impact of self-reported attributes on the likelihood of receiving e-mail messages, it was most pronounced for underweight women (they received 77% more e-mail messages than women who reported that they were overweight) and "blondes." For men, the "penalty" for overweight was less severe. In contrast, stated income and "shortness" had a strong impact on men's e-dating "success" while they had a marginal impact on women's "success."

Sociology

One of the most seminal sociological investigations of e-dating resulted in the book, "Double click: Romance and commitment among online couples" by Andrea Baker (2005). The uniqueness of this publication is that it is not just empirical (based on a scientifically designed survey of 89 couples that met online) but also predictive. The major goal of this research was to use survey and interview data to discover the keys for success in e-dating. Based on this research, the author outlined the POST model, a four factor model that purports to predict success in e-dating. The four factors include:

1. **Place** (where the e-daters met on- and offline): This research defined quite a number of different "places" for e-daters to meet. The findings indicated that the more specifically

related to relationship building the meeting place was (such as a chat room dedicated to a topic that is of interest to both parties or an e-dating service) the more successful the relationship that resulted from meeting there was likely to be.
2. **Obstacles** (the number and types of obstacles that the e-daters faced and had to overcome): The researcher defined two types of obstacles. Those that resulted from pre-existing relationships (marriage, cohabitation of one or more of the parties) and those that resulted from geographical distance. The findings indicated that more such obstacles existed at the beginning of the relationship the less likely was the relationship to be stable and long-term.
3. **Self-presentation** (self disclosure versus secrecy, deception versus truth, and appearances versus truth): The findings from the above research indicated that self-disclosure, honesty and appearance that was congruent with the other party's expectations were associated with the relationship's positive outcome. Holding back information about one's self, particularly once the online relationship moves on the face-to-face stage, pretending to be someone other than who one really is and providing information about one's appearance that proved to be untrue once the e-daters met face-to-face were all predictors of failure of the e-dating relationship.
4. **Timing:** The findings from this research were that the longer the e-daters waited to meet each other face-to-face and the longer they waited to initiate a sexual relationship, the more likely they were to establish a successful relationship. Indeed, cybersex was one of the best predictors of an e-dating relationship's failure. These authors interpret this finding to suggest that starting a sexual relationship too early on- or off-line hinders the development of other areas of shared interest and could potentially subdue other areas of potential conflict a couple encounters, resulting in the couple not knowing if they are truly compatible with each other.

The above study has a number of direct implications to our research. First, it confirms the fact that daters engage in different types of behavior throughout their e-dating career. It proposes the idea that some e-dating strategies are more successful than others and that successful male and female e-daters tend to use different strategies. This research also suggests through its in-depth interviewing with e-daters that strategies are not constant—they may change throughout the e-dating career, as e-daters become more aware of the constraints of the e-dating environment.

The strength of the Baker (2005) study is that unlike most of its predecessors, it was based on primary sources—the reports of e-daters in response to an open-ended questionnaire administered to them. However, as noted by the author, this methodology also had its weaknesses. E-daters who participated in this study reported on events that happened weeks and sometimes months before the survey was administered to them. This introspective methodology could have introduced distortions into the data that a more "real-time" approach, like the one employed in our study, does not have. Also, because the e-daters reports were made after the fact, it was difficult to use the data to describe the actual developmental process of e-dating, which is the main focus of our research.

TOWARD A THEORY OF E-DATING DEVELOPMENT

As indicated in the previous sections and based on the literature that we reviewed, our theory of e-dating development is based on the following premises:

1. The typical e-dating career consists of six stages or steps: (a) *Construction* of a profile; (b) *Searching* for appropriate matches; (c) *Sending* winks, e-mail messages, text messages, and so forth; (d) *Responding* to winks, e-mail messages, text messages, and so forth; (e) *Setting* up face-to-face dates; (f) *Conducting* face-to-face dates; and (g) *Concluding* the process by either starting a new cycle of the e-dating process, establishing a relationship with one date, or quitting the process all together without finding a desirable match.
2. Because of primarily cultural reasons, even though the steps in the e-dating process are identical for males and females, the *sequence* of these stages and the amount of time and energy that each gender spends on some of the stages differs. Thus, because males are expected to be the initiators in the dating game, they tend to start "searching" activities earlier and spend more time and energy on these activities than females do. Similarly, males also tend to engage in contact initiation activities (sending winks and e-mails) earlier in the e-dating process and they invest more time and energy in these activities throughout their e-dating career than females do.
3. Because males are more numerous on e-dating services than females are (and possibly because females are less inclined to initiate contact with males), the males' attempts to establish contact with females are less likely to be successful than the females' attempts to establish contact with males.
4. This reality, results in males *increasing* their investment of time and effort in initiating contact over time more than females do. Thus, while the two genders might start the e-dating process with similar behaviors and expectations, as the process unfolds and as a result of input from the environment (lack of response from females), males increase their contact initiation activities, while females decrease their engagement in such activities.
5. The end result from this process is that males are less successful (and possibly less satisfied) with their e-dating experiences relative to females.

Table 1 shows the typical e-dating development stages for each gender. Please note that the six stages can conclude after one "round" or repeat for a longer period of time, with new rounds initiated repeatedly. Also, the theory assumes that while some males might behave initially "like females" (initiating few contacts with females and expecting females to contact them) and some females might initially behave "like males" (initiating contact with males), as the e-dating process unfolds and because of the different input from the environment that males and females get, both genders "converge" into the "typical" sequence proposed by our theory.

Table 1. The typical female and male e-dating career

Females	Males
A. **Constructing** or revising a profile	A. **Constructing** or revising a profile
B. **Searching** for appropriate matches	B. **Searching** for appropriate matches
D. **Receiving** winks, e-mail messages or instant messages	C. **Sending** (and eventually receiving) winks, e-mail messages, or instant messaging
E. **Setting** up face-to-face dates (usually by telephone)	E. **Setting** up face-to-face dates (usually by telephone)
F. **Conducting** dates	F. **Conducting** dates
G. **Concluding** the process by either starting a new cycle, establishing a committed relationship or quitting all together	G. **Concluding** the process by either starting a new cycle, establishing a committed relationship or quitting all together

Figure 1. Type of activity engaged in by gender.

Males (n=10)	Week 1	Week 2	Week 3	Week 4	Week 5
	Constructing or revising a profile (70%)	Conducting a search for matches (50%)	Conducting a search for matches (50%)	Initiating contact with matches (57%)	Responding to contact from others (57%)
	Conducting a search for matches (20%)		Constructing or revising a profile and Responding to contact from others (40%)	Conducting a search for matches (28%)	
	Initiating contact with matches (10%)				

	Week 6	Week 7	Week 8	Week 9	Week 10	Week 11
	Initiating contact with matches (43%)	Responding to contact from others (33%)	Initiating contact with matches (33%)	Initiating contact with matches (33%)	Setting up dates (100%)	Responding to contact from others (100%)
	Responding to contact from others (43%)	Initiating contact with matches (33%)	Meeting dates (25%)	Setting up a dates (33%)		
		Setting up dates (33%)	Setting up dates (33%)	Meeting dates (25%)		

Females (n=12)	Week 1	Week 2	Week 3	Week 4	Week 5
	Constructing or revising a profile (100%)	Constructing or revising a profile (50%)	Constructing or revising a profile (50%)	Conducting a search for matches (45%)	Responding to contact from others (82%)
			Responding to contact from others (25%)	Responding to contact from others (36%)	

	Week 6	Week 7	Week 8	Week 9	Week 10	Week 11
	Conducting a search for matches (36%)	Responding to contact from others (80%)	Responding to contact from others (63%)	Responding to contact from others (67%)	Responding to contact from others (60%)	Setting up dates (50%)
	Responding to contact from others (27%)	Conducting a search for matches (20%)	Conducting a search for matches (25%)	Setting up dates (16%)	Setting up dates (20%)	Meeting dates (25%)
				Meeting dates (16%)	Meeting dates (20%)	Responding to contact from others (25%)

Based on the above assumptions, we have outlined below a number of research questions that we explored empirically. For our initial investigation, we decided to restrict our research to the very first three assumptions in the model, namely, (1) the e-dating process consists of stages and that these differ for males and females; (2) the behavior of males and females throughout the e-dating process is different, and (3) the input that male and female e-daters receive from the environment (namely, from other e-daters) is different. (Figure 1)

RESEARCH QUESTIONS

First Research Question

- **Does the e-dating process consist of distinct steps that are characterized by different behaviors?**

As indicated in the previous sections, the steps or stages in the e-dating process include: (1) constructing a profile, (2) searching, (3) initiating communication, (4) receiving communication, (5) setting face-to-face dates, (6) conducting dates, and (7) concluding the e-dating process.

- **Do males and females follow a different sequence of steps in the e-dating process?**

As mentioned in the previous sections, our model expects males to start the searching behaviors earlier and engage in more activities that involve establishment of contact with others than females do. Females, on the other hand, are expected to start the search behavior later and engage in it less frequently. Females are also expected to initiate less contact and respond to more contact from males.

Second Research Question

- Do male and female e-daters behave differently throughout the e-dating process?

In this preliminary research, we analyzed differences between male and female e-daters in a number of areas including: the intensity of search behavior, including the number of profiles searched, the establishment of a favorites list and the number of profiles included in it. We also explored the number of "winks," e-mail messages and text messages sent, the number of rejection messages sent, and the number of individuals barred from having contact with our respondents.

Our expectations were that while males will engage in more search and contact initiation behaviors (such as sending "winks," e-mail messages and text messages), females were expected to engage in more contact-limiting-behaviors such as sending rejection messages and barring other e-daters from having access to their profile. As for the setting and holding of dates, which are also active e-dating behaviors, our expectation was that since these require the cooperation of two parties, females will engage in more of these activities than males.

Third Research Question

- Do male and female e-daters receive different inputs from the environment to their e-dating behavior?

Our expectations were that male and female e-daters do indeed receive different input from the environment, with males getting fewer "winks," e-mail messages and text messages than female do. Also, as indicated in relation to the previous research question, because setting and holding dates is an activity that depends on the behavior of the other party (who needs to agree to the date), we expected males to have fewer dates than females.

EMPIRICAL INVESTIGATION

Sample and Instrument

Based on the literature in the previous sections, we constructed an inventory of e-dating activities that followed the six stages in our e-dating development theory (see Appendix 1 for the research inventory). We administered the inventory to a group of 22 volunteers from two undergraduate classes in a private Midwestern university in the U.S. during the first semester of the 2006-2007 academic year. The average age of the participants was 20.4 and the number of males and females was about equal (12 females and 10 males). All participants were white and the majority were of upper middle class background.

Data Collection Process

The participants in the study were not paid for their participation. Instead, they received a free three-month subscription to Match.com (paid by the researchers), the largest e-dating service in the industry. In addition, they received credit for one assignment in the two courses that they were doing with one of the investigators. Participants were informed that participation was not compulsory and if they chose not to participate they were provided with an alternative assignment to do for the course.

Participants were also informed that because of the small sample and the need to keep participants as "similar to each other as possible," only heterosexual students that were not in a committed relationship were invited to participate. Also, even though we did not require participants to be e-dating novices, all participants indicated that they had no previous e-dating experience prior to joining the study.

The participants were assured by the researchers that the data that they volunteered was only going to be analyzed in aggregate. No personal data about participants was to be made public.

They did not receive information about our e-dating theory, but they were told that our major focus was their behavior, thoughts and feelings throughout the e-dating experience.

To establish a baseline of knowledge on e-dating for all participants, the researchers subscribed the participants to Match.com. This was followed with a two-hour lab tutorial on the various features of the Match.com service, including detailed information on how to create a profile, conduct and save searches, initiate contact and respond to contacts from others, and how to manage a personal Match.com account.

The management features of the account, which included a tally of: messages sent, messages received, winks sent, winks received, members rejected, members barred, members entered into favorites list, and so forth, were particularly important for our study because the participants were expected to report this information throughout the study.

Participants received 12 copies of the inventory and were asked to fill up and submit one copy per week, starting the third week of the semester. The majority of the participants submitted between 8 and 12 completed inventories. Even though participants were told that they could exit the study at any time and without an explanation, not one of the participants did so.

Findings

To address the above research questions, we conducted a number of analyses.

First Research Question

To address the first research question which addressed the issue of whether stages could be discerned in the activities of e-daters and whether these stages were different for males and females, descriptive statistics was used to tabulate the activities of the respondents each week. Only activities that the respondents referred to as their "main" activity for the week were tabulated.

Table 1 shows the distribution of the respondents' main activity each week by gender. As demonstrated in the table, there were observable differences in the activities that males and females engaged in each week during the test period.

Thus, for week 1 the main activity engaged in by the majority of males (70%) and females (100%) was to construct or revise their Match.com profile. In week 2 and 3, half (50%) of the females were still constructing or revising their profile while half (50%) of the males had moved on and were conducting a search for matches.

Interestingly, in week 4, 57% of the males were initiating contact with others, an activity that a few of the females (19%) started at the ninth week, and by the eleventh week only half (50%) of the females were initiating contact.

During week 5, most men (and women) were back to responding to contacts. In weeks 6 through 9, men were continuing to initiate contact with matches, an activity that women never engaged in over the eleven week period. Indeed, most women in weeks 7 through 10 were engaged in responding to contact from others.

To complete our analysis for the first research question, we used a chi square to explore whether gender (male vs. female) was related to the main activity e-daters engaged in each week. Even though differences were observed in the percentages, only weeks 3 and 8 were significant at the 0.05 level, and weeks 4, 9, and 10 were significant at the 0.10 level.

Second Research Question

To address the second research question which focused on e-daters' initiating contact with other e-daters, we examined e-daters' "active behaviors" such as the number of winks sent, e-mails sent, instant messages sent, rejection messages sent, and barring of other e-daters from communicating with the e-dater.

Table 2. Differences in e-daters active behaviors

Active Behaviors	Averages for Males and Females Over The Study Duration (12 Weeks)	T-test statistics (P-value) (Equal variances not assumed)
The number of winks sent	Females = 0.208 Males = 9.57	T= -11.42 (p=0.000)
The number of e-mails sent	Females = 0.207 Males = 9.56	T= -11.41 (p=0.000)
The number of instant messages (IMs) sent	Females = 1.43 Males = 7.87	T= -7.96 (p=0.000)
The number of rejection messages sent	Females = 0.495 Males = 1.59	T= -4.89 (p=0.000)
The number of people barred from accessing one's profile	Females = 2.89 Males = 0.000	T= 6.429 (p=0.000)
The number of dates held	Females = 1.16 Males = 0.101	T= 8.92 (p=0.000)

Using a t-test for independent samples analysis, we compared the active behaviors of males and females across all twelve weeks of the study period.

The results of this analysis are presented in Table 2. As indicated in the table, the differences between male and female e-daters were significant for all active behaviors. Thus, females were sending significantly fewer winks, e-mails, instant messages, and rejection messages. However, females barred more people from accessing their profile than did males. Paradoxically (but in line with the e-dating theory), the last "active" behavior shown in Table 2, the number of dates held over the twelve week study period, was significantly higher for females than males. This is a paradoxical finding because females were less active on every one of the behaviors in this category.

Third Research Question

To address the third research question which focused on the input that e-daters received from the environment, we examined the input that the e-daters received from others, such as the number of "hits" on their Match.com profile, the number of winks made to them, the number of e-mail messages sent to them, and the number of instant messages sent to them.

Using a t-test for independent samples analysis, we compared the behaviors of males and females. Table 3 shows these differences. As indicated in Table 3, the male and female e-daters differed significantly ($p < 0.05$) on three of the four behaviors in the table. Thus, females received significantly more hits, winks and e-mail messages. Also, although no significant difference was found between male and female receiving text messages, females still received instant messages more frequently than males.

DISCUSSION AND CONCLUSION

As indicated in the previous sections, our goal was to describe a theory of e-dating development and to explore it empirically. The findings from our preliminary research suggest that:

1. Six distinct stages can be identified empirically in the e-dating career of male and female e-daters. These, as our theory predicted, include: (a) constructing a profile, (b) searching, (c) initiating communication, (d)

Table 3. Differences in responses from the environment to d-daters' behaviors.

Input from the Environment	Averages for Males and Females Over The Study Duration (12 Weeks)	T-test statistics (P-value) (Equal variances not assumed)
The number of visits ("hits") received to their profile	Females = 157.03 Males = 15.98	T= 7.43 (p=0.000)
The number of winks sent to them	Females = 71.44 Males = 6.63	T= 7.93 (p=0.000)
The number of e-mails sent to them	Females = 23.43 Males = 5.43	T= 7.07 (p=0.000)
The number of instant messages (IMs) sent to them	Females =0.376 Males = 0.130	T= 1.89 (p=0.061)

receiving communication, (e) setting face to face dates, (f) conducting dates, and (g) concluding the e-dating process.

2. The sequence of stages for males and females is not the same, with males initiating contact with females earlier and investing more time and energy in initiating contact activities throughout the e-dating process than females do.
3. Even though some males and females exhibit behaviors that are typical of the other gender (some males establish a profile and wait for females to contact them and some females start contacting males right away instead of waiting for males to contact them), both genders tend to exhibit the behavior pattern outlined in our theory; namely, males tend to initiate more contact and females tend to initiate less.
4. The inputs that each gender receives from the environment are different and can possibly explain the differences in their behavior initially and over time. Thus, males tend to get fewer approaches from females across all "passive" behavioral categories (e.g., they receive fewer hits on their profile, winks, e-mail messages and text messages).

Obviously, this exploratory research can be extended in a number of different directions:

First, our e-dating theory proposed a number of hypotheses that we did not test in the preliminary investigation that we are reporting on in this chapter. Future investigations may extend the scope of this research to include questions like: Do the differences between male and female behavior increase over time? Does the input that the two genders receive from the environment change over time? Can the changes in behavior exhibited by e-daters be explained by changes in expectations? Obviously, answering each of the above questions in the affirmative would lend more support to the model that we proposed here. Thus, if the differences between the behavior of male and female e-daters increase over time, it would strengthen our contention that e-dating is, indeed, a developmental process in which behaviors change in response to changes in perceptions.

Second, our sample of undergraduate, white, upper middle class, heterosexual students can be extended to include participants of different ages, races, socio-economic classes, and social orientations. Furthermore, the number of participants in each of the above categories can be increased to allow for more generalizable conclusions.

Third, in this exploratory study, we used an open-ended questionnaire to elicit as much information about a range of topics related to e-dating. Future research might triangulate the research methodology by employing, on one hand, more quantitative measures of some of the variables that we identified here, and on the other, in-depth interviewing to identify additional variables that our inventory did not fully account for.

Fourth, if indeed e-dating services can be seen as different environments for male and female daters, this may have far-reaching implications. Thus, future research might explore the extent to which different e-dating services cater to the unique dating needs of males and females, by categorizing e-dating services as male or female "friendly."

Following this line of reasoning, male friendly services can be expected to create an environment that is friendlier to males by establishing rewards for females for responding to male contacts. An e-dating service that is already doing it is eHarmony. One of the unique features of the eHarmony service is that it correlates the number of "good matches" that an e-dater receives from the service with whether the e-dater responded to contacts from previously provided matches. This system is applied equally to males and females. Since females are less inclined to initiate contact and/or to respond to male contact than males are, this principle of rewarding e-daters for responding to contacts helps males more than it helps females.

Similarly, a female friendly service would establish an environment that is even friendlier to females than current services are by providing females with additional input on prospective male matches, such as through a ranking of all e-daters. Again, even though ranking would be applied equally to both males and females, since females are more often in a position to screen a large number of prospective matches, they would benefit more from ranking than males will. If a service offers this feature to its customers, the result would be a more attractive environment for females than for males.

Another implication from our research has to do with perceptions. If indeed the reality of e-dating for males and females is so different, are both genders aware of it? Does this awareness affect male and female inclination to use e-dating services? Is this awareness affected by age, social status, marital status or culture? Future research into e-daters perceptions might explore whether they can be modified. For example, if older females were aware that in their age group the ratio of males to females is even greater than in the younger age groups, would they be more inclined to become e-daters than they currently are? Will this awareness remove the stigma that is currently associated with e-dating, particularly in the older age groups?

Another related issue has to do with the strategies that e-daters employ to become successful in this process. Our study indicated that males and females tend to use different strategies. An interesting set of questions that relates to this finding is whether the strategies that each gender uses are affected by individual differences. For example, can a male who is particularly attractive (physically or thanks to a higher level of education or income) use more feminine strategies (initiate less contact) and still be successful? Are e-daters aware of how their individual attributes affect their strategies and their success rate? Do they modify their strategies based on this awareness?

Another related issue has to do with education. Given that e-dating is becoming ubiquitous, should the educational system invest resources in training young people to be efficient e-daters? Should males and females get *different* training on how to conduct themselves on an e-dating Web site given that the environment in which they operate is not the same? Should the educational system consider e-dating skills as important for young adults as driving, cooking, and other "survival skills"?

Another set of questions that the discussion in the previous section raises is related to the exposure that e-dating may involve and the risks that are associated with this exposure. Recent popular publications discuss the fact that the exposure of e-daters might lead (particularly in the case of female e-daters) to risks of violence (Loviglio, 2007; Moraski, 2007). The popular literature is replete with prescriptions on how to conduct "safe" e-dating.

Some e-dating services (e.g., True.com) have based their business model on providing their customers with background checks on other e-daters' criminal records and matrimonial status, lobbying for such searches to be enforced by law on all services or for services that do not conduct such searchers to acknowledge it on their Web sites (Heydary, 2006). Following this line of reasoning, an interesting direction for future research would be to explore the extent to which different e-dating business models are perceived by e-daters as more or less secure, the extent to which service providers can enhance the sense of security of their customers and the degree to which security enhancing features result in higher revenues for e-dating services.

Privacy concerns and the possible invasion of privacy that some e-dating services involve is another important direction for future research. Thus, ranking or categorizing e-dating services on how "invasive" they are could be an interesting future line of research on e-dating. Such categorization may reveal that some services provide more information to e-daters about other e-daters than is desirable. Indeed, some users of e-dating services (particularly, females) might experience the real-time features of e-dating services (e.g., provision of information to other daters on whether an e-dater is active in real time) as surveillance.

Given the variance in exposure between different e-dating services, it would be interesting to empirically explore how e-daters feel about the invasion of privacy that the various e-dating business models entail. Are males more comfortable with high levels of exposure than females? Are younger e-daters more comfortable with high levels of exposure than older e-daters are? Do demographic variables (such as level of education, ethnicity or income) impact the degree to which e-daters are willing to tolerate different levels of self-exposure or invasion of their privacy?

Another possible direction for future research is to explore how different "types" of e-dating services utilize different business models and how this impacts users. The e-dating arena consists of services that cater to marriage-oriented, friendship-oriented, or sex-oriented users (as well as to users who combine these orientations). Future research might explore the features that differentiate these types of e-dating services from each other, the degree of exposure of daters to each other is involved in each type, the extent to which e-daters are aware of the differences between the "types" and the degree to which the combination of features that each type represents affects the success of the business model used by the service.

Given that this chapter does not focus only on the behavior of e-daters, there are other, wider societal implications that follow from the discussion in the previous sections.

One such implication is the possible scope for abuse of information that e-daters make public. There are references in the popular literature to the use of information in social networking services by employers to spy on their employees (Lavallee, 2007; White, 2007). These reports suggest that many employers use information that individuals have posted on social networking and e-dating services (possibly a long time before the individual joined the labor force) as a basis for selection of candidates for jobs, promotion of employees, and even for harassment of employees on the job. Future research might explore the extent to which employers do indeed engage in spying on their employees by using data from social networking and e-dating services and the impact that this may have on employees' life at work.

The potential for abuse of social networking and e-dating services raises a set of other issues that involve the political and legal system. If indeed the social networking and e-dating sector poses potential dangers, should society regulate the industry to make sure that customers are more protected than they currently are? Should e-dating services be required to check their customers' criminal record or marital status? Should they be required to acknowledge on their Web site, in a manner similar to pharmaceutical companies, if they do *not* conduct such searches? Should e-dating services that involve "ranking" of e-daters by other e-daters in a manner similar to eBay's rating of buyers and sellers be outlawed because this practice involve the potential for defamation of customers? Should e-dating services be barred from discriminating against groups of e-daters that they do not wish to serve (e.g., gays)?

Only the future will tell how many of these issues will be addressed by researchers and/or by society as a whole and how this will lead to a transformation of e-dating as we know it today.

REFERENCES

Baker, A. (2005). *Double click: Romance and commitment among online couples.* Cresskill, NJ: Hampton Press.

Ben-Ze'ev, A. (2004). *Love online: Emotions on the Internet.* Cambridge: Cambridge University Press.

Berry, D. (2005). *Romancing the Web: A therapist's guide to the finer points of online dating.* Manitowoc, WI: Blue Waters Publications.

Bowlby, J. (1980). *Attachment and loss: Vol. 3, Loss, sadness and depression.* New York, NY: Basic Books.

Collins, N. (1980). Working models of attachment: Implications for explanation, emotion and behavior. *Journal of Personality and Social Psychology, 71,* 810–832. doi:10.1037/0022-3514.71.4.810

Consumer Search. (2005). Online dating sites best rated dating sites, services. Retrieved May 26, 2006, from http://www.consumersearch.com/www/Internet/online-dating/fullstory.html.

Culbreth, J. (2005). *The boomers' guide to online dating.* U.S.: Rodale Inc.

Davis, K. (1941). Intermarriage in caste societies. *American Anthropologist, 43,* 376–395. doi:10.1525/aa.1941.43.3.02a00030

Edgar, H., Jr., & Edgar, H., II. (2003). *The ultimate man's guide to Internet dating: The premier men's resource for finding, attracting, meeting and dating women online.* Aliso Viejo, CA: Purple Bus Publishing.

Fraley, R., & Shaver, P. (2000). Adult romantic attachment: Theoretical developments, emerging controversies, and unanswered questions. *Review of General Psychology, 2*(2), 132–154. doi:10.1037/1089-2680.4.2.132

Greenwald, R. (2003). *Find a husband after 35: Using what I learned at Harvard Business School.* Random House Publishing Book.

Gubernick, D. (1994). Bi-parental care and male female relations in mammals. In: S. Parmigiani & F. Vom Saal (Eds.), *Infanticide and parental care* (pp. 427-463). Chur, Switzerland: Harwood.

Hazan, C., & Shaver, P. (1987). Romantic love conceptualized as an attachment process. *Journal of Personality and Social Psychology, 52,* 511–524. doi:10.1037/0022-3514.52.3.511

Heer, D. (1974). The prevalence of black-white marriage in the United States, 1960 and 1970. *Journal of Marriage and the Family, 36,* 246–258. doi:10.2307/351151

Heydary, J. (2006, September, 26). Regulation of online dating services sparks controversy. *Wall Street Journal,* p. A5.

Hitsch, J., Hortacsu, A., & Ariely, D. (2005). *What makes you click.* Paper presented at the AEA Meeting, Choice Symposium, Northwestern University. Estes Park.

Johnson, R. (1980). *Religious assortative marriage in the United States.* New York, NY: Academic Press.

Kalmijn, M. (1991). Shifting boundaries: Trends in religious and educational homogamy. *American Sociological Review, 56*, 786–800. doi:10.2307/2096256

Kalmijn, M. (1993). Trends in black/white intermarriage. *Social Forces, 72*, 119–146. doi:10.2307/2580162

Kalmijn, M. (1998). Intermarriage and homogamy: Causes, patterns, trends. *Annual Review of Sociology, 24*, 395–421. doi:10.1146/annurev.soc.24.1.395

Kennedy, R. (1952). Single or triple melting pot? Intermarriage in New Haven, 1870-1950. *American Journal of Sociology, 58*, 56–59. doi:10.1086/221073

Kirkpatrick, L., & Davis, K. (1994). Attachment style, gender and relationship stability: A longitudinal analysis. *Journal of Personality and Social Psychology, 66*, 502–512. doi:10.1037/0022-3514.66.3.502

Lavallee, A. (2007, June 13). Firms tidy up clients' bad online reputations. *Wall Street Journal,* p. B1.

Lieberson, S., & Waters, M. (1988). *From many strands: Ethnic and racial groups in contemporary America.* New York, NY: Russell Sage.

Loviglio, J. (2007, June 16). Two sex convictions in online dating case. *Wall Street Journal,* p. B5.

Mare, R. (1991). Five decades of educational assortative mating. *American Sociological Review, 56*, 15–32. doi:10.2307/2095670

Merton, R. (1941). Intermarriage and the social structure: Fact and theory. *Psychiatry, 4*, 361–374.

Moraski, M. (2007, June 16). Beware of digital Don Juans. *Wall Street Journal,* p. A5.

Orr, A. (2004). *Meeting, mating and cheating: Sex, love, and the new world of online dating.* New Jersey: Reuters.

Pietromonaco, P., & Carnelley, K. (1994). Gender and working models of attachment: Consequences for perceptions of self and romantic relationships. *Personal Relationships, 1*, 63–82. doi:10.1111/j.1475-6811.1994.tb00055.x

Pistole, C. (1995). College students ended love relationships: Attachment style and emotion. *Journal of College Student Development, 1*, 53–60.

Qian, Z. (1997). Breaking racial barriers: Variations in interracial marriage between 1980 and 1990. *Demography, 34*, 263–276. doi:10.2307/2061704

Rose, D. (1999). *Internet soul mates: Finding the love of your life through the Internet.* Phoenix, AZ: Productiones Deanna, LLC.

Rosenfeld, M. (2005). A critique of exchange theory in mate selection. *American Journal of Sociology, 110*(5), 1284–2027. doi:10.1086/428441

Silverstein, J., & Lasky, M. (2004). *Online dating for dummies.* Hoboken, NJ: Wiley Publishing Inc.

TruDating. (2006). Online dating service directory. Retrieved May 25, 2006, from the http://www.trudating.com/.

White, E. (2007, January, 1). Employers reach out to recruit with Facebook. *Wall Street Journal,* p. D3.

APPENDIX: E-DATING RESEARCH INVENTORY

Demographic Information:

Gender: M   F   Age: _____   Major:_____

This study is about e-dating. In particular, its purpose is to discover the decision and learning process through which students evolve during their e-dating experience using Match.com.

In order to participate in this study, you must meet the following criteria:

1. You have to be new to the e-dating experience. If you have extensive experience in e-dating, you should exclude yourself from participating in this study.
2. You should not be in a committed relationship. If you are in such a relationship, you should exclude yourself, too.
3. You are interested in establishing a committed relationship and willing to invest the time and energy that it will take to do so, as well as to invest the time and energy that it will take to document the process through this research.

If you are participating for extra credit, your responses will not be anonymous, as we will need your identity to add the extra credit to your grade.

Your responses may be kept anonymous if you do not request extra credit for your course.

You may drop out of the study at any time, but will forfeit the extra credit.

If you meet these criteria and accept these conditions, please sign below:

_____
Name

_____
Student Signature   Date

Date _____
ID Number_____

Questions for Students' Weekly Journals

Please fill in the answers to the following questions and submit every Friday.

Referencing your Match.com Web site activities this week, please answer the following questions:

1. **In which major activity did you engage in this week?**

    a. Constructing or revising a profile
    b. Conducting a search for matches
    c. Initiating contact with matches
    d. Responding to contacts from others
    e. Setting up dates
    f. Meeting dates
    g. Concluding the process

   Please circle the letter that corresponds to this week's major activity.

If there was a close second, what was it?

Please describe in detail what you did this week in relation to each of the following activities:

Activity A - Constructing or Revising a Profile
2. **If you created or revised a profile this week, please print and attach it.**
    - Did you consult (i.e., show your profile and ask for feedback, or otherwise seek advice) anyone on the process? Please list those with whom (e.g., parent, sibling, roommate, friend, etc.) you consulted.
    - What other resources did you use (e.g., books or online references) in constructing your profile?
    - Did you find this difficult?
    - Did you exaggerate, minimize or knowingly omit any aspect of your profile?
    - If so, which aspects did you exaggerate, minimize or omit?
    - If you revised your profile, what did you change? Why did you change it?

Activity B - Searching
3. **If you conducted searches this week, please tell us:**
    - Did you create your own customized search?
    - If so, what attributes did you add/change beyond the basic attributes provided by the service?
    - Did you save your search for future reference?
    - Did you receive a list of "matches" from the service?
    - If so, did you follow up on it?
    - Did you follow up on one single match that you received from the service (Is it fate?)?
    - Did you search the profiles of matches who approached you?
    - Please estimate how many profiles you visited this week as part of your search process?
    - Did you keep any profiles that you found in a "favorites" list?
    - If so, how many did you place in "favorites" this week?

Activity C - Initiating Contact
4. **If you initiated any contact with others this week, please tell us:**
    - How many winks did you send this week?
    - How many e-mail messages did you send this week?
    - How many instant messages did you send this week?

- How many rejection messages did you send this week?
- How many matches did you decide to "bar" from contacting you this week?
- How many of the people that you contacted this week do you think will respond?
- How many of the people that you contacted this week do you think you will meet?

Activity D - Receiving and Responding to Contact from Others

5. If you received any communication from others this week, please tell us:
   - How many "hits" did your profile get this week? (report the current number of hits)
   - How many winks did you get this week?
   - How many e-mail messages did you get this week?
   - How many instant messages did you get this week?
   - Of those contacting you this week, with how many will you communicate in the future?
   - How many of the people that contacted you this week do you think you will meet?

Activity E - Setting up Dates

6. If you scheduled any face-to-face dates this week, please tell us:
   - How did you set the date(s) (telephone, e-mail, instant messaging, text-messaging, etc)?
   - In each case, how many messages did you exchange with each match before scheduling the date(s)?
   - In each case, did you initiate (ask the other person for) a date(s)?
   - If any of the date(s) that you set were cancelled, why?
   - How many people with whom you have scheduled dates this week do you think you will actually meet?
   - With how many people that you scheduled dates this week do you think you will have a long-term relationship?

Activity F - Conducting Dates

7. If you conducted any face-to-face dates, please tell us:
   - How many dates (list only those from Match.com) did you have this week?
   - Did all your Match.com dates show up as planned?
   - If they didn't show up, please tell us why they didn't (if you know):

Activity G - Concluding the dating process

8. If you concluded the e-dating process this week, please tell us:
   - Do you intend to turn off your profile?
   - If so, why?

This work was previously published in Social Networking Communities and E-Dating Services: Concepts and Implications, edited by Celia Romm Livermore & Kristina Setzekorn, pp. 292-313, copyright 2009 by ISR Publishing (an imprint of IGI Global).

Chapter 15
A Trination Analysis of Social Exchange Relationships in E-Dating

Sudhir H. Kale
Bond University, Australia

Mark T. Spence
Bond University, Australia

ABSTRACT

More than half a billion users across the globe have availed themselves of e-dating services. This chapter looks at the marketing and cross-cultural aspects of mate-seeking behavior in e-dating. We content analyzed 238 advertisements from online matrimonial sites in three countries: India (n=79), Hong Kong (n=80), and Australia (n=79). Frequencies of mention of the following ten attribute categories in the advertiser's self-description were established using post hoc quantitative analysis: love, physical status, educational status, intellectual status, occupational status, entertainment services, money, demographic information, ethnic information, and personality traits. Past research on mate selection using personal ads and the three countries' positions on Hofstede's dimensions of culture were used in hypotheses generation. The results support several culture-based differences in people's self-description in online personal ads; however, some anticipated differences were not realized, suggesting that some cultural differences may not be as strong as Hofstede (2001) suggests.

DOI: 10.4018/978-1-60960-759-3.ch015

INTRODUCTION

Family researchers and psychologists have investigated the attributes people desire in their life partners for almost 70 years (cf. Neely, 1940; Smith & Monane, 1953). However, cross-cultural differences in attribute preferences did not receive much attention until about 15 years ago (Buss, 1989). While several researchers have used personal ads to identify the qualities heterosexual males and females are looking for in a potential mate (Harrison & Saeed, 1977; Hirschman, 1987; Goode, 1996), transporting this discourse to the domain of online personals is a somewhat recent development.

Arvidsson (2006) observes that Internet dating is an aspect of a more general trend to construct a common social world through communicative interaction. The "common social world" constructed through Internet personals will be impacted by the culture permeating the advertiser and the target audience (Barnlund, 1989; Kale, 1991). The core ideas and norms of a culture contribute toward an individual's internal representation of the self, and how that self is related to important others (Fiske, Kitayama, Markus, & Nisbett, 1998). Perceptions of the internal self will impact what the individual advertiser says about one's self when seeking a potential mate. Since individual cultures across the globe show considerable differences along several important dimensions (e.g., Hofstede, 1991), these differences should be reflected in search behavior and interpretations of romantic love and intimacy across cultures (Dion & Dion, 1996). In a recent investigation of online ads, Ye (2006) observed that significant cultural differences can be observed in mate selection between Chinese and Americans.

The present study focuses on how cultural differences impact people's external self-representation in e-dating. E-dating is of interest for two reasons: to gain insights into its explosive, widespread adoption throughout the World; and— what would seem to be a source of *resistance* to adopting this communication channel—is that it is an "impoverished" medium (Walther, 1996): it is devoid of face-to-face interaction, which prompts unanswered questions concerning how individuals present themselves in a "faceless" situation.

According to social exchange theory, the sustainability of a relationship is determined by satisfaction with the rewards vis-à-vis the costs in that relationship as compared to available alternatives (Bagozzi, 1975; Hirschman, 1987). To enhance the odds of initiating and maintaining a viable intimate relationship, advertisers in personals ads are likely to offer and emphasize those aspects of self they believe a potential date or partner would find rewarding (Gonzales & Meyers, 1993). Which characteristics are deemed rewarding would be impacted by the culture of the target audience for e-dating (Hall & Hall, 1990; Kale, 1991).

This chapter explores differences in e-dating ads across three culturally diverse countries— India, Hong Kong, and Australia. The choice of countries was based on two considerations: first, to facilitate the content analysis, we wanted to choose countries where the use of English was widespread, and we wanted three countries which exhibited considerable across-country cultural diversity. Using Hofstede's (1991) dimensions of national culture, we have generated a series of hypotheses on how cultural differences will impact self-presentation in online personal ads. The hypotheses are then tested through a post hoc quantitative analysis of ads on e-dating sites in these countries. The results suggest important cultural differences in self-presentation as well as some significant interactions between sex and country in accounting for variance in self- presentation.

STUDY BACKGROUND

[O]nline dating systems have begun to influence not only individual lives but also cultural notions

A Trination Analysis of Social Exchange Relationships in E-Dating

of love and attraction ... But despite the incredible number of people using these services, we know little about how users perceive each other... It's possible, too, that different subpopulations of users within the site are seeking entirely different things and using different evaluative techniques. (Fiore, 2007)

Research on people's dating preferences is not new. Previous studies have explored differences in preferences with regard to an ideal partner between men and women (see Rajecki, Bledsoe & Rasmussen, 1991), reasons for placing personal ads (see Jason, Moritsugu, & De Palma, 1992), evidence of evolutionary influences in dating preferences (see Symons, 1979; Sadalla, Kenrick, & Venshure, 1987), motivations for interracial dating (see Yancey & Yancey, 1998), effects of forewarnings on evaluation of target profiles (see Leon, Rotunda, Sutton, & Schlossman, 2002), applications of the marketing exchange theory to lonely hearts ads (see Hirschman, 1977), and cross-cultural differences in desired mate attributes (see Buss, 1989; Parekh & Beresin, 2001; Ye, 2006).

The two research streams of male-female differences in partner expectations and applying evolutionary perspective to explain these differences are closely related. Several studies on personal ads and dating behavior have found that behaviors of men and women in the dating context are consistent with traditional sex-role stereotypes (Nevid 1984; Urberg, 1979; Davis, 1990). For instance, Davis (1990) looked at 328 personal advertisements sampled from a major Canadian newspaper. His findings suggest that the men were more likely to desire a particular physical attribute than women, and the women were more likely to stipulate that their companion be employed, possess intelligence, have a profession, and be financially well-off. In their study of 800 "lonely hearts" advertisements, Harrison and Saeed (1977) concluded that women were more likely than men to advertise their attractiveness, seek financial security, and look for someone older than them. Men, more than women, were in search of attractiveness and youth, and in return offered financial security and professed their interest in marriage.

Cosmides and Tooby (1987) and Buss (1987) have been largely credited for spearheading the evolutionary psychology paradigm in mate selection. Cosmides, Tooby, and Barkow (1992, p. 3) define evolutionary psychology as "simply psychology that is informed by the additional knowledge that evolutionary biology has to offer, in the expectation that understanding the process that designed the human mind will advance the discovery of its architecture." The evolutionary perspective suggests that the ideal mate for a male is a female possessing high reproductive capacity, which mostly equates with a young female (Thornhill & Thornhill, 1983; Buss, 1987). Males, therefore, find females with relatively youthful facial characteristics attractive (Symons, 1979). For a female, the ideal male is someone who will successfully compete for resources and can provide for their offspring (Sadalla, Kenrick, & Venshure, 1987). Attributes such as robust health, clear skin, and strong muscles in a male are likely to be more attractive to females than average health, skin, and muscles (Alley & Cunningham, 1991). Consistent with the evolutionary perspective, men place more emphasis on physical attractiveness when choosing partners for sex or marriage, and women place relatively more emphasis on socio-economic status, earnings potential, and college education in choosing their mates (Barscheid & Walster, 1974; Townsend & Wasserman, 1997).

Leon et al. (2003) suggest that as the computer age has progressed, people's lives have become increasingly busy and the ability to meet potential mates is reduced. Personal ad dating, they write, is ubiquitous because it fulfills a social need and a niche. In their study of individuals who had placed newspaper personal ads, Jason, Moritsugu, and DePalma (1992) concluded that those placing ads were well educated and financially successful. Their reasons for seeking the newspaper as a channel for mate seeking were high mobility and lack of access to traditional modes of meeting others (i.e., friends, family, and work). Around 85% said they were new to the area, 69% reported difficulty in meeting people through social activities, 61% felt uncomfortable in meeting people in singles bars, and 59% reported lack of familial contacts for introduction to potential mates. In conclusion, the authors state that the self-advertisers either did not have other ways to explore potential relationships, or that they may have tired of the more conventional channels and means of introduction.

CONCEPTUAL FRAMEWORK

People hold a certain view of themselves on the basis of what is perceived as culturally appropriate by others in the culture or within the in group (Usunier, 1996). Building on Foa and Foa's (1974) social exchange theory and the status characteristics theory (see Berger, Cohen, & Zelditch, 1966; Bereger & Fisek, 1974; Humphreys & Berger, 1981), Hirschman (1987) identified ten resource categories that comprehensively encapsulate the resources offered and sought in exchange between men and women using personal advertisements. These categories are: love, physical status, educational status, intellectual status, occupational status, entertainment services, money, demographic information, ethnic information, and personality trait information.

Love comprises emotional commitment, companionship, warmth, and emotional/affective personality traits. *Physical status* involves those physical characteristics that are valued within a society. *Educational status* refers to formal education and college degrees. *Intellectual status* relates to characteristics typically associated with high intelligence. *Occupational status* refers to those occupations that are held in high regard within a society. *Entertainment services* refer to non-sexual activities to be done with another person. *Money* is an expression of wealth, affluence, or financial well-being. *Demographic information* comprises general descriptive characteristics such as age, marital status, and place of residence. *Ethnic information* concerns specifics such as race, religion, nationality, and caste. *Personality trait* information refers to statements about one's personality but "does not include traits related to sexual or emotional characteristics" (Hirschman, 1987; p. 101).

We posit that the frequency of mentions of characteristics belonging to each of the ten categories described above would be associated with culture of the advertiser and the target audience (Kale, 1991). Operationalizing and explaining differences arising out of culture requires a suitable cultural framework (Kale & Barnes, 1992). Previous literature presents several frameworks with which to operationalize culture. Key among these are the classification schemata proposed by Kluckhohn and Strodtbeck (1961), Hall and Hall (1990), Inkeles and Levinson (1969), Trompenaars and Hampden-Turner (1998), Schwartz (1994), and Hofstede (1991; 2001). While there exists some disagreement among researchers as to which cultural framework is most appropriate, for the purposes of this study, we chose the one proposed by Geert Hofstede (1991). Of all the frameworks, Hofstede's (1991) seems to have the most overall acceptance, an intuitive appeal, and the advantage of quantification. Hofstede proposes five dimensions of culture: power distance, individualism, masculinity, uncertainty avoidance, and long-term orientation.

Power distance (PDI) is the extent to which the less powerful members of organizations and institutions (like the family) accept and expect that power is distributed unequally. This dimension represents the amount of inequality in the distribution of power, status, and wealth. It suggests that a society's level of inequality is endorsed by the followers as much as by the leaders. Power and inequality across humans are extremely fundamental facts of any society. All societies are unequal, but some are more unequal than others.

Countries are scored on the power distance dimension, receiving a number between 1 and 100. For this dimension, higher numbers means there is greater distance (inequality) in the distribution of power.

Individualism (IDV) assesses the bond between an individual and his or her fellow individuals. Individualist societies are characterized by loose ties across people; thus, in a country receiving a higher score everyone is expected to look after him/herself and his/her immediate family. On the collectivist side, we find societies in which people are integrated into strong, cohesive in groups, often extended families (with uncles, aunts and grandparents), clans, or tribes which continue protecting them in exchange for unquestioning loyalty.

Masculinity (MAS) versus femininity refers to the stereotypical sex-role differences across societies. In masculine cultures sex roles are sharply differentiated and traditional masculine values such as achievement, assertiveness, and competition are relatively more valued. In feminine cultures sex roles are less sharply distinguished and attributes such as nurturing and caring are relatively more valued. Masculine societies tend to be hero worshippers whereas feminine societies tend to sympathize with the underdog. The women in feminine countries have the same modest, caring values as the men. In masculine countries (those with high scores), women tend to be somewhat assertive and competitive, but not as much as men; masculine countries, therefore, show a considerable gap between men's values and women's values.

Uncertainty avoidance (UAI) reflects a society's level of tolerance for uncertainty and ambiguity. This dimension indicates to what extent a culture programs its members to feel either uncomfortable or comfortable in unstructured situations. Unstructured situations are novel, unknown, surprising, and ambiguous. Strong uncertainty-avoiding cultures try to minimize the possibility of such situations by enacting strict laws and rules, safety and security measures, and through their belief in an absolute truth. People in strong UAI countries are also more emotional, and motivated by inner nervous energy. The opposite type, uncertainty accepting cultures (those with *low* UAI scores), are more tolerant of opinions different from what they are used to; they try to have as few rules as possible, and allow multiple truths to exist.

Long-term orientation (LTO) versus short-term orientation: Also called "Confucian Dynamism," this dimension assesses a society's capacity for patience and delayed gratification. Long-term oriented cultures—reflected in high LTO scores—tend to save more money and exhibit more patience in reaping the results of their actions. Short-term-oriented societies want to maximize the present rewards and are relatively less prone to saving or anticipating long-term rewards. Hofstede (2001, p. 359) writes, "Long Term Orientation stands for the fostering of virtues oriented towards future rewards, in particular perseverance and thrift. Its opposite pole, Short Term Orientation, stands for the fostering of virtues related to the past and present, in particular, respect for tradition, preservation of 'face' and fulfilling social obligations."

Scores for the three countries chosen for this research along the five cultural dimensions are shown in Table 1. Our starting position was to select Australia, our country of citizenship for further analysis. We then endeavored to select two other countries that exhibited considerable differences

Table 1. Scores of India, Hong Kong, and Australia on Hofstede's cultural dimensions (Hofstede, 2001)

	PDI	IDV	MAS	UAI	LTO
India	77	48	56	40	61
Hong Kong	68	25	57	29	96
Australia	36	90	61	51	31
World Average	56.5	43	51	65	48

in scores along the dimensions of PDI, IDV, UAI, and LTO; although as can be seen, there is little variance on masculinity, with all three countries scoring slightly above the world average.

Research Hypotheses

Based on earlier research on male-female representation of the self (Hirschman, 1987; Feingold, 1990; Gonzales & Meyers, 1993) and on the three countries' scores on Hofstede's cultural dimensions, we advance the following hypotheses, starting with gender effects.

H1: Women, relative to men, are:
   a. More likely to make love-related references
   b. More likely to mention physical characteristics
   c. More likely to mention entertainment services
   d. Less likely to mention money
   e. Less likely to mention educational status
   f. Less likely to make intellect-related references
   g. Less likely to mention occupational status

H2: Relative to Australia, India and Hong Kong are:
   a. Less likely to make love-related references
   b. Less likely to mention physical characteristics
   c. Less likely to mention entertainment services
   d. Less likely to mention personality traits
   e. More likely to mention money
   f. More likely to mention educational status
   g. More likely to make intellect-related references
   h. More likely to mention occupational status
   i. More likely to provide ethnic information
   j. More likely to provide demographic information

THE STUDY

The present study is based on a sample of Australian, Indian, and Hong Kong online personal advertisements. By creating a personal profile on dating Web sites, the partner seeker is packaging an offering, a product if you will, as well as specifying to whom this offering should ideally appeal. Hirschman (1987) investigated male- and female-placed personal advertisements in magazines as examples of complex, heterogeneous marketing exchanges and stated "personal advertisements represent the offering of people as products, as a set of marketable resources in search of an appropriate buyer" (p. 101).

Dowd and Pallotta (2000) propose that people approach mate selection by looking at partners who share interests, beliefs, and economic potential and other areas of similarity. According to them, people think of love as an investment in their future well-being that must be approached carefully and rationally, much as consumers do when purchasing products and services.

The current study content analyzed Australian, Indian, and Hong Kong personal advertisement posted on dating Web sites in each respective country. A terse perusal of personal ads on the Hong Kong Web site revealed that many were from ex-pats who may not share that culture's dominant values and norms, thus Hong Kong specific hypotheses are guardedly advanced. A priori we anticipate seeing the biggest differences between Australia and India.

RESEARCH METHODOLOGY

Sample

A sample of 238 personal online advertisements (121 females and 117 males) were downloaded and printed from three dating Web sites: 79 personal ads were from the Australian Web site, www.rsvp.com, 79 from the Indian Web site, www.shaadi.com, and 80 from Hong Kong (www.singlesofhongkong.com). To select individual ads, after logging on to each Web site the third personal ad posted was printed and every third ad thereafter. There is no reason to believe that this sample selection method would result in response bias; for example, personal ads are not organized by age, surname or religious affiliation. We limited our sample to heterosexuals that fell into the age range of 21-35.

Coding Scheme

All three of the Web sites had two sections. The first portion was pre-formatted with fill-in-the-blank headers such as education, occupation, hobbies, desired level of commitment and religion/ethnicity. This section was not analyzed for two reasons: (1) The headers were not consistent across the Web sites, and more importantly; (2) Our interest was in the self-presentation section. It is in this unstructured section that participants can draw attention to any aspects about themselves they desire to project—in other words, to engage in impression management. In many cases these narratives included repeating and/or elaborating upon information from the first section. As would be expected, in most cases these narratives also provided additional insights into the person's activities, interests and lifestyle. It was this section that was coded and content analyzed.

For our purposes, we counted words within the narratives that mapped onto 10 pre-determined categories, for example, love, personality traits, entertainment and physical characteristics (see hypotheses H2a-j above for the complete list of categories). We acknowledge that words interpreted as signaling each of the 10 categories of interest can be debated. For our purposes, words categorized as 'love' had to have nurturing or bonding connotations, examples of which would include *family* or *caring*; personality trait descriptors would include comments like *up for a challenge* and *likes sports*; entertainment services referred to capabilities they had with which they could indulge a partner, such as *play the guitar* and *accomplished dancer*; physical characteristics was in reference to words that served to signal one's beauty or health, such as *not that good looking*, *I have all my own teeth*, and *charming smile*, and so forth.

Four students enrolled in a consumer behavior course who were not informed as to the purpose of the study categorized the narratives. A practice coding test run took place during which students sat separately and then compared their categorizations. The inter-rater reliability was unacceptably low. The source of the low inter-rater reliability was twofold. The first source of discrepancy that was quickly resolved was to limit the analysis to self-descriptions. Thus, coders were informed to not count any portion of a narrative that was directed at someone other than the person writing the narrative. Comments not directed at their self were typically about desired qualities in a mate, but on occasion included comments about family members or friends. The second source of disagreement stemmed from the fact that often narratives had seemingly redundant references, making it difficult to agree on the appropriate number of times to count a comment. Consider, for example, the following verbatim extract from an Australian male:

I have quite a hectic lifestyle ... busy most nights of the week ... and getting busier. ... I love having a good time as well as love going nuts and having fun ... I love having a good time ...

Arguably this could be counted as just two personality traits, those comments related to a 'hectic lifestyle' and those that referred to 'having a good time'. However, it was decided that the person writing this narrative wanted to emphasize these points and therefore that the redundancies should be counted—in other words, we endeavored to be as liberal in our interpretation of word meaning as was feasible. This narrative was therefore counted as seven personality traits (three of which relate to their hectic lifestyle and four to having fun).

In light of the frequency with which redundancies like the aforementioned appeared, it was decided that the appropriate way to categorize words was by having two research assistants sit together to code each narrative and resolve any differences concerning how to categorize a word at that time. Given that each narrative is an independent sample it is reasonable to assume that should the narratives be categorized by different individuals there would be disparities, but these differences should approximate a normal distribution about the true mean.

For illustrative purposes, two verbatim narratives are presented complete with spelling and grammar errors along with the resultant coding counts for four of the categories of interest: love, personality trait, entertainment and physical characteristics.

Australian Male

"OK i know im **not that good looking** but maybe ive got more to offer than looks. I'm generally a *quiet* sort of person when I meet new people but as I get to know them I come out of my shell. I'm caring, *honest*, have a *good sense of humour* maybe even a bit warped sometimes, value friends and family. I think i'm pretty *easy going* and *like to get out* and *have a bit of fun* every now and then. Not much into clubbing more a pub with a few friends or backyard and a few drinks with friends person. All in all im just hoping to find that someone special.

Indian Female

"I am a *well mannered*, *happy* and *easy going* type of girl. I am very *hard working* and i love to support my partner. I am *very kind* and i love to be treated with kindness and love. *Down to earth* and i love to be with a big family and relations."

DATA ANALYSIS AND FINDINGS

For each of the ten categories of interest a 3 x 2 ANOVA was run, the dependent variables being country and gender. In all but two cases (money and ethnic information), the overall model's F-statistic was significant at $p = .05$, hence only the significance of the main and two-way interaction effects are reported. In the case of money and ethnic information, the number of subjects reporting zero counts (i.e., made no reference to these issues) were n=232 and n=210, respectively, hence these categories were excluded from further analysis. Therefore, H1d, H2e and H2i are reported as having insufficient data, an interesting insight in itself. (Box 1 and Box 2)

We fully elucidate the analysis pertaining to love-related references, the first category of interest. Analyses for the other categories were similar, but the findings are presented more succinctly.

Box 1. Australian male coding counts

Love: 3 (appropriate words underlined for illustrative purposes)
Personality trait: 6 (in *italics*)
Entertainment: 4 (including its obverse, 'Not much into clubbing')
Physical characteristic: 1 (in **bold**)

Box 2. Indian female coding counts

Love: 4 (underlined)
Personality trait: 6 (in italics)
Entertainment: 0
Physical characteristic: 0

A Trination Analysis of Social Exchange Relationships in E-Dating

Based on the 3 x 2 ANOVA with love-related references the dependent variable, there was a main effect due to country (p <.001) as well as a significant two-way interaction between country and gender (p<.001); however, there was no main effect due to gender (p=.865). Cell means appear in the figure below. Australians were significantly more likely to make love-related references (x = 1.92) than were either individuals from Hong Kong (x = 1.09) or India (x =.71). Pairwise comparisons revealed that the main effect of country was due to a significant difference between India and Australia (p <.001), as well as a significant difference between Hong Kong and Australia (p =.009); however, there was no significant difference between India and Hong Kong (p =.123). Findings related to country are therefore consistent with H2a. (Figures 1, 2, 3, 4, 5, 6, 7, & 8)

To examine the two-way interaction, t-tests by gender within each country were conducted. Within India there was no significant difference in the number of love-related references between men and women (p =.133), whereas there was a significant difference between genders in both Hong Kong and Australia (p <.001). Surprisingly, men were more likely to make love-related references than were women in Hong Kong (means equal 1.73 and 0.45, respectively), whereas the reverse was true in Australia (mean equals 1.12 versus 2.79). We therefore cannot support H1a.

Figure 1.

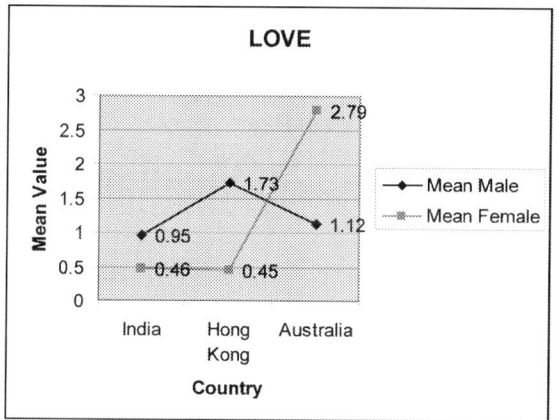

Figure 2.

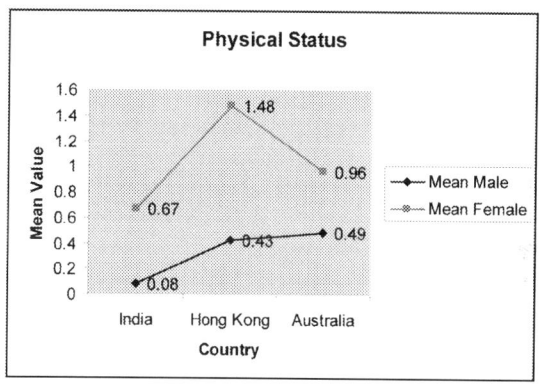

Figure 3.

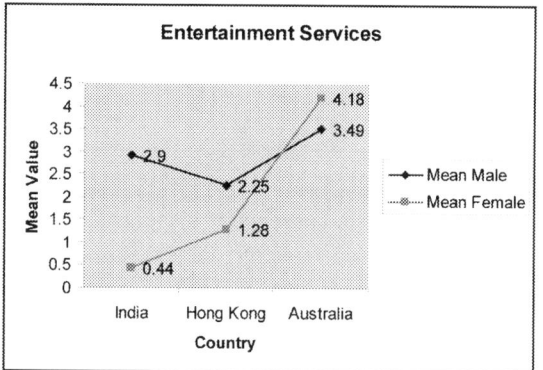

Figure 4.

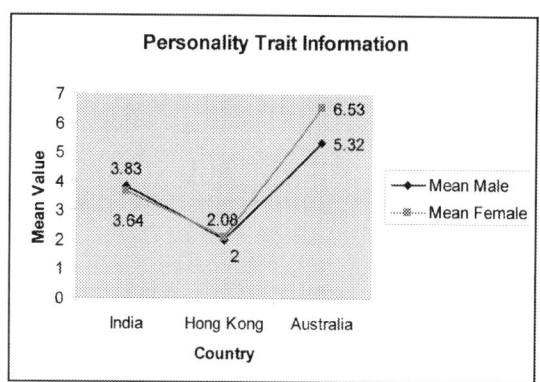

Figure 5.

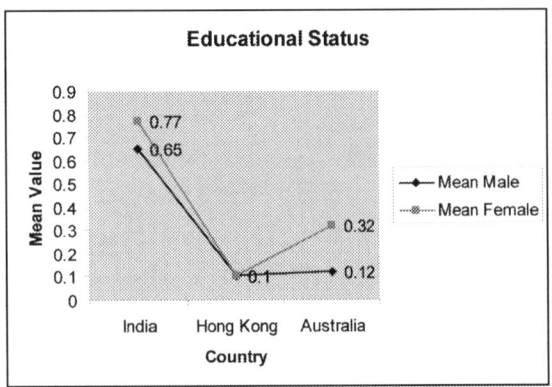

Figure 6.

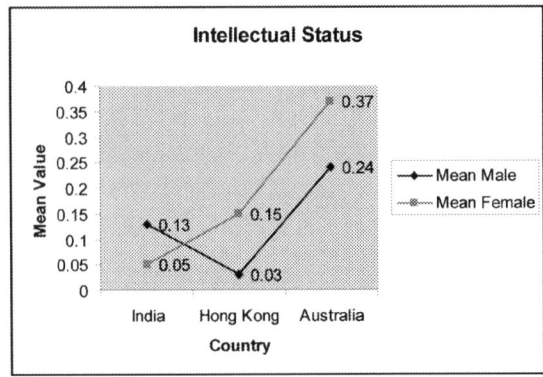

Figure 7.

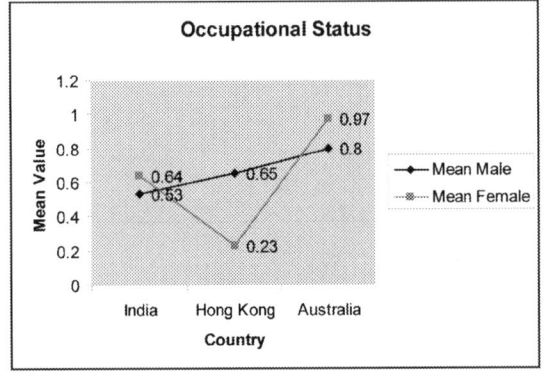

Figure 8.

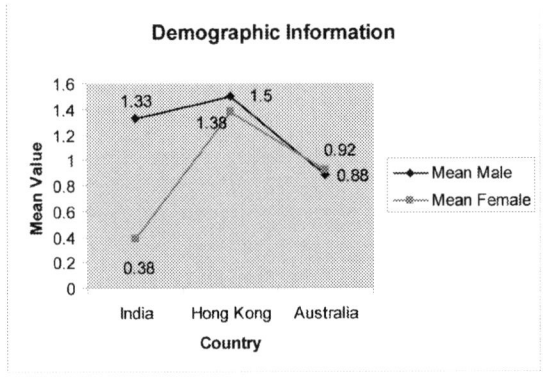

MEAN VALUES FOR LOVE-RELATED REFERENCES

There is a significant difference in the number of references to physical characteristics by country (p = 0.029) and sex (p < 0.001); there is no two-way interaction (p = 0.163). Gender differences are as predicted (H1b); however, H2b receives only partial support. Australians were more prone to mention physical characteristics than were Indians, but those from Hong Kong had the greatest proclivity overall.

With respect to entertainment services mentioned, there is a significant difference due to country (p < 0.001); and the predicted gender effect only approaches significance (p = 0.058).

There is, however, a significant gender by country interaction (p = 0.029). Overall, and consistent with H2c, Australians are more likely to mention entertainment services than are individuals from India or Hong Kong. However, an interesting pattern emerges with respect to gender: India and Hong Kong males make more entertainment-related comments than do women, whereas the reverse is true in Australia. H1c is therefore not supported.

Concerning comments about personality traits, there was a main effect due to country (p < 0.001), but no significant difference due to gender (p = 0.392) or the two-way interaction (p = 0.367). These data support H2d: Australians are more prone to mention personality traits than are individuals from India or Hong Kong (and, as it turns out, Indians are more prone than are those from Hong Kong).

There is a significant difference in the number of references made about educational status by country ($p < 0.001$); however, there is no significant gender effect ($p = 0.358$) nor two-way interaction ($p = 0.778$); H1e is therefore not supported. Indians made more education-related references than did individuals from Hong Kong or Australia, although there was no significant difference between the latter two; H2f therefore receives only partial support.

With respect to comments about one's intellect, there was as a significant effect due to country ($p = 0.011$), but not a significant effect due to gender ($p = 0.384$) or the two-way interaction ($p = 0.384$), H1f is therefore not supported. Furthermore, contrary to expectations the main effect of country was opposite to the expected pattern: Australians were more likely to make intellect-related references than were individuals from either India or Hong Kong; H2g is therefore not supported.

There is a significant difference in references to occupational status by country ($p = 0.004$); but no significant difference due to sex ($p = 0.662$) or the sex by country interaction ($p = 0.054$); H1g is therefore not supported. Contrary to expectations, Australians were more prone to make occupation-related references than were individuals from India or Hong Kong; H2h is therefore not supported.

With respect to references to demographic information, there is a significant main effect due to country ($p = 0.006$) as well as gender ($p = 0.041$); there is also a significant two-way interaction ($p = 0.036$), the latter driven by India. While individuals from Hong Kong were more likely to mention demographic information than were individuals from Australia, the same cannot be said about individuals from India; thus, H2j receives partial support.

DISCUSSION

Findings of this study are interesting in that they reinforce some of the earlier findings and fail to support others. For example, it was expected that women will make more love-related offerings in their ads than would men. In this study, the support for sex-related differences between men and women in their self-descriptions related to love was not statistically significant, which is contrary to studies by Hirschman (1977), Gonzales and Myers (1993), and Koestner and Wheeler (1988). On the other hand, significant differences were reported in references to physical characteristics across both gender as well as country. Hong Kong residents made the most number of comments related to physical characteristics, possibly a reflection of the racial and ethnic diversity of advertisers from this region. (Recall that a perusal of Hong Kong ads suggested that many were posted by ex-pats.)

In keeping with evolutionary psychology, it was postulated that women would be more attracted to someone with a higher education, occupational status and intelligence. Consequently, men, to enhance the odds of initiating and maintaining a viable relationship were expected to stress their educational attributes, occupational status, and intellect more than women (see Gonzales & Myers, 1993). The current study did not support this evolutionary paradigm. This is a counter-intuitive finding that deserves more research.

We hypothesized that Australians would make more love-related references than Indians or Hong Kong residents. This was supported by our data analysis. In individualistic countries, people rank love as an important ingredient in a marriage whereas in collective societies, marriage is typically an arrangement that includes two families, not just individuals. It was also expected that individuals from individualistic societies would value novelty, variety, and pleasure (Hofstede, 1991). Consequently, individualistic Australians would mention entertainment services more than would the collectively oriented advertisers from India and Hong Kong. This was supported by our content analysis. Also supported was the expectation that Australians were more likely to mention their personality traits than Indians or Hong Kong residents.

An intriguing finding to emerge from this study was the insufficient mentions of money and financial status across all advertisers. It was anticipated that large power distance societies such as India would overtly mention financial status compared to small power distance societies such as Australia—this was not the case. For all intents and purposes, no one mentioned money.

Finally, two counter-intuitive findings deserve mention. In keeping with the large difference in power distance between Australia on the one hand, and India and Hong Kong on the other, it was expected that Australians were relatively less likely to make overt references to their intellect and occupational status. Content analysis of the ads suggested the exact opposite. Advertisers from India and Hong Kong made significantly less references to their intelligence and occupation as compared to the partner seekers from Australia. Other contributing factors such as the relatively low-context culture in Australia may have been responsible for these findings (see Hall & Hall, 1990).

CONCLUSION AND MANAGERIAL IMPLICATIONS

The Internet serves as a new and massive laboratory for sociological and psychological research. Leon et al. (2003) observe that the Internet can be utilized to conduct research efficiently due to the ease of implementation and the relatively few resources that are required. They further contend that the methodologies inherent in administering studies on the Internet or using samples from the Web appear as valid as those produced in lab settings. We consider this post hoc quantitative analysis of e-dating narratives a successful means of revealing cross-cultural differences.

Meeting others or dating through personals is very similar to Internet dating because in both cases the advertisers describe themselves in writing before any face-to-face encounter occurs. In order to get what they desire, advertisers typically project an image that would be deemed as attractive by the target audience. Underlying cultural values as well as gender differences are therefore expected to influence the self-presentation of lonely hearts advertisers. This study embraced Hofstede's (2001) widely used framework for classifying countries. Based on differences in each country's scores on Hofstede's five dimensions, we posited and subsequently confirmed differences in how potential mates present themselves, in particular in the number of references they made about love, physical status, entertainment services, personality traits, educational status and intellectual status. However, and importantly, support was lacking for several other characteristics, most notably references to 'money' and 'ethnicity', which were rare and non-significantly different across both genders and countries. The lack of money-related references seems particularly odd, given the myriad of studies suggesting that women seek men who can provide financial security (e.g., Barscheid & Walster, 1974; Harrison & Saeed, 1977; Davis 1990; Townsend & Wasserman, 1997). Apparently, commenting about one's money is not an opening gambit for men, a finding that is reflected in a casual perusal of women's sites, many of which do not include a photo or photos, a means of reflecting one's physical attractiveness, a characteristic widely acknowledged as sought after by men. We can only speculate, but protecting one's privacy is a plausible explanation for both gender's behaviors.

Despite instances of non-significant differences due to gender or country, the general trend is that there *are* differences across countries; thus, we should anticipate a proliferation of country specific Web sites as well as sub-culture-specific

Web sites within a country. Consider the latter. In reference to targeting niche audiences, in the USA there is Jdate.com that caters to Jewish singles, largeandlovely.com, a site wherein those overweight can feel at ease, and for those with advanced degrees—a proxy for income—there are BrainDates.com and DocDates.com. In these cases, e-dating sites are "qualifying" participants on one attribute, presumably one that is non-compensatory to those seeking partners. Like any number of.com growth areas, we should first witness a proliferation in e-dating Web sites—a trend well underway—followed by a consolidation in Web sites. A rule of thumb is that the value of a social network—whether a mobile phone service provider, an auction site, a multiplayer online role playing game (MORG) or a dating service—is proportional to the square of the number of participants. In the e-dating realm, it will therefore behoove the less popular sites to consolidate: if you are an individual wanting to advertise, ceteris paribus, advertise in the medium that appeals to the largest, *appropriate* audience.

This study is not a definitive treatise that confirms or disproves cultural differences. Further studies should consider using more culturally diverse populations and perhaps a tighter taxonomy of the (ten) attributes comprising the advertisers' self-descriptions advanced by Hirschman (1987). What we can say is that this study does qualify as one of the many successful applications of 'netology'—using the Internet to study social behaviors.

REFERENCES

Alley, T., & Cunningham, M. (1991). Average faces are attractive, but very attractive faces are not average. *Psychological Science*, 2, 123–125. doi:10.1111/j.1467-9280.1991.tb00113.x

Arvidsson, A. (2006). Quality singles: Internet dating and the work of fantasy. *New Media & Society*, 8(4), 671–690. doi:10.1177/1461444806065663

Bagozzi, R. (1975). Marketing as exchange. *Journal of Marketing*, 39(October), 32–39. doi:10.2307/1250593

Barnlund, D. (1989). *Communicative styles of Japanese and Americans*. Belmont, CA: Wadsworth.

Barscheid, E., & Walster, E. (1974). Physical attractiveness. In: L. Berkowitz (Ed.), *Advances in experimental psychology* (pp. 157-215). New York, NY: Academic Press.

Berger, J., Cohen, B., & Zelditch, M., Jr. (1966). Status characteristics and expectation states. In: J. Berger, M. Zelditch, Jr., & B. Anderson (Eds.), *Sociological theories in progress* (vol. 1, pp. 29-46). Boston, MA: Houghton Mifflin.

Berger, J., & Fisek, H. (1974). A generalization of the theory of status characteristics and expectation states. In: J. Berger, T. Conner, & M. Fisek (Eds.), *Expectation states theory* (pp. 163-205). Englewood Cliffs, NJ: Winthrop.

Buss, D. (1987). Sex differences in human mate selection criteria: An evolutionary perspective. In: C. Crawford, D. Crebs, & M. Smith (Eds.), *Sociobiology and psychology: Ideas, issues, and applications* (pp. 335-352). Hillsdale, NJ: Erlbaum.

Buss, D. (1989). Sex differences in human mate preferences: Evolutionary hypotheses in 37 cultures. *The Behavioral and Brain Sciences*, 12, 1–49.

Cosmides, L., & Tooby, J. (1987). From evolution to behavior: Evolutionary psychology as the missing link. In: J. Dupre (Ed.). *The latest and the best: Essays on evolution and optimality* (pp. 227-306). Cambridge, MA: MIT Press.

Cosmides, L., Tooby, J., & Barkow, J. (1992). Introduction: Evolutionary psychology and conceptual integration. In: J. Barlow, L. Cosmides, & J. Tooby (Eds.), *The adapted mind: Evolutionary psychology and the generation of culture* (pp. 3-15). New York, NY: Oxford University Press.

Davis, S. (1990). Men as success objects and women as sex objects: A study of personal advertisements. *Sex Roles, 23,* 43–50. doi:10.1007/BF00289878

Dion, K., & Dion, K. (1996). Cultural perspectives on romantic love. *Personal Relationships, 3*(1), 5–17. doi:10.1111/j.1475-6811.1996.tb00101.x

Dowd, J., & Pallotta, N. (2000). The end of romance: The demystification of love in the postmodern age. *Sociological Perspectives, 43,* 549–581.

Fiore, A. (2007). Online dating research at Berkley. http://people.ischool.berkeley.edu/~atf/dating.

Fiske, A., Kitayama, S., Markus, H., & Nisbett, D. (1998). The cultural matrix of social psychology. In: D. Gilbert, S. Fiske, & G. Lindzey (Eds.), *Handbook of social psychology* (4th ed., pp. 915-981). New York, NY: McGraw Hill.

Foa, U., & Foa, E. (1974). *Societal structures of the mind.* Springfield, IL: Charles C. Thomas.

Gonzales, M., & Meyers, S. (1993). "Your mother would like me": Self-presentation in the personal ads of heterosexual and homosexual men and women. *Personality and Social Psychology Bulletin, 19,* 131–142. doi:10.1177/0146167293192001

Goode, W. (1996). Gender and courtship entitlement: Responses to personal ads. *Sex Roles, 34,* 141–170. doi:10.1007/BF01544293

Hall, E., & Hall, M. (1990), *Understanding cultural differences.* Yarmouth, ME: Intercultural Press.

Harrison, A., & Saeed, L. (1977). Let's make a deal: An analysis of revelations and stipulations in lonely hearts advertisements. *Journal of Personality and Social Psychology, 35*(4), 257–264. doi:10.1037/0022-3514.35.4.257

Hirschman, E. (1987). People as products: Analysis of a complex marketing exchange. *Journal of Marketing, 51*(January), 98–108. doi:10.2307/1251147

Hofstede, G. (1991). *Culture and organizations: Software of the mind.* London: McGraw Hill.

Hofstede, G. (2001). *Culture's consequences: Comparing values, behaviors, institutions, and organizations across nations.* Thousand Oaks, CA: Sage.

Humphreys, P., & Berger, J. (1981). Theoretical consequences of the status characteristics formulation. *American Journal of Sociology, 86*(5), 953–983. doi:10.1086/227350

Inkeles, A., & Levinson, D. (1969). National character: The study of modal personality and sociocultural systems. In: L. Gardner & E. Aronson (Eds.), *The handbook of social psychology* (vol. 4, pp. 418-516). Reading, MA: Addison Wesley.

Jason, L., Moritsugu, J., & DePalma, D. (1992). Advertisements as a strategy for meeting people. *Psychological Reports, 71,* 1311–1314. doi:10.2466/PR0.71.8.1311-1314

Kale, S. (1991). Culture-specific marketing communications: An analytical approach. *International Marketing Review, 8*(2), 18–30. doi:10.1108/02651339110004078

Kale, S., & Barnes, J. (1992). Understanding the domain of cross-national buyer-seller interactions. *Journal of International Business Studies, 23*(1), 101–132. doi:10.1057/palgrave.jibs.8490261

Kluckhohn, F., & Strodtbeck, F. (1961). *Variations in value orientations.* Westport, CT: Greenwood Press.

Koestner, R., & Wheeler, L. (1988). Self-presentation in personal advertisements: The influence of implicit notions of attraction and role expectations. *Journal of Social and Personal Relationships*, *5*, 149–160. doi:10.1177/026540758800500202

Leon, D., Rotunda, R., Sutton, M., & Schlossman, C. (2003). Internet forewarning effects on ratings of attraction. *Computers in Human Behavior*, *19*, 39–57. doi:10.1016/S0747-5632(02)00017-1

Neely, W. (1940). Family attitudes of denominational college and university students, 1929 and 1936. *American Sociological Review*, *4*, 512–522. doi:10.2307/2084426

Nevid, J. (1984). Sex differences in factors of romantic attraction. *Sex Roles*, *11*, 401–411. doi:10.1007/BF00287468

Parekh, R., & Beresin, E. (2001). Looking for love? Take a cross-cultural walk through the personals. *Academic Psychiatry*, *25*, 223–233. doi:10.1176/appi.ap.25.4.223

Rajecki, D., Bledsoe, S., & Rasmussen, J. (1991). Successful personal ads: Gender differences and similarities in offers, stipulations, and outcomes. *Basic and Applied Psychology*, *12*, 457–469. doi:10.1207/s15324834basp1204_6

Sadalla, E., Kenrick, D., & Venshure, B. (1987). Dominance and heterosexual attraction. *Journal of Personality and Social Psychology*, *52*, 730–738. doi:10.1037/0022-3514.52.4.730

Schwartz, S. (1994). Beyond individualism/collectivism: new cultural dimensions of value. In: U. Kim, H. Triandis, C. Kagitcibasi, S. Choi, & G. Yoon (Eds.), *Individualism and collectivism: Theory, method and applications* (pp. 85-119). Thousand Oaks, CA: Sage.

Symons, D. (1979). *The evolution of human sexuality*. New York, NY: Oxford University Press.

Thornhill, R., & Thornhill, N. (1983). Human rape: An evolutionary analysis. *Ethology and Sociobiology*, *4*, 137–173. doi:10.1016/0162-3095(83)90027-4

Townsend, J., & Wasserman, T. (1997). The perception of sexual attractiveness: Sex differences in variability. *Archives of Sexual Behavior*, *26*, 243–268. doi:10.1023/A:1024570814293

Trompenaars, F., & Hampden-Turner, C. (1998). *Riding the waves of culture: Understanding cultural diversity in global business* (2nd ed.). New York, NY: McGraw-Hill.

Urberg, K. (1979). Sex role conceptualization in adolescents and adults. *Developmental Psychology*, *15*, 90–92. doi:10.1037/h0078082

Usunier, J. (1996). *Marketing across cultures* (2nd ed.). Hertfordshire, UK: Prentice Hall.

Walther, J. (1996). Computer-mediated communication: Impersonal, interpersonal, and hyperpersonal interaction. *Communication Research*, *23*, 3–43. doi:10.1177/009365096023001001

Yancey, G., & Yancey, S. (1998). Interracial dating: Evidence from personal advertisements. *Journal of Family Issues*, *19*(3), 334–348. doi:10.1177/019251398019003006

Ye, J. (2006). Seeking love online: A cross-cultural examination of personal advertisements on American and Chinese dating Web sites. *Global Media Journal*, *5*(8).

Chapter 16
Online Matrimonial Sites and the Transformation of Arranged Marriage in India

Nainika Seth
University of Alabama in Huntsville, USA

Ravi Patnayakuni
University of Alabama in Huntsville, USA

ABSTRACT

Online personals have been a remarkably successful in the Western World and have been emulated in other cultural contexts. The introduction of the Internet can have vastly different implications on traditional societies and practices such as arranged marriages in India. This chapter seeks to investigate using an ethnographic approach the role of matrimonial Web sites in the process of arranging marriages in India. It seeks to explore how these Web sites have been appropriated by key stakeholders in arranging marriage and how such appropriation is changing the process and traditions associated with arranged marriage. The key contributions of this study are in that it is an investigation of complex social processes in a societal context different from traditional western research contexts and an exploration of how modern technologies confront societal traditions and long standing ways of doing things. Our investigation suggests that the use of matrimonial Web sites have implications for family disintermediation, cultural convergence, continuous information flows, ease of disengagement, virtual dating and reduced stigma in arranged marriages in India.

DOI: 10.4018/978-1-60960-759-3.ch016

Online Matrimonial Sites and the Transformation of Arranged Marriage in India

INTRODUCTION

Online personals have been a remarkable success story in the United States, attracting as many as 40 million unique visitors at their peak in 2003 (Mulrine, 2003). At a time when e-commerce ventures were being viewed with suspicion by investors and as the stock market hit new lows subsequent to its run up in 1999-2000, this was a significant phenomenon. Online personals typically cater to singles, providing them an opportunity to find mates or dates beyond their traditional social networks of friends, school, work, neighborhood or place of worship. Adapting to a different societal context, one that is more conservative and traditional, Web sites that assist in brokering marriages have emerged in India. In 2006, some 7.5 million users used their services, increasing from 4 million in 2004 (Lakshman, 2006). As in the case of online personals in U.S., which have the potential to affect how we arrange our social selves, online matrimonial sites can influence the process of arranging marriages with wider implications for family structure and relationships.

Marriage is viewed differently in India as compared to the West where it is largely a matter of individual choice. In India, marriage is viewed not so much as a union between two individuals as the beginning of an enduring relationship between two families. Weddings are usually protracted events that mark the end of lengthy negotiations between two extended families including aunts, uncles, and even cousins once step removed (Seymour, 1999). Referred to as 'arranged marriage', they are rarely based purely on individual preference, choice or love. Marriage symbolizes and affirms the collective nature of family and larger kinship units in which the families are embedded. In contrast, the western notion of marriage labeled as 'love marriage' is frowned upon by the more traditional family elders (Dion & Dion, 1996).

Globalization of the economy, urbanization and the increased influence of western popular culture from books to movies and television shows, have brought about changes in the society. 'From joint family to nuclear family' is an oft repeated phrase that is used to summarize changes in the family in India during modern times. The decline in the influence of extended and joint family ties has resulted in structural holes in family networks, making it difficult for families to find suitable life-partners for their children. This led to the emergence of matchmaking services and classified advertisements (referred to as matrimonials) in newspapers. With the advent of the Internet, a new channel in the form of matrimonial Web sites has emerged as an alternative way to find partners for marriageable members of the family. The introduction of technology in the form of matrimonial Web sites in an otherwise socially-enabled process provides the setting for a fascinating exploration of changing social mores and the interaction of technology and society.

Research on electronic dating, online personals, matchmaking and social networks is limited (Close & Zinkhan, 2003; Fiore & Donath, 2004), more so in the type of societal context provided by India. This chapter investigates the impact of matrimonial Web sites on the process and practices associated with arranged marriage in India. Specifically, it seeks to answer the research questions: (1) how are the affordances provided by matrimonial Web sites appropriated by stakeholders in the process of arranging marriage; (2) what is the impact of such appropriation on the process; and (3) how does the use of such technologies shape traditions and norms associated with marriage. The investigation is informed by the theory of social construction of technology where the central premise is that technology as designed provides users with a range of possibilities which shape usage and are in turn shaped by users. The intent of the study is not to propose and validate hypotheses but to gain a deeper insight into the phenomena and an understanding of the how technology is shaping and in turn shaped by users in such complex social processes. An ethnographic approach to data collection and analysis is deployed for this purpose in this investigation. The purpose of an ethnographic approach is not so much to show that technology is used but to show how it is socially appropriated. The key contributions of this study thus are an investigation of complex social processes in a societal context different from traditional western research contexts and the introduction of modern technologies where technology confronts with traditions and long standing ways of doing things. It will provide the platform for a wider exploration of the impact of modern computing and communication technologies on traditional societies.

THEORETICAL FOUNDATION

Early research in the adoption and use of IT approached the phenomenon from a technology deterministic perspective (Markus & Robey, 1988) that focused on the impact of IT, treating it as an exogenous, invariant and monolithic artifact. Researchers then argued that IT innovations are not necessarily adopted passively as standard templates of an idea, rather it undergoes a "developmental process in adoption" involving the redefinition of specific sub-components of the new technology and their interaction with local user context (Rice & Rogers, 1980).

Introduction of new technologies invariably exerts pressure on individuals, organizations and society to change, adjust, or adapt to the new technology. However, the effects of new IT are more a function of how they are used by people rather than a function of the technology itself. Actual behavior in the context of new technologies may often differ from intended use (Markus & Robey, 1988). People adapt systems to their particular work needs, or they may resist them or not use them at all.

Structuration theory has been proposed as a theoretical lens for developing a better understanding of the interaction between organizations, technology, and people (Orlikowski & Robey, 1991; Orlikowski, 1992b). Central to structuration theory is the concept of "duality of structure" which is used to theorize that structures that are inherent in new technologies are different from the structures that emerge in human action as people interact with these technologies. Further, theoretical extension of this approach has been put forth in the form of adaptive structuration theory that has been used for studying organizational change that accompanies usage of new technologies (DeSanctis & Poole, 1994). Drawing upon structuration (Giddens, 1984) and appropriation (Bijker & Law, 1992), Desanctis and Poole (1994) propose adaptive structuration theory for explaining the process of incorporating new technologies into work practices. Appropriation holds that people actively select how functionality and social structures embedded in the technology are used, and that a given feature may be deployed in different ways depending on how it is appropriated. Thus a key concept that emerges from organizational research in adoption and implementation is that structures, rules and resources provided by technologies and institutions are subject to appropriation by users.

Underlying the notion of social construction of technological systems (Bijker, Hughes, & Pinch, 1987) is a similar set of premises, albeit embedded in a larger sociological context. Social constructionist theory argues that just as technology is shaped by political, economic, social and technical factors, its use will be shaped by individual and societal influences (Bijker et al., 1987; Bijker & Law, 1992). Technology-as-designed provides a range of possibilities for appropriation by users. When technology is deployed by users, they appropriate technology in different ways so that technology-in-use is different from technology-in-design (Carroll, Howard, Vetere, Peck, & Murphy, 2002). As a result technology is shaped and reshaped over time and may eventually reach a state of equilibrium where it becomes embedded in users' lives. Its continued use will depend on recurring reproduction and reinforcement of appropriated use, failing which the technology may be disappropriated by users. Thus, according to social constructionist theory, the way in which a technology is used cannot be understood without understanding the context in which the technology is embedded. Rather than use that is faithful to design, instrumental uses and dominant attitudes influence the incorporation of new technologies by users. Moreover, the manner in which technology is appropriated further influences the design of technology which in turn shapes use and users indicating that there is no linear path between technology adoption, its use and its impact on society.

The advent of matrimonial Web sites represents the introduction of a new technology into the complex social process of arranged marriage. Their design, largely modeled upon the design of online personals in U.S., is likely to have features that mirror the more western societal context of finding dates and partners. The Web sites offer a number of affordances to users such as content rich personal profiles, more choice, ability to search and filter, many-to-many communication, direct communication and disintermediation among others. However, the adoption of such online services would not depend simply on the characteristics and availability of technology, but on how users appropriate and repurpose the technology artifact for their use. In addition, societal factors such as the image of these Web sites, testimonials and references from the close circle of family friends that bear influence in such matters will also play a critical role in the use of matrimonial Web sites. Context-specific technical characteristics associated with arranged marriages in India such as caste, sub-caste and 'dowry' will play a role in determining if these services are used and how they are used to accommodate such considerations. The social construction of technology perspective allows us to investigate the adoption and use of matrimonial Web sites for arranging marriages to provide a rich and deep insight into the interplay between technology and complex socially embedded roles, relationships and rituals.

ARRANGED MARRIAGE IN INDIA

A nation of over one billion people, India is a country of many contrasts and contradictions. A visitor may witness signs of a progressive economy in its infrastructure, media and use of mobile communication devices. At the same time, institutions such as marriage and the role of women continue to be dominated by traditions. Sociologists categorize the Indian family structure as 'patrifocal' in nature (Seymour, 1999). Prevalent norms and values emphasize the interdependent nature of family relationships in contrast to independence and personal autonomy. From a very young age, children are socialized to identify with the family as a whole and discouraged from developing an autonomous self. They are conditioned to place the interests of the family ahead of their own. Alienating and confronting parents and family is still an anathema to most young people, especially in important decisions such as career selection and marriage. Furthermore, cultural mores frown upon the socialization among men and women in the form of dating and relationships. As a result, arranged marriage is still the dominant way for families and individuals to find partners for marriage. Once married, norms dictate that as a wife, a woman should put the needs of her husband and his extended family above her own needs. In a majority of instances, the newly married couple takes up residence at the bridegroom's parents' home.

The typical western view of arranged marriages tends to be biased by its own traditions and values which emphasize individual choice and responsibility. For many in the West, an arranged marriage represents women being treated as property. Their wishes subordinated to patriarchy's desire for property and power. They find it difficult to comprehend that women (or even men) could be pushed into marriage, sight unseen. Although some of these views are well-justified for sections of the society that are socially and economically handicapped and vulnerable. For many others, arranged marriages represent a lifetime of commitment to family and mutual goals. Arranged marriages can provide a degree of emotional security and economic stability that most people in the West would not expect from marriage. Even when raised in a western culture, Indians prefer arranged marriages. The practice has left Indians with the lowest rate of intermarriage of any major immigrant group in the United States with fewer than 10% marrying outside their ethnic group (Bellafante, 2005).

The process of arranging marriage can be a long and elaborate process involving the extended family and friends that culminates in elaborate wedding ceremonies that extend over several days. Traditionally, parents start the process when their children are considered to be of marriageable age, which in the case of women tends to be 22 or 23 years of age and for men around 26 years. The process may be put on hold if they are pursuing higher education or hastened when they start drawing a regular paycheck. The need for suitable alliances is broadcast to the extended family as well as friends. Biographic information about potential matches is exchanged using formal (résumé) and/or informal (oral description) communication. The process of selection is layered and nuanced involving many different considerations. Traditionally, the caste of a prospective match would be a major consideration. More recently, anecdotal evidence suggests that even though caste and sub-caste play a role, primacy is often awarded to level of education, profession, economic background (and potential) and the family of the prospective match. Informal background checks are performed to assess nature, character, prior relationships, habits (such as smoking and drinking which are frowned upon) and reputation of the family. These checks are usually conducted through the informal network of friends and relatives. The screening process also involves astrologers (who may also perform the role of a priest for the family) who evaluate the horoscope of the prospective bride or groom for compatibility.

Once suitable matches are screened, the prospective groom and his family visit the prospective bride's family for a face-to-face meeting to assess compatibility of both the families and the prospective partners. Usually, the prospective partners are allowed to spend a brief amount of time to talk to each other. At this stage the process has progressed closer to a likely successful arrangement and has greater stakes for the families involved. Too many rejections (especially of the prospective bride) can create tension in the family and are considered to be a stigma on the family. More likely, one of a few of the handpicked matches results in satisfactory agreement among the families. What follows the agreement of marriage are complex negotiations about the logistics of marriage—where, when (at an auspicious time determined by astrologers), who will attend, number of guests, dowry if applicable, involvement of priests and rituals to be performed, among other myriad details. In most instances these negotiations exclude the bride and groom. The agreement is formalized with an engagement ceremony which can be relatively simple ceremony marked by an exchange of rings or in other instances as ceremonial as the wedding itself.

The brief description here summarizes the traditional arranged marriage, one that is still largely prevalent in the Indian society. However, over the past several decades, social and geographical mobility have weakened the extended family structure and increasingly replaced it with a more nuclear family structure. As a consequence, social networks provided by the extended family structure are no longer available to parents for finding suitable partners for their marriageable offspring. The absence of such social networks is felt even more by the large, growing, mobile and educated middle class of India. This led to the emergence of matchmaking services, classified advertisements, and more recently online matrimonial services. Rao and Rao (1982) indicate that such anonymous channels of matchmaking as matrimonial ads are more prevalent in urban India where a majority of the middle class reside. As the nuclear family structure becomes more prevalent, the trend in arranged marriage is to allow greater participation of the prospective bride and groom. Some argue that this is changing arranged marriage to one that is more of an 'assisted marriage' (Bellafante, 2005). Perhaps the most important change is the granting of 'veto' power by parents to their offspring on any marriage proposal introduced by them. Prospective partners often go on 'arranged' dates, which may be supervised under the watchful eyes of an elder relative. Sometimes there is an extended period of dating prior to a formal agreement to the marriage.

Earlier studies of arranged marriages in India have looked at role of dowry (Anderson, 2003), status of women (Rao & Rao, 1982) and application of Markov decision making models to the marriage decision making process (Batabyal, 1998). The advent of online matrimonial services introduces a technological artifact into the process of arranged marriages bringing technology to the forefront, amidst changing social practices. At the very least, the technology provides a number of affordances to families seeking partners for marriageable family members. In an otherwise information-sparse environment that consists of either the limited and cryptic information of a newspaper classified or the filtered and often embellished information provided by a brokerage service, online services allows their users to post extensive information about potential partners. No longer dependent on the social network, users have access to a significantly larger pool of prospective partners. The richer information in each profile enables users to perform more complex searches and use a variety of criteria to filter and screen potential partners. Depending on the online service, further communication may be facilitated with online chat and/or e-mail allowing for further exchange of information and interest. Finally, since the technology does not distinguish between parents or prospective partners, it has the potential to completely disintermediate the role of family members and emulate the more western model of individuals finding their own partners for marriage. Thus, the use of matrimonial Web sites in India provides a fascinating setting for examining how these affordances provided by the technology are appropriated. It provides an opportunity to examine changes in power and control structures and the relationship between technology and social institutions.

RESEARCH METHOD

The objective of this research is to examine the adoption and use of technology in situ in a complex social process involving numerous stakeholders in order to develop a grounded understanding of the phenomena. Our aim here is to observe what people actually do rather than what they say they do or what they say they should be doing. We rely on ethnography to observe, document and interpret the appropriation of technology in arranged marriages in India.

Ethnography as a research method was developed by social and cultural anthropologists where the researcher spends a considerable amount of time in the field observing the phenomenon within its social and cultural context (Myers, 1999). In recent years, an increasing emphasis has been placed by researchers on the social and organizational contexts of information systems and ethnographic research has emerged as an important tool for studying these contexts (Myers, 1999; Schultze & Leidner, 2002). Early IS research that used the ethnographic approach focused on human-computer communications (Suchman, 1987) and was the basis for the widely known *In the Age of the Smart Machine* by Zuboff (1988). More recently, ethnography has been used to study management of information systems (Davies & Nielsen, 1992), development of information systems (Orlikowski, 1991; Myers & Young, 1997), their implementation (Orlikowski, 1992a), knowledge work (Schultze & Leidner, 2002), and their impact (Randall et al., 1999).

With its emphasis on participant observation over extended periods of time, ethnography is considered to be one of the most in-depth research methods possible (Myers, 1999). The method places primacy over first-hand observations made by researchers who are immersed in the social and work lives of their subjects (Atkinson & Hammersley, 1994; Myers, 1999). By focusing on socially-situated observations, we develop rich descriptions of the how participants in arranged marriage engage with each other, adopt and appropriate technology and analyze the role of technology in shaping the social context to generate theoretical insights. As ethnographers we adopt a sense-making and learning role as compared to the more conventional scientific approach of formulating and testing hypotheses. The approach deploys a flexible and somewhat unstructured research design where the actual progression of the phenomenon (e.g., an arranged marriage) and study participants drive the data collection process.

As ethnographers, researchers act as their own research instrument; as a result they are driven by their unique identity, knowledge, experience and subjectivity. The researcher has to rely on his/her personal experience in engaging with the research phenomenon to develop an understanding and generate theoretical insights. The ethnographic narrative arising from the study then become experiences of shared subjectivity. In writing ethnography, researchers often engage in writing and rewriting their own identities (Chawla, 2006). Given that ethnography is often associated with observing cultural context as an outsider (Atkinson & Hammersley, 1994), we as natives[1] of the culture are in some ways 'insiders' to social setting in which we perform our investigation. However, it offers the advantage of being readily accepted, having a shared history, understanding of the context and related experiences. Moreover, participants are less likely to view us as outside observers because we look native, speak the language, and are to some degree (albeit loosely) embedded in the social fabric of their daily lives. Chawla (2006) argues that as an ethnographer, native or otherwise, researchers enter the field entrenched with degrees of outsiderness that instills a certain amount of objectivity and distance into their observation and analysis. Moreover, just like other research that adopts a more scientific approach, ethnographic research is expected to meet standards of objectivity (Schultze & Leidner, 2002). As scientists, ethnographic researchers have to balance subjectivity and objectivity in a manner that convinces the academic community of the generalizability and reliability of their inferences.

DATA COLLECTION AND ANALYSIS

Data for the study was collected over a span of fifteen months which included two visits, 63 days in 2006 and 49 days in 2007, studying Web sites and follow-up conversations over telephone. During the first visit, the authors spent time talking to people and collecting secondary data about matrimonial Web sites, investigating the sites and the success stories posted on these sites. In the second visit, secondary data was used as a basis for identifying broad issues and research questions for primary data collection. A majority of the time on both visits was spent in Mumbai, and supplemented by data collected from other major metropolitan cities over brief visits. As in studies of this nature, the emphasis is not as much on using a representative sample as it is to develop a deep understanding of the phenomenon. At the same time, we did wish to get a sampling of different families and the roles played by different members of the family in using online matrimonial services. The data collection process involved conversations with prospective partners, parents, siblings and close relatives which ranged from informal interviews to just observing conversations as they took place in households.

Table 1. SMI services offered by matrimonial Web sites in India

Search (Information Gathering)	Matching (Decision Making)	Interaction (Relationship Formation)
• Religious • Social background (caste, sub-caste, Gotra[3], Manglik) • astrological information (horoscopes and sin signs) • lifestyle(smoking, alcohol consumption, vegetarian), • culture (languages spoken and values-liberal, traditional, modern, etc.) • complexion (fair or 'wheatish' rarely dark) • body type (slim or average never heavy) • living conditions (income, living with parents, nationality, citizenship and work status in different countries such as U.S.	• Horoscope based matching • Push (results delivered in the mailbox) and pull matching (filtering based on user criteria) • Can pursue multiple matches simultaneously	• Contact through the service • Phone and e-mail addresses • Built-in chat services

Since the focus of our investigation was online matrimonial services, a considerable amount of time was spent in understanding the technology itself. This involved two aspects; the first was to understand the nature and type of services provided by different Web sites[2]. We document these using the search, matching, and interaction framework (SMI) proposed by Ahuvia and Adelman (1992) in Table 1. The SMI framework is based on the primary roles performed by any market intermediaries namely, searching, matching and transacting. Ahuvia and Adelman (1992) developed the SMI framework to categorize the processes that are involved in the marriage market and proceeded to describe the marriage market intermediaries in terms of how they performed these processes. As marriages do not happen in a vacuum, the search-matching-interaction framework integrates the context in which the relationships dyads are embedded with the interpersonal processes involved in the formation of the relationship. The second aspect involved observing the ongoing appropriation of the online service by users as reflected in the profiles and success stories documented on these Web sites. These success stories are obviously intended as testimonials by other users for the service, but they provide an additional source of information and details on how the partners decided to adopt the service, how they used the service, and the role of other family members. While these secondary sources of data are not central to this investigation, they helped us in understanding of the context and develop a more complete understanding of the phenomenon.

During our visits, we spent a considerable amount of time talking to different families that were actively engaged in the process of finding a suitable partner for a family member. They were in different stages of the process; while some had just begun to test the waters, others were actively evaluating candidates. In one instance, we were able to follow the process right up to the actual wedding ceremony itself, which was attended by one of the authors. Our time in the field was spent initially in identifying families that would be suitable candidates for collecting data from the social network of our relatives in India and introductions made through this network. Conversations about arranged marriage took place in a variety of settings. A considerable amount of time was spent in participating in day-to-day activities of the participants, many of which involved shopping, eating out, or simply sharing a ride with them as they commuted either for work or social engagements. When a search is active, it was not too difficult to get a family talking as it would invariably be at the top of their minds. While families were observed as such, discussions took place over afternoon tea or a meal; there were many opportunities where there were one-to-one conversations. Apart from group settings, the cultural context and topic are such that women are more apt to discuss and share their feelings, emotions and thoughts on the subject in depth in a one on one conversation.

Ethnographic research suffers from unique issues of validity and reliability (LeCompte & Goetz, 1982); replication of these studies pose problems of variation in context. Collectively, we spoke to about 39 individuals during our stay and during subsequent follow-up telephone calls. Of these, 23 were women and 16 were men. Of the men, six were fathers of the women, four were brothers of the women, twq were uncles of the men and four were the men in the marriage market. Of the women, 12 were the women who were candidates for marriage, six were mothers of the women, four were sisters of the women, and one was the mother of a man in the marriage market. About three-fourths of the respondents were from the northern part of India and the remaining were from the southern part of India. All of our study participants were from metropolitan cities and because of the nature of our sampling process, we did not have access to people in small towns and rural India. The average household income of the participants likely ranged from 25,000 rupees per month (about 600 dollars) to about more than 100,000 rupees a month (2,400 dollars). We did not ask direct questions about income because such questions were not appropriate in the social milieu in which we were interacting with the participants. Thus the participants constituted members of the Indian middle and upper middle class. It is also our assumption that many lower-middle class families do not use the Internet for matrimonial matchmaking. Almost all of our participants belonged to the upper three castes in the Indian caste system, and our social network limited our access to lower castes. The first few of the participants were members of the authors' extended family, friends and larger circle of acquaintances who then directed us to others who were participating in matrimonial Web site-based matchmaking. This limitation in our sampling limits the generalizability of our study. Our study is limited in scope to urban middle class families participating in the Web-based matchmaking and is influenced by the authors' perceptions of traditional arranged marriages in India, as well as what our study participants, especially the fathers and mothers, recounted about how arranged marriages used to take place in their time.

The nature of our conversations focused on information included in the profiles, how the process of arranged marriage was conducted using the Web sites, the role of the family versus the partners themselves and interactions between families and partners before and during the decision making process. The questions were woven into the conversation, sometimes requiring repeat interactions and were transcribed at the end of the day. Due to the nature of immersion in the field, it is not possible to precisely draw a boundary on how many hours of actual conversation form the pool of field experience. In order to keep track of our conversations, we kept a daily log individually that included both field notes and our own reflections. Once every week during our visits, both authors would spend a couple of hours going over their own and each other's logs to fill gaps and discuss progress.

Data coding and categorization was done manually by the each of the authors at the end of the study. The next step was discussion and synthesis of the variations in coding and categorization schemes used by the authors. At this step, where needed, further data was collected through follow-up phone conversations. Or coding and categorization schemes centered around our key research questions, the role of the family in initiation and the decision making process, the degree and length of courtships, the sequence of courtship in the matchmaking process, and the preferences of the participants and family members about the chosen characteristics of the potential partner.

CHANGING ROLES, SHIFTING TRADITIONS AND CULTURAL CONVERGENCE: ONLINE MATRIMONIAL SERVICES AND ARRANGED MARRIAGE

There are a growing number of Web sites that are dedicated to providing matrimonial services in India. Major players from U.S. in Internet-related businesses view this as a potential market and have tied up with Indian firms, for example, Yahoo! along with a venture capital firm has taken up a stake in BharatMatrimony.com. Microsoft has ties with Shaadi.com, another popular Web site. The number of users of these services has grown from about 4 million in 2004 to 7.5 million in 2006 according to estimates provided by Internet & Mobile Association of India (Lakshman, 2006). Although the online market in these services is only about 4% of the estimated $500 million spent on off-line matrimonial services, it is expected to continue to grow at the rate of 40%-50% every year. As the bulk of the off-line market consists of print classifieds, if trends in the U.S. newspaper industry are anything to go by, online matrimonial services could soon overtake print media. A growing educated middle-class and sustained economic growth is only likely to further fuel the growth in these services.

Popular beliefs, especially in urban India, indicate that the popularity of these sites reflects the changing face of India. Traditionally, the family plays a very important role in arranged marriages in India. It begins with announcing the entry of the prospective bride or groom into the marriage market. The family influences the matching and selection process to preserve traditional notions of compatibility in terms of age, family culture, caste, community and horoscopes. Like most cultures, the bride is usually given away by the father, but unlike western culture not to the groom but to the groom's family. Once married, the bride becomes a member of the groom's family and is expected to have only weak ties to her own family.

The fact that most grooms stay in their parents' homes, which is the norm, further reinforces the symbolic nature of this transition from one family to another for the bride. For this reason, the compatibility of the bride with the groom's family and not just the groom is considered to be of greater importance than the relationship between the groom and bride's family. To understand how families are using online matrimonial services for arranging marriages, we look at the matchmaking process in terms of search, matching and interaction. As we look at how technology is appropriated, we address the research questions: how do features presented by matrimonial Web sites used for arranging marriage, how are they changing the nature of the marriage process, and the norms and traditions associated with arranged marriage. We primarily focus on the role of family in the process against the backdrop of social and cultural changes permeating the Indian Diaspora.

SEARCH

The decision that a son or daughter should enter the marriage market is usually made by the parents and as a consequence the process is initiated by parents or a trusted family elder. For somebody to raise the matter of their own marriage would be considered bold and indicate that they are self-absorbed rather than thinking in terms of the interests of the family. Many young people in fact choose to avoid or postpone discussion of marriage using education or career as excuses because they believe they will be heading off confrontation between their expectations of a partner in marriage with those of the family. Some view it as a battle they will eventually 'loose' as the family will eventually 'force' them to compromise and make a choice. Others view it just the opposite, using delay as a tactic to wear down their parents as they start worrying about the window of marriageable age slipping away and are then ready to allow the son or daughter to have an upper hand in the selection process. All this suggests the absence of open communication within the family as individual members try to conform to their expected roles while engaging in signaling and power play.

Traditionally, the parents and family elders would start the process by raising the matter when they meet relatives and friends that they have started 'looking' for a match for their son or daughter. With the changing family structure and weakening social networks, this increasingly poses a challenge for families. For expatriates, who have been disconnected from these networks in India, the challenges are only greater. Choices may be few, not match the desired profile and the family also opens itself to pressure from the extended family. Before online matrimonial services, families would rely on third-party matchmaking services and classified advertisements. Finding the right matchmaking service presented its own challenge as most would not disclose the demographic nature of their pool of prospects. They might be skewed towards a particular community that the family was not interested in or avoid a community altogether thereby reducing the efficacy of that brokerage service. Family elders indicate that until recently using matrimonial ads in newspapers was stigmatized and considered to be the last resort. Research in different national contexts indicates that till recently the users of matchmaking services were stigmatized (Darden & Koski, 1988). Classifieds were resorted to when families had problems finding matches the traditional way. Anonymous and charged by the word, parents would craft cryptic classifieds in attempt to compress all their myriad criteria in addition to posting basic biographic information into 50 words or less.

On the Web the restrictions posed by classifieds on the amount of information disappears. Whereas photographs would generally be exchanged only after the initial responses to classified were filtered, on the Web, most users post photographs along with the profile. This suggests that the stigma associated with using impersonal and anonymous methods has eroded in most cases. Many families view this as the preferred method because they feel that they have a better chance of finding suitable matches in view of the more detailed information and larger pool of prospects. By browsing the Web site, users can make a quick judgment about the demographics of the pool and its suitability to their selection criteria.

As it is customary in Hindu families for parents to search for their children's life partner, my parents were doing the same by speaking with relatives and also circulating our 'bio-data' via various electronic resources[4]. [1]

Even with online services, parents continue to perform the role of initiating, searching and filtering potential partners. Gender stereotypes persist in the new medium. This is consistent with research on dating ads in other countries which suggest that gender and role stereotypes and expectations likely persist and change very gradually over time (Peres & Meivar, 1986; Koestner & Wheeler, 1988). Women indicate that in the Indian societal context even if they were actively involved in the process or actually posted their own profile, they wanted to maintain the appearance that the process was being initiated and managed by their parents. In all of our conversations, we seldom found the roles reversed or even an attempt to convey the appearance that they had been reversed. Women, in most cases, did not post their own profiles because they were afraid of being considered as 'fast and easy'. They would recount the experience of their friends who posted their own profiles had ended up in situations where the men were exchanging messages with them without any interest in marriage and were just looking to have a good time. Parents also perceived a greater responsibility in having their daughter married at the right time with the right match because they perceived that they were custodians of their daughters until they joined their 'true' family. This put greater pressure on parents to initiate and start the search for a suitable partner. Parents also believe they can pre-empt their daughters finding their own partners or avoid the issue especially in the case of career-oriented and ambitious women who were not interested in getting married in their twenties. One 28-year-old woman who is a consultant in a multinational firm explained:

*I have to agree to put up with the ad because my parents keep pressurizing me.... at least this way it appears to them that I am interested in getting married and at the same time I can keep rejecting the matches they get for me.*₂

Her parents said:

*the girls today want a very specific kind of husband.. he should be liberal and modern⁵ in his thinking.. she does not like anyone we find in our community.. by posting this ad perhaps we can find such a person in our specific caste and subcaste somewhere else.*₃

Another woman stated:

*My profile was created by my elder brother and every day he checked for a partner for me.*₄

While parents conceded to their daughters' desire for a like-minded partner, they also sought to preserve traditional notions of the primacy of caste and community.

Interestingly, a greater proportion of the profiles of grooms were posted by prospective grooms themselves. Traditional gender stereotypes hold that men should be allowed greater independence (as long as they stayed or had their parents live with them). The perception that men are more technologically savvy created the circumstances for unmarried men to play a bigger role in online matchmaking. Parents felt they could deal with the low-tech nature of classifieds, but often felt ill equipped to deal with computers and the World Wide Web.

The shift from classified to online services has created increased opportunity for communication that was otherwise absent or predicated on non-verbal gestures and behavior. Once a classified is published, the action is followed with responses that come in batches and decline rapidly with time. The batched nature resulted in periods of waiting followed by some uncomfortable discussions among parents and their children. With online services, the profile has a greater shelf life yielding a steadier stream of responses. Moreover, there was the opportunity to engage in a more continuous as opposed to episodic search for potential partners. This created increased opportunity for communication within the family and more importantly for a tacit convergence of expectations.

Caste and sub-caste continue to be major considerations in search. In each of the online matrimonial services, a user could select bride or groom by caste, sub-caste, religion, language, state and age among other criteria. Almost everyone we talked to were very specific about religion and caste as filtering criteria. This was also reflected in the profiles that were posted on the Web sites. For example, a Hindu (religion), Brahmin (caste) female would want a Hindu, Brahmin male for a husband. In another posting, a 24-year-old woman from Mumbai sought a Hindu from the same caste speaking a specific language (Malayalam), but having moderate and liberal attitudes.

The profiles also reflected social taboos associated with suitable partners. Many users specifically mentioned smoking and drinking. Partners who did not smoke or drink are preferred. Social taboos associated with smoking and drinking persisted in the new medium. In talking to women, drinking alcohol was not preferred for men and absolutely taboo for women. Other criteria for search that persist in online profiles are horoscope and skin color. Most men sought partners who had a fair or 'wheatish' complexion demonstrating the continued belief in India that fair skin is associated with beauty.

MATCHING

Traditionally, arranged marriages have been brokered by family and friends, an elaborate process laced with social nuances that involves matching candidates on the basis of caste, community, religion and horoscopes. The family plays two important gate-keeping roles; the first is that of controlling the entry of new members into the family, especially the bride, and ensuring that they are compatible with the family's values and traditions. In its second role, the family perpetuates the caste, community and religious divisions in the society. These societal divisions are viewed as surrogates for compatibility and for ensuring that traditions are carried on from one generation to the next. Our conversations with families showed that parents continued to perform this important gate-keeping role. In most cases, members of the family first screened the responses and made the first contact before allowing the prospective partners to meet each other.

We met in coffee day after our parents talked to each other.[5]

Our parents then arranged a meeting for us.[6]

My father showed an interested in his profile and give him contact no. and e-mail ID. He accepted and forward my profile to his parents .... then he talked to my father and said he want to come to my parent's home in Dehradun. Within a week he came and finalized the matter. Then my sister and Jijaji (Brother-in-law) went to his home at NOIDA and found suitable and said ok from our side. His parents and my parents talked to each other on the phone and after that the marriage was fixed.[7]

After reading my profile and showing it to her father, her second cousin phoned my father (who was the contact listed with my profile) requesting more information about me.[8]

Firstly, both my parents and my partner's parents contacted each other to make sure that all necessary requirements were suitable in order for the marriage proposal to materialize.[9]

I approached to her and received a response from her father.[10]

With online services, there is potential for disintermediation of the role played by parents and their status as gatekeepers starts to diminish. The process of screening and matching is no longer solely dependent on parents. Family could control information as either they went and met intermediaries or received the responses to classifieds. Now the information is always there, only a few clicks away. It is not surprising that sometimes the initiative towards matchmaking is now being taken by the prospective partners themselves. Even if they found their own partner, given the dominant role of parents and the strong bond with family, they would still seek the approval of their families. Whether it was the family that acted as a gatekeeper or a final consenting authority, the family is ever present in the matchmaking process.

My thanks to ... for bringing the two families together[11]

We interacted about our family details and the kind of partner being sought. We both were satisfied with each others' families, culture, background etc.[12]

With the **consent** *and blessings of our families.*[13]

It is interesting that even if partners took initiative, they would mimic the very criteria that would be applied by their families, such as caste, community and religion. It did not matter whether the profiles were posted by family elders or the candidates themselves; religion, language, community, caste and sub-caste were always a consideration. Almost all profiles mentioned their own religion and caste as well as their preferred religion and caste of the partner.

marrying a Muslim is out of question. [14]

our daughter-in-law is a Hindu ... we still wish our son would have chosen a Muslim girl. [15]

If I marry a Bengali (language).. it will be easier for her to interact with my grandparents and extended family. [16]

Brahmins (caste group) are very particular about who they marry. [17]

The use of online matrimonial services in fact seems to make it easier to find someone within the sub-caste of your choice. In the absence of these matrimonial services, the ability to find someone within one's caste group depended on the reach of your extended family and the resources available. Families in the past would often compromise by marrying in the community outside the caste group because of limitations in the pool of applicants available. However, with geographical barriers removed by online services, it has become possible to find someone belonging to the exact same caste or sub-caste as that being sought by the family. This same someone (invariably the bride) is also willing to move halfway across the world to live with her newly wedded partner. Thus online services not only perpetuate traditional notions of an acceptable partner, but also provided increased choice. Sometimes, this would also create decision delay, as there was always hope that someone more perfectly matching the filtering criteria could come along in the future.

This is great! The girl of my dreams could be on the other side of the world. [18]

I can't believe that I can find someone who matches my exact profile needs thousands of miles away. [19]

Matching the horoscopes was also an important concern for many families. Some online services even provided this as an additional feature of their services.

the gotras should not be common. We would also ask for your details for purpose of horoscope matching [20]

Our families met and even the horoscopes matched well [21]

Sticking with our family tradition, we matched horoscopes and got elders consent. [22]

A cursory analysis would suggest that online matrimonial services simply replicate the off-line process of arranging marriages. It is evident that the criteria used for matching partners are largely carried over online. At the same time these services provide greater transparency and access that is loosening the grip of the family over information and eroding their role as gatekeepers. At times, the role is completely disintermediated by the presence of online services. Using online services increases the pool of potential partners and provides greater choice by breaking down geographical barriers and filling the gap created by weakening social networks. The increased choice does come at a cost—that of information overload.

INTERACTION

Perhaps the biggest change that online matrimonial services have introduced to arranged marriage is opportunity for interaction. In the traditional off-line mode, once a potential match was identified, the next step would be for the prospective groom's family (with or without the groom) to visit the prospective brides family at home. In more traditional families this may even be preceded by several visits by other relatives and/or meetings of both families at another relative's home. Setting up the visit was often a complex negotiation. The prospective partners are usually allowed to spend a brief amount of time alone to talk to each other. Other than that, it was rare for them to get an opportunity to get to know their future partner in marriage better. The bride and groom would also not have much say in any of the decisions about the wedding negotiations and arrangements.

With online services, many of these restrictions are being lifted. Prospective partners are often allowed to interview each other online without parental supervision. This represents an interesting shift in the role of the family in determining the compatibility of both partners. By allowing such communication, the responsibility of assessing compatibility and selection are gradually moving away from the family elders to the prospective candidates. Since the joint family structure is slowly disappearing, the parents often seem content to apply the broad criteria of religion, caste, age, economic background, and so forth, on the choice of a partner and leave the rest (personality, likes, dislikes, etc.) to their children. With the help of online services, a period of online courtship has emerged. Potential partners are allowed to get to know each other by exchanging e-mails, talking to each other over the phone or chatting online. In almost all instances, this was long-distance without any face-to-face meetings. Although most families still do not allow the prospective partners to meet alone or go out on a date, they did not seem to have a problem with electronic interaction. This step of getting to know a potential partner better seems to have emerged with modern communication technologies and precedes any formal meeting between the two families.

we chatted for 3 months[23]

she sent me an e-mail.. and the next day everything was finalized.[24]

We dated ....(electronically).. for six months.[25]

We chatted and spoke on the phone for hours and found that we are a perfect match for each other.[26]

We chatted on (name of Web site) and thereafter we decided we r going to marry even before our families could meet.[27]

Earlier her family was not at all interested in me but slowly with our persistence they were forced to get us married.[28]

The norms of matchmaking in India have traditionally been different. After a meeting of the families, the bride's family would patiently wait for a response from the groom's family. If the days stretched into weeks and there was no response from the family, it indicated that they are not interested in pursuing the match further. The implicit rejection when a marriage proposal is turned down could also carry a social stigma. There are no such unspoken rules on online interaction. Chatting and dating can go on for a long time before there is any discussion of marriage. Online interaction allows users to disengage easily without any stigma associated with such rejection. This can at times create problems and emotional issues for some users who are not used to having such extended relationships.

we chatted for many months but he never wanted to take it forward.[29]

The traditional way of finding marriage partners through family and friends provides a certain amount of accountability; there is tacit trust which when violated can have social implications. With the search for partners going online, the process is taken out of a social context. While the search can cut across traditional social networks to find potential partners they are otherwise unlikely to reach, it makes judging the credibility of the information online even harder. Many families indicated that conducting a background check was very difficult with online matrimonial services. In some cases, the whole thing seemed to fall apart, after the families met as they did not approve of each other, or the family was different from what they had expected based on the descriptions provided online.

the girl was educated and pretty but she turned out to be crazy..$_{30}$

the family did not seem as reputable as they claimed to be..$_{31}$

In many instances, it was still possible to get some background checks done through family and friends. While the Web sites themselves did not provide any easy methods to facilitate a background check, some families went out of their way to do a background check by trying to find mutual contacts in the community who could help them with more information about the family. Also used in lieu of the background check was applying the traditional filters of religion, language, community, caste, and so forth, with the implicit assumption that users similar in background also bore a level of trustworthiness. The inability to validate the information from traditional networks is another reason why families found it useful to allow the potential partners to communicate. Families acquiesced to online and long-distance interaction so that prospective partners could sort out values, norms and beliefs. As long as the potential partner fit the traditional filters, families let the partners do further selection themselves.

In my profile, I had mentioned ... who happens to be my dad's uncle and his neighbor in (city name), and a common link between the two families. At that point of time he was in UK and I was in Bangalore. He called me and we had a conversation

... We exchanged all the details about us and both the families through.. uncle$_{32}$

I received an e-mail stating that she was interested in speaking with me. Soon after ... we began to chat on the Internet via Yahoo! Messenger. As soon as we started to communicate ... our sessions lasted well into the wee hours of morning.$_{33}$

Then we started chatting over msn. Which was on for 3 months$_{34}$

It is evident that online matrimonial services have introduced new elements into the process of arranging marriage that are made possible by technology. The perception of relative anonymity and informal nature of the medium along with the absence of context provided by traditional social networks has allowed potential partners to play a greater role in the process. With the process shielded from the view of the immediate social network of the family and therefore any possibility of social sanction or stigma, new forms of interactions and steps in the process are emerging. Families do not have to engage in elaborate orchestration of the interaction between families prior to settling on a choice for partner. The informal nature of interaction and absence of face-to-face communication make it easier to engage and disengage and as an interesting consequence obviate the need for signaling and the ambiguity associated with signaling (are they interested or not?).

DISCUSSION

Our analysis of the use of online matrimonial services for arranged marriages reveals the possibilities created by technology and how they are appropriated by users. To many, the change may appear to be glacial in pace, but against the backdrop of a society that has a history and traditions dating back several millennia, they show shifting roles, changing traditions and convergence with the more modern view of marriage and family held by western culture. Information technology in the form of matrimonial Web sites creates both online and off-line possibilities for users as they go about the traditional process of finding life partners and new members for their families.

Families seeking partners for their marriageable son or daughter typically operate in an information-sparse environment. Reliance on the friends and family network meant that much of the information was subjective, by word of mouth but embedded in the social context. The cryptic nature of information from classifieds and the commercial nature of third-party services created many challenges for families. The digital world, where space is not at a premium, creates a more *information rich environment* in which the search can be conducted. The relative *anonymity* and *privacy* provided by using the service at home or an Internet café reduces concerns about any social stigma that may be associated with having to rely on such services rather than social networks. Moreover, our data suggests that as the institution of joint family recedes from family life, any negative association about the use of such services is disappearing.

With online services, the role of immediate social network, classifieds and third-party brokerage services is *disintermediated*. Technology allows users to *cut across boundaries* created by distance and social networks allowing creating a larger pool of potential candidates. To ease the process of selection, all Web sites provide *search tools* for users to specify their criteria. All Web sites capture and allow search across traditional criteria of religion, caste, community, language in addition to age and economic background. This reflects the influence of social context on the design of Web sites. At the same time, the technology is reducing the need for this information. In an information-sparse environment, the application of these criteria as filters served as surrogate indicators of compatibility, ensuring that the families of the bride and groom had similar values and traditions. In an information-rich environment where all kinds of information about habits, likes and dislikes is available, the need for surrogate indicators diminishes. This allows families and potential partners to express their subjective preferences that are more direct and reflect more rational criteria. As many of the Web sites are modeled after online personals popular in western cultures, these design elements in the form of such content and ability to communicate directly using e-mail and chat appear to engender a *cultural convergence* between western and Indian notions of partner selection. Some of the Web sites, as an indicator of such cultural convergence, provide dating services in addition to matchmaking services. Video profiles are also becoming an option that is used by some users on these sites.

The roles played by different family members in the process are also adjusting with the use of technology. Parents and/or other family act as gatekeepers to control the information flow from various sources as well as information about prospective matches. In the digital world, the information is (persistent) always available online and easily *accessible* to all members of the family. This weakens the role of parents as gatekeepers. It also changes the nature of the flow, which is more continuous as compared to the more episodic flow of classified advertisements and meeting with relatives or brokerage services. The *continuous flow* of information creates more opportunities for interaction among family members over casual conversations. The increased communication surfaces concerns of different family members, bridges role boundaries and generation gap, and helps creates a greater consensus within the family.

Perhaps one of the more significant changes made possible with online services is the ability to have *direct* communication with potential partners and their family. Online communication does not bear the same credence as face-to-face meeting in formal social settings. The relatively anonymous and informal nature of interaction reduces the

Figure 1. Change in SMI processes with the use of matrimonial Web sites

Search	Matching	Interaction
Information rich environment	Matchmaking based on traditional criteria	Email, chat and phone interaction
Provided anonymity and privacy	Information access reduced Family role as gatekeeper to merely specifiers of criteria	Extended courtship
Distance and geography no bar in finding the perfect match		Virtual dating may precede matchmaking
Gender, role and cultural stereotypes persist in matrimonial ads		Limited parental supervision
		Interaction brings to the fore credibility issues in online information
		Background checks may still be performed offline
		Family interaction reduced
		Easier to disengage if match does not work out
		Reduced stigma in disengaging

- Disintermediation of Family
- Cultural Convergence
- Information Persistence and Accessibility
- Continuous rather than episodic information flow and communication
- Ease of engagement and disengagement
- Reduced Stigma
- Virtual Dating

perceived risk of adverse social consequences of allowing prospective partners to communicate directly. Furthermore, devoid of the context provided by traditional social networks, families feel the need to allow more extensive communication between prospective partners to assess compatibility. This also allows the prospective partners to play a greater role in assessing compatibility and making choices further disintermediating the role of other family members in matching and selection. Figure 1 graphically presents the results of our study.

LIMITATIONS AND DIRECTIONS FOR FUTURE RESEARCH

Our study investigates the influence of technology and its use on arranged marriage in India with the use of online matrimonial services. The study has several limitations in its present form. The nature of the research method creates limitations on replicability and generalizability of the study. As natives of the culture we were studying, we carried with us a tacit understanding of the social backdrop of our study. While the additional insight allows us to see subtle differences and changes and a more nuanced understanding of the phenomenon, at the same time, it creates the potential for subjectivity and bias.

Study subjects were Indians residing in India and did not include Indians residing in other countries who represented a significant proportion of users of online matrimonial services. The immersive nature of field study in ethnography made it difficult to study this segment as they visited India only for short periods of time and it was difficult to talk to them and observe them in a natural setting. However, among the families we studied, several had considered potential partners who were non-resident Indians.

Our study was conducted primarily in Mumbai and Delhi, which are large metropolitan cities. Culture, social mores and traditions differ significantly across urban and rural India. As a result, the narrative cannot be considered as reflective of India in general but primarily urban families. However, it should be noted that a majority of users of online matrimonial services reside in larger cities.

In our attempt to focus on the use of online matrimonial services, we did not study families that arranged marriages using more traditional methods. A more insightful comparison and contrast of the two different routes to arranged marriage would have helped to isolate the underlying changes driving the two processes.

CONCLUDING REMARKS

Online personals, e-dating and matrimonial Web sites are changing the rules of how relationships are formed and maintained in communities all over the world. In societies where dating itself is taboo according to social and religious norms, online matrimonial services are filling the gap left by the absence of social networks in societies transitioning to urban and modern culture. Since there are no established mores about using online media, the online matrimonial services mirror existing social practices. As the technology is used and appropriated by users, both social practices and the technology evolve. The use of online matrimonial services provides an interesting illustration of the social construction of technology.

The use of technology demonstrates a tension created by the affordances provided by technology and entrenched social traditions and practices. The use of online services has diluted the societal norms about socializing among opposite sexes but at the same time preserved traditional notions of compatibility by providing easy access to information about religion, caste and community. Although it is still not acceptable to go out on dates, online relationships are considered acceptable and allowed to continue over extended periods of time without parental supervision. Men and women who are seeking life partners are playing bigger roles in arranged marriage but still consider parents to be the final arbiters. Gender stereotypes continue to persist as women do not wish to give the appearance of driving the process but feel comfortable using the technology to actively participate in the process.

Online matrimonial services are not adopted as an instrument to bring about social change. To be accepted by families, they need to reflect and perpetuate societal and religious traditions and values. As they are recurrently used, the possibilities created by technology and its appropriation by users, creates a new equilibrium that reflects the new social reality created by technology and users. In the case of online matrimonial services, the subtle influence of technology cannot be overlooked as the use of online content, instant messaging and e-mail is expanding the influence of the younger generation over their elders in arranged marriage to create a new social norm that bears closer resemblance to western notions of marriage.

This has implications for the social construction of similar technologies in different societal and cultural contexts. When introducing new technologies with social implications such as cell phones and wireless services, the initial adoption of these technologies is only likely to succeed if on the surface these technologies mirror the traditional norms of behavior and social interaction. However, the appropriation of these technologies by users over a period of time brings about changes in social relationships and interactions. These changes, in their own way, change the structure and features of the technology, further driving social change. For example, India is one of the largest growth markets for the use of mobile phones and the phones were first used in India by affluent families and for business uses; over a period of time as prices came down, they found their way to the lower middle class, the rural areas and the self-employed such as street vendors and maids. In line with the new social classes that have adopted mobile phones, the technologies themselves changed to encourage usage, such as the predominance of pre-paid phone plans, long battery life and the use of am/fm radio as a standard feature, as these phones are used as music players by a majority of the population. This has further fueled the growth in the market of these phones in India.

REFERENCES

Ahuvia, A., & Adelman, M. (1992). Formal intermediaries in the marriage market: A typology and review. *Journal of Marriage and the Family*, *54*(2), 452–463. doi:10.2307/353076

Anderson, S. (2003). Why dowry payments declined with modernization in Europe but are rising in India. *The Journal of Political Economy*, *111*(2), 269–310. doi:10.1086/367679

Atkinson, P., & Hammersley, M. (1994). Ethnography and participant observation. In: N. Denzin & Y. Lincoln (Eds.), *Handbook of Qualitative Research* (pp. 248-261). Thousand Oaks, CA: Sage.

Batabyal, A. (1998). Aspects of arranged marriages and the theory of Markov decision processes. *Theory and Decision*, *45*(3), 241–253. doi:10.1023/A:1004998730922

Bellafante, G. (2005, August 23). Courtship ideas of South Asians get a U.S. touch. *The New York Times*.

Bijker, W., Hughes, T., & Pinch, T. (1987). *The social construction of technological systems: New directions in sociology and history of technology*. Cambridge, MA: MIT Press.

Bijker, W., & Law, J. (1992). General introduction. In: W. Bijker & J. Law (Eds.), *Shaping technology/building society: Studies in socio-technical change*. Cambridge, MA: MIT Press.

Carroll, J., Howard, S., Vetere, F., Peck, J., & Murphy, J. (2002). *Just what do the youth of today want? Technology appropriation by young people*. Paper presented at the The 35th Hawaii International Conference on System Sciences. Hawaii.

Chawla, D. (2006). Subjectivity and the "Native" ethnographer: Researcher eligibility in an ethnographic study of urban Indian women in Hindu arranged marriages. *International Journal of Qualitative Methods*, *5*(4), 1–13.

Close, A., & Zinkhan, G. (2003). Romance and the Internet: The emergence of e-dating. In: B. Kahn & M. Luce (Eds.), *Advances in consumer research* (vol. 31, pp. 153-157). Valdosta, GA: Association for Consumer Research.

Darden, D., & Koski, P. (1988). Using personal ads: A deviant activity? *Deviant Behavior*, *9*(3), 383–400.

Davies, L., & Nielsen, S. (1992). An ethnographic study of configuration management and documentation practices in an information technology centre. In: K. Kendall, K. Lyytinen, & J. De Gross (Eds.), *The impact of computer supported technology on information systems development*. Amsterdam: Elsevier/North Holland.

DeSanctis, G., & Poole, M. (1994). Capturing the complexity in advanced technology use: Adaptive structuration theory. *Organization Science*, *5*(1), 121–147. doi:10.1287/orsc.5.2.121

Dion, K., & Dion, K. (1996). Cultural perspectives on romantic love. *Personal Relationships*, *3*(1), 5–17. doi:10.1111/j.1475-6811.1996.tb00101.x

Fiore, A., & Donath, J. S. (2004). *Online personals: An overview*. Paper presented at the CHI. Vienna, Austria.

Giddens, A. (1984). *The constitution of society: Outline of the theory of structuration*. Cambridge: Polity press.

Koestner, R., & Wheeler, L. (1988). Self-presentation in personal advertisements: The influence of implicit notions of attraction and role expectations. *Journal of Social and Personal Relationships, 5*(1), 149–160. doi:10.1177/026540758800500202

Lakshman, N. (2006). Here come the bride sites. *Business Week*, 42.

LeCompte, M., & Goetz, J. (1982). Problems of reliability and validity in ethnographic research. *Review of Educational Research, 52*(1), 31–60.

Markus, M., & Robey, D. (1988). Information technology and organizational change. *Management Science, 34*(5), 583–598. doi:10.1287/mnsc.34.5.583

Mulrine, A. (2003, September 29). Love.com: For better or for worse, the Internet is radically changing the dating scene in America. *U.S. News and World Report*.

Myers, M. (1999). Investigating information systems with ethnographic research. *Communications of the Association for Information Systems, 2*(23), 1–20.

Myers, M., & Young, L. (1997). Hidden agendas, power, and managerial assumptions in information systems development: An ethnographic study. *Information Technology & People, 10*(3), 224–240. doi:10.1108/09593849710178225

Orlikowski, W. (1991). Integrated information environment or matrix of control? The contradictory implications of information technology. *Accounting. Management and Information Technologies, 1*(1), 9–42. doi:10.1016/0959-8022(91)90011-3

Orlikowski, W. (1992a). CASE tools as organizational change: Investigating incremental and radical changes in systems development. *MIS Quarterly, 17*(3), 309–340. doi:10.2307/249774

Orlikowski, W. (1992b). The duality of technology: Rethinking the concept of technology in organizations. *Organization Science, 3*(3), 398–427. doi:10.1287/orsc.3.3.398

Orlikowski, W., & Robey, D. (1991). Information technology and the structuring of organizations. *Information Systems Research, 2*(2), 143–169. doi:10.1287/isre.2.2.143

Peres, Y., & Meivar, H. (1986). Self-presentation during courtship: A content analysis of classified advertisements in Israel. *Journal of Comparative Family Studies, 17*(1), 19–31.

Randall, D., Hughes, J., O'Brien, J., Rodden, T., Rouncefield, M., & Sommerville, I. (1999). Banking on the old technology: Understanding the organizational context of 'Legacy' issues. *Communications of the Association for Information Systems, 2*(8), 1–27.

Rao, V., & Rao, V. (1982). *Marriage, the family and women in Indi*. Delhi, India: Heritage Publications.

Rice, R., & Rogers, E. (1980). Reinvention in the innovation process. *Science Communication, 1*(4), 499–514. doi:10.1177/107554708000100402

Schultze, U., & Leidner, D. (2002). Studying knowledge management in information systems research: Discourses and theoretical assumptions. *MIS Quarterly, 26*(3), 213–242. doi:10.2307/4132331

Seymour, S. (1999). *Women, family and child care in India: A world in transition*. Cambridge: Cambridge University Press.

Suchman, L. (1987). *Plans and situated actions: The problem of human-machine communication*. Cambridge: Cambridge University Press.

Zuboff, S. (1988). *In the age of the smart machine*. New York, NY: Basic Books.

ENDNOTES

[1] Both authors were born and raised in India and have resided in the U.S. for the past 15 years.
[2] A list of Web sites is listed in Appendix A
[3] A gotra is the lineage or clan assigned to a Hindu at birth based on an astrological condition.
[4] Each quote is replicated in Appendix B to translate the colloquial intent in conventional English. The quotes are identified by the number in subscript at the end of the quote.
[5] Being liberal and modern for most women in India means that the men do not subscribe to traditional stereotypes of women as homemakers and are subservient to the husband and his family. It has implications for everything from how they can dress to their careers.

This work was previously published in Social Networking Communities and E-Dating Services: Concepts and Implications, edited by Celia Romm Livermore & Kristina Setzekorn, pp. 329-352, copyright 2009 by ISR Publishing (an imprint of IGI Global).

APPENDIX A

List of Matrimonial Web sites:

1. **BharatMatrimony.com:** www.bharatmatrtimony.com
2. **Hindumatrimony.com:** www.hindumatrimony.com
3. **PunjabiMatrimony.com:** www.punjabimatrimony.com
4. **iMilap.com:** www.imilap.com
5. **Jeevansathi.com:** www.jeevansathi.com
6. **A1 Indian Matrimonials:** www.a1im.com
7. **Shaadi.com:** www.shaadi.com
8. **eMatrimonials:** www.geocities.com/ematrimonials
9. **Godbless Matrimonials:** www.godblessmatrimonials.com/india/
10. **Matrisearch.com:** matrisearch.com
11. **Falguni Mehta's Marriage Bureaus:** www.falgunimehta.com/.

APPENDIX B

1. *As is customary in Hindu families, parents search for their children's life partner. My parents are doing the same by speaking to relatives and circulating my 'bio-data' using various electronic resources.*
2. *I have to agree to put up with the ad because my parents keep putting pressure me…. at least this way it appears to them that I am interested in getting married while at the same time I can keep rejecting the matches they propose.*
3. *The girls today want a very specific kind of person as a husband.. he should be liberal and modern in his thinking.. she does not like anyone we find from our community.. by posting this ad perhaps we can find such a person from our specific caste and subcaste.*
4. *My elder sister created my profile and every day she searched for a life partner for me.*
5. *We met for coffee the day after our parents felt positive about taking the next step.*
6. *Our parents then arranged for us to meet.*
7. *My father showed an interest in his profile and give him (prospective groom) our contact no. and e-mail ID. He accepted and forwarded my profile (prospective bride) to his parents .... He (prospective groom) then talked to my father and said he wanted to visit my parents at their home in Dehradun. He (prospective groom) visited my parents within a week and expressed his desire to marry me (prospective bride). Then my sister and Jijaji (Brother-in-law) visited his home at NOIDA and found the family acceptable and gave their consent for the proposal. His parents and my parents talked to each other over the phone and finalized the marriage."*
8. *After reading my bio-data and showing it to her father, she contacted her cousin, who then phoned my uncle (who happened to be the contact listed with my bio-data) to request more information about me.*
9. *First, both my parents and my partner's parents contacted each other to make sure that all necessary requirements were met in order for the marriage proposal to proceed further.*
10. *I approached to her and received a response from her father.*

11. *My thanks to ... for bringing the two families together*
12. *We talked about the details of our families and the kind of partner we sought. Both of us were satisfied with each others' families, culture, background etc.*
13. *With the consent (emphasis mine) and blessings of both our families.*
14. *marrying a Muslim is out of question*
15. *our daughter-in-law is a Hindu ... we still wish our son had chosen a Muslim girl (in marriage).*
16. *If I marry a Bengali (one who speaks Bengali and by implication from the state of West Bengal).. it will be easier for her to interact with my grandparents and the extended family.*
17. *Brahmins (caste group) are very particular about who they marry.*
18. *This is great! The girl of my dreams could be on the other side of the world.*
19. *I can't believe that I can find someone who matches my exact needs, who may be thousands of miles away.*
20. *The gotras should not be the same. We would also ask for your horoscope for the purpose of matching it with ours.*
21. *Our families met and even our horoscopes matched well.*
22. *In keeping with our family tradition, we exchanged horoscopes and obtained parental consent.*
23. *we chatted for 3 months*
24. *She sent me an e-mail.. and the next day everything was finalized.*
25. *We dated ....(electronically).. for six months.*
26. *We chatted (online) and spoke over the phone for hours. We found that we are a perfect match for each other.*
27. *We chatted on (name of Web site) and then decided that we are going to get married even before our families met.*
28. *Earlier her family was not at all interested in m,e but slowly with our persistence they were forced to let us get married.*
29. *We chatted for many months but he never wanted to take it forward..*
30. *The girl was educated and pretty but she turned out to be crazy*
31. *The family did not seem as reputable as they claimed to be..*
32. *In my profile, I had mentioned ... who happens to be my dad's uncle and their neighbor in (city name), and a common link between the two families. At that point in time he was in UK and I was in Bangalore. He called me and we had a conversation ... We exchanged all the information about us and both the families through.. uncle*
33. *I received an e-mail stating that she was interested in speaking to me. Soon after ... we began to chat on the Internet via Yahoo! Messenger. As soon as we started to communicate ... our sessions lasted well into the wee hours of morning.*
34. *Then we started chatting over msn. Which went on for 3 months.*

Compilation of References

(2002). *Goodbody Economic Consultants*. Dublin, Ireland: Entrepreneurship in Ireland.

(2002). *U.S. Census Bureau*. Washington, D.C.: Survey of Women Business Owners.

Acker, J. (1990). Hierarchies, Jobs, Bodies: A Theory of Gendered Organizations. *Gender & Society*, *4*(2), 139–158. doi:10.1177/089124390004002002

Adam, A., Richardson, H., Tattersall, A., & Keogh, C. (2004). *WINWIT: Women in North West Information Technology, ESF Report*. Salford University of Salford, Informatics Research Institute.

Adams, A., Buckingham, C. D., Lindenmayer, A., & McKinlay, J. B. (2008). The influence of patient and doctor gender on diagnosing coronary heart disease. *Sociology of Health & Illness*, *30*(1), 1. doi:10.1111/j.1467-9566.2007.01025.x

Adams, R. B., & Ferreira, D. (2008). *Women in the Boardroom and Their Impact on Governance and Performance*. Retrieved January 2010 from http://ssrn.com/abstract=1107721

Adogbeji, B., & Akporhonor, B. A. (2005). The impact of ICT (Internet) on research and studies: The experience of Delta State University students in Abraka, Nigeria. *Library Hi Tech News*, *10*, 17–21. doi:10.1108/07419050510644347

Ahl, H. (2006). Why Research on Women Entrepreneurs Needs New Directions. *Entrepreneurship: Theory & Development*, *30*(5), 595–622. doi:10.1111/j.1540-6520.2006.00138.x

Ahlström, S., Bloomfield, K., & Knibbe, R. (2001). Gender differences in drinking patterns in nine European Countries: Descriptive findings. *Substance Abuse*, *22*(1), 69–85. doi:10.1080/08897070109511446

Ahuja, M. (2002). Women in the information technology profession: A literature review, synthesis, and research agenda. *European Journal of Information Systems*, *11*, 20–34. doi:10.1057/palgrave/ejis/3000417

Ahuvia, A., & Adelman, M. (1992). Formal intermediaries in the marriage market: A typology and review. *Journal of Marriage and the Family*, *54*(2), 452–463. doi:10.2307/353076

Akhter, S. H. (2003). Digital divide and purchase intention: Why demographic psychology matters? *Journal of Economic Psychology*, *24*, 321–327. doi:10.1016/S0167-4870(02)00171-X

Akman, I., & Mishra, A. (2010). Gender, age and income differences in internet usage among employees in organizations. *Computers in Human Behavior*, *26*(3), 482–490. doi:10.1016/j.chb.2009.12.007

Alapack, R., Blichfeldt, M., & Elden, A. (2005). Flirting on the Internet and the hickey: A hermeneutic. *Cyberpsychology & Behavior*, *8*(1), 52–61. doi:10.1089/cpb.2005.8.52

Aldrich, H. E., & Elam, A. B. (1997). Strong Ties, Weak Ties and Strangers: Do Women Owners Differ from Men in Their Use of Networking to Obtain Assistance? In Birley, S., & Macmillan, I. C. (Eds.), *Entrepreneurship in a Global Context*. London: Routledge.

Aldrich, H. (1989). Networking Among Women Entrepreneurs. In Hagan, O., Rivchun, C., & Sexton, D. (Eds.), *Women-Owned Business* (pp. 103–132). New York: Praeger.

Compilation of References

Aldrich, H., & Zimmer, C. (1986). Entrepreneurship Through Social Networks. In Sexton, D. L., & Smilor, R. W. (Eds.), *The Art and Science of Entrepreneurship* (pp. 3–23). Cambridge, MA: Ballinger.

Aldrich, H. E., Carter, C., et al. (2002). *With Very Little Help From Their Friends: Gender and Relational Composition of Nascent Entrepreneurs' Startup Teams*. 22nd Annual Entrepreneurship Research Conference, Babson College, MA.

Aldridge, A., Forcht, K., & Pierson, J. (1997). Get linked or get lost: Marketing strategy for the internet. *Internet Research: Electronic Networking Applications and Policy*, *7*(3), 161–169. doi:10.1108/10662249710171805

Ali, K.A.M., Jemain, A.A., Yusoff, R.Z., & Abas, Z. (2007). efficient cost management through excellence quality management practices among local authorities in Malaysia. *Top Quality Management and Business Excellence*, *18*(1-2, 99.

Alic, M. (1986). *Hypatia's Heritage: A history of women in science from Antiquity through the Nineteenth Century*. Boston: Beacon Press.

Allen, M. W., Armstrong, D. J., Riemenschneider, C. K., & Reid, M. F. (2006). Making sense of the barriers women face in the information technology work force: Standpoint theory, self-disclosure, and casual maps. *Sex Roles*, *54*, 831–844. doi:10.1007/s11199-006-9049-4

Allen, M. W., Reid, M., & Riemenschneider, C. (2004). The role of laughter when discussing workplace barriers: Women in Information Technology jobs. *Sex Roles*, *50*(3/4), 177–189. doi:10.1023/B:SERS.0000015550.92555.7e

Alley, T., & Cunningham, M. (1991). Average faces are attractive, but very attractive faces are not average. *Psychological Science*, *2*, 123–125. doi:10.1111/j.1467-9280.1991.tb00113.x

Almack, J. C. (1922). The Influence of Intelligence on the Selection of Associates. *School and Society*, *16*(410), 529–530.

Alreck, P., & Setle, R. B. (2002). Gender effects on internet, catalogue and store shopping. *Journal of Database Marketing*, *9*(2), 150–162. doi:10.1057/palgrave.jdm.3240071

Amarsaikhan, D., Lkhagvasuren, T., Oyun, S., & Batchuluun, B. (2007). online medical diagnosis and training in rural Mongolia. *Distance Education*, *28*(2), 195–211. doi:10.1080/01587910701439241

Ammenwerth, E., Mansmann, U., Iller, C., & Eichstadter, R. (2003). factors affecting and affected by user acceptance of computer-based nursing documentation: Results of a two-year study. *Journal of the American Medical Informatics Association*, *10*(1), 69–84. doi:10.1197/jamia.M1118

Anastasopoulos, V., Brown, D. A. H., & Brown, D. L. (2002). *Women on Boards: Not Just the Right Thing... But the 'Bright' Thing*. Conference Board of Canada.

Anderson, A. R., & Miller, C. J. (2003). "Class Matters": Human and Social Capital in the Entrepreneurial Process. *Journal of Socio-Economics*, *32*, 17–36. doi:10.1016/S1053-5357(03)00009-X

Anderson, S. (2003). Why dowry payments declined with modernization in Europe but are rising in India. *The Journal of Political Economy*, *111*(2), 269–310. doi:10.1086/367679

Andersson, A., Vimarlund, V., & Timpka, T. (2002). Management demands on information and communication technology in process-oriented health-care organisations. *Journal of Management in Medicine*, *16*(2/3), 159–169. doi:10.1108/02689230210434907

Antony, P., & Gayathri, V. (2008). Ricocheting gender equations: Women workers in the call centre industry. In Saith, A., Vijayabaskar, M., & Gayathri, V. (Eds.), *ICTs and Indian social change: diffusion, poverty, governance* (pp. 291–381). Los Angeles: Sage Publications.

Anvari, M. (2007). Impact of information technology on human resources in healthcare. *Healthcare Quarterly (Toronto, Ont.)*, *10*(4), 84–88.

Apfelbaum, D. (2009). Wer verdient wie viel? Ergebnisse der c't-Gehaltsumfrage 2008. *c't*, *6*, 92-99.

ΑΡΙΣΤΟΤΕΛΕΙΟ ΠΑΝΕΠΙΣΤΗΜΙΟ ΘΕΣΣΑΛΟΝΙΚΗΣ, (2008). *Μελέτη σπουδών και επαγγελματικής σταδιοδρομίας στο Τμήμα Ηλεκτρολόγων Μηχανικών & Μηχανικών Υπολογιστών Α.Π.Θ. σε σχέση με τα χαρακτηριστικά φύλου*. Report in Greek, 2008, Retrieved January, 29th, 2010, from http://newton.ee.auth.gr/genderIssues/docs/THMMY_GenderNew_vFINAL.pdf

Arenius, P., & Kovalainen, A. (2006). Similarities and Differences Across the Factors Associated with Women's Self-employment Preference in the Nordic Countries. *International Small Business Journal*, *24*(1), 31–57. doi:10.1177/0266242606059778

Armstrong, D. J., Riemenschneider, C. K., Allen, M. W., & Reid, M. F. (2007). Advancement, voluntary turnover and women in IT: A cognitive study of work-family conflict. *Information & Management*, *44*, 142–153. doi:10.1016/j.im.2006.11.005

Arvidsson, A. (2006). Quality singles: Internet dating and the work of fantasy. *New Media & Society*, *8*(4), 671–690. doi:10.1177/1461444806065663

Ash, S., Gorman, P. N., Seshadri, V., & Hersh, W. R. (2004). Perspectives on CPOE and patient care. *Journal of the American Medical Informatics Association*, *11*(2), 95–99. doi:10.1197/jamia.M1427

Ashcraft, C., & Blithe, S. (2009). *Women in IT: The facts*. Boulder, CO: National Center for Women & Information Technology. Retrieved from http://www.ncwit.org/pdf/NCWIT_WomenInITFacts_FINAL.pdf

Atkinson, P., & Hammersley, M. (1994). Ethnography and participant observation. In: N. Denzin & Y. Lincoln (Eds.), *Handbook of Qualitative Research* (pp. 248-261). Thousand Oaks, CA: Sage.

Babaeva, L., & Chirikova, A. (1997). Women in business. *Russian Social Science Review*, *38*(3), 81–92. doi:10.2753/RSS1061-1428380381

Bagozzi, R. (1975). Marketing as exchange. *Journal of Marketing*, *39*(October), 32–39. doi:10.2307/1250593

Baker, T., Aldrich, H. E., & Liou, N. (1997). Invisible entrepreneurs: The neglect of women business owners by mass media and scholarly journals in the USA. *Entrepreneurship and Regional Development*, *9*(3), 221–238. doi:10.1080/08985629700000013

Baker, P. M. A., & Ward, A. C. (2002). Bridging temporal and spatial "gaps": The role of information and communication technologies in defining communities. *Information Communication and Society*, *5*(2), 207–224. doi:10.1080/13691180210130789

Baker, A. (2005). *Double click: Romance and commitment among online couples*. Cresskill, NJ: Hampton Press.

Baldwin, L. P., Clarke, M., & Jones, R. (2002). Clinical ICT systems: Augmenting case management. *Journal of Management in Medicine*, *16*(2/3), 188–198. doi:10.1108/02689230210434925

Ballington, J., & Karam, A. (Eds.). (2005). *Women in Parliament: Beyond Numbers*. IDEA.

Bargh, J., McKenna, K., & Fitzsimons, G. (2002). Can you see the real me? Activation and expression of the "true self" on the Internet. *The Journal of Social Issues*, *58*, 33–48. doi:10.1111/1540-4560.00247

Barkham, R., Gudgin, G., Hart, M., & Hanvey, E. (1996). *The Determinants of Small Firm Growth: An Inter - Regional Study in the United Kingdom 1986-1990. Regional Policy and Development Series*. London: Jessica Kingsley Publishers.

Barnlund, D. (1989). *Communicative styles of Japanese and Americans*. Belmont, CA: Wadsworth.

Barrett, M., & Davidson, M. J. (2006). *Gender and Communication at Work*. Aldershot, UK: Ashgate.

Barringer, B., & Ireland, D. (2006). *Entrepreneurship-Successfully Launching New Ventures*. Pearson Education Prentice Hall.

Barscheid, E., & Walster, E. (1974). Physical attractiveness. In: L. Berkowitz (Ed.), *Advances in experimental psychology* (pp. 157-215). New York, NY: Academic Press.

Batabyal, A. (1998). Aspects of arranged marriages and the theory of Markov decision processes. *Theory and Decision*, *45*(3), 241–253. doi:10.1023/A:1004998730922

Bates, T. (2002). Restricted Access to Markets Characterizes Women-Owned Businesses. *Journal of Business Venturing*, *17*, 313–324. doi:10.1016/S0883-9026(00)00066-5

Baum, J. A. C., & Silverman, B. S. (2004). Picking Winners or Building Them? Alliance, Intellectual and Human Capital as Selection Criteria in Venture Financing and Performance of Biotechnology Startups. *Journal of Business Venturing*, *19*, 411–436. doi:10.1016/S0883-9026(03)00038-7

Compilation of References

Beard, L., Wilson, K., Morra, D., & Keelan, J. (2009). A survey of health-related activities on second life. *Journal of Medical Internet Research. 11*(2). Retrieved on March 19, 2010 from http://www.jmir.org/2009/2/e17/HTML

Becta - British Educational Communications and Technology Agency. (2008). *How do boys and girls differ in their use of ICT?* Retrieved January, 29th, 2010, from http://partners.becta.org.uk/upload-dir/downloads/page_documents/research/gender_ict_briefing.pdf

Bellafante, G. (2005, August 23). Courtship ideas of South Asians get a U.S. touch. *The New York Times*.

Belliveau, M. A. (2005). Blind Ambition: The Effects of Social Networks and Institutional Sex Composition on the Job Search Outcomes of Elite Coeducational and Women's College Graduates. *Organization Science, 16*(2), 134–150. doi:10.1287/orsc.1050.0119

Bem, S. L. (1981). Gender schema theory: A cognitive account of sex typing. *Psychological Review, 88*(4), 354–364. doi:10.1037/0033-295X.88.4.354

Bem, S. L. (1993). *The lenses of gender: Transforming the debate on sexual inequality*. New Haven: Yale University Press.

Ben-Ze'ev, A. (2004). *Love online: Emotions on the Internet*. Cambridge: Cambridge University Press.

Berg, B. (2001). *Qualitative Research Methods for the Social Sciences*. Boston: Allyn & Bacon.

Berger, J., & Fisek, H. (1974). A generalization of the theory of status characteristics and expectation states. In: J. Berger, T. Conner, & M. Fisek (Eds.), *Expectation states theory* (pp. 163-205). Englewood Cliffs, NJ: Winthrop.

Berger, J., Cohen, B., & Zelditch, M., Jr. (1966). Status characteristics and expectation states. In: J. Berger, M. Zelditch, Jr., & B. Anderson (Eds.), *Sociological theories in progress* (vol. 1, pp. 29-46). Boston, MA: Houghton Mifflin.

Berry, D. (2005). *Romancing the Web: A therapist's guide to the finer points of online dating*. Manitowoc, WI: Blue Waters Publications.

Bhide, A. (1993). Bootstrap Finance: the art of start-ups. *Harvard Business Review*, (Nov-Dec): 109–117.

Bijker, W., & Law, J. (1992). General introduction. In: W. Bijker & J. Law (Eds.), *Shaping technology/building society: Studies in socio-technical change*. Cambridge, MA: MIT Press.

Bijker, W., Hughes, T., & Pinch, T. (1987). *The social construction of technological systems: New directions in sociology and history of technology*. Cambridge, MA: MIT Press.

Bimber, B. (2000). Measuring the gender gap on the internet. *Social Science Quarterly, 81*(3), 868–876.

Bird, S. R., & Sapp, S. G. (2004). Understanding the Gender Gap in Small Business Success. *Gender & Society, 18*(1), 5–28. doi:10.1177/0891243203259129

Birley, S. (1985). The role of Networks in the Entrepreneurial Process. *Journal of Business Venturing, 1*(1), 107–117. doi:10.1016/0883-9026(85)90010-2

Blake, M. K. (2006). Gendered Lending: Gender, Context and the Rules of Business Lending. *Venture Capital, 8*(2), 183–201. doi:10.1080/13691060500433835

Blickenstaff, J. (2005). Women and science careers: Leaky pipeline or gender filter? *Gender and Education, 17*, 369–386. doi:10.1080/09540250500145072

Blisson, D., & Kaur Rana, B. (2001). *The Role of Entrepreneurial Networks: the Influence of Gender and Ethnicity in British SMEs*. The 46th International Conference for Small Business, Taipei, Taiwan.

BLK – Bund-Länder-Kommission für Bildungsplanung und Forschungsförderung (1987). Gesamtkonzept für die informatische Bildung. *Materialien zur Bildungsplanung und Forschungsförderung, 16*.

Blossfeld, H.-P., Bos, W., Hannover, B., Lenzen, D., Müller-Böling, D., Prenzel, M., & Wößmann, L. (2009). *Geschlechterdifferenzen im Bildungssystem. Jahresgutachten 2009*. Wiesbaden: VS Verlag für Sozialwissenschaften. Retrieved March 23, 2009, from http://www.aktionsrat-bildung.de/fileadmin/Dokumente/Geschlechterdifferenzen_im_Bildungssystem__Jahresgutachten_2009.pdf

BMBF – Bundesministerium für Bildung und Forschung (Ed.). (2008). *Studiensituation und studentische Orientierungen*. 10. Studierendensurvey an Universitäten und Fachhochschulen. Bonn. Retrieved March 11, 2009, from http://www.bmbf.de/pub/studiensituation_studentetische_orientierung_zehn_lang.pdf.

BMBF – Bundesministierium für Bildung und Forschung. (2009). *CeBIT-Umfrage: Frauen wollen MINT!* Retrieved March 11, 2009, from http://www.komm-mach-mint.de/Startseite/News/CeBIT-Umfrage-Frauen-wollen-MINT!

Boellstorff, T. (2008). *Coming of age in Second Life an anthropologist explores the virtually human*. New York: Princeton UP.

Boerma, W.G.W., & van den Brink-Muinen. (2000). Gender-related differences in the organisation and provision of services among general practitioners in Europe: A signal to Health Care Planners. *Medical Care, 38*(10), 993–1002. doi:10.1097/00005650-200010000-00003

Bonneville, L., & Pare, D. J. (2006). Socioeconomic stakes in the development of telemedicine. *Journal of Telemedicine and Telecare, 12*(5), 217–219. doi:10.1258/135763306777889073

Borooah, V. K., & Collins, G. (1997). Women and Self-Employment: An Analysis Of Constraints and Opportunities in Northern Ireland. In Deakins, D., Jennings, P., & Mason, C. (Eds.), *Small Firms: Entrepreneurship in the Nineties* (pp. 72–88). London: Paul Chapman Publishing.

Bortoluzzi, M., & Piergiorgio, P. (2009). Multimodal Analysis of Virtual Learning Environments: A University Campus in Second Life. In Sapio, B. (Eds.), *The Good, the Bad and the Challenging. The User and the Future of Information and Communication Technologies* (Vol. 1, pp. 443–453). Koper, Slovenia: ABS-Center.

Bowker, N., & Tuffin, K. (2002). Disability discourses for online identities. *Disability & Society, 17*(3), 327–344. doi:10.1080/09687590220139883

Bowker, N. I., & Tuffin, K. (2007). Understanding positive subjectivities made possible online for disabled people. *New Zealand Journal of Psychology, 36*(2), 63–71.

Bowker, N., & Tuffin, K. (2003). Dicing with deception: People with disabilities' strategies for managing safety and identity online. *Journal of Computer-Mediated Communication, 8*(2). Retrieved on January 15, 2009 from http://jcmc.indiana.edu/vol8/issue2/bowker.html

Bowlby, J. (1980). *Attachment and loss: Vol. 3, Loss, sadness and depression*. New York, NY: Basic Books.

Bowlin, W. F., & Renner, C. J. (2008). Assessing gender and top-management-team pay in the mid-cap and small-cap companies using data envelopment analysis. *European Journal of Operational Research, 185*(1), 430. doi:10.1016/j.ejor.2007.04.022

Braet, O., & Ballon P (2007). Business model scenarios for remote management. *Journal of Theoretical and Applied Electronic Commerce research, 2*(3), 62–79.

Brass, D. J. (1985). Men's and Women's Networks: A Study of Interaction Patterns and Influence in an Organization. *Academy of Management Journal, 28*(2), 327–343. doi:10.2307/256204

Braten, I., & Stromso, H. I. (2006). Epistemological beliefs, interest, and gender as predictors of internet-based learning activities. *Computers in Human Behavior, 22*, 1027–1042. doi:10.1016/j.chb.2004.03.026

Bretts, M. (1993). She shall overcome. *Computerworld, 27*, 67–70.

Briedis, K. Egorova, T. Heublein, U. Lörz, M., Middendorff, E., Quat. H. & Spangenberg, H. (2008). *Studienaufnahme, Studium und Berufsverbleib von Mathematikern. Einige Grunddaten zum Jahr der Mathematik*. HIS: Forum Hochschule, 9. Retrieved March 17, 2009, from http://www.his.de/pdf/pub_fh/fh-200809.pdf

British Telecom. (2008). *Mum Magnates- A study into entrepreneurial mums and how they keep their new born businesses thriving*. The Red Consultancy-www.bizmums.yell.com

Britt, H., Bhasale, A., & Miles, D. A. (1996). The sex of the general practitioner: A comparison of characteristics, patients and our conditions managed. *Medical Care, 34*(1), 403–415. doi:10.1097/00005650-199605000-00003

Broder, J. (2007, February 7). *Edwards bloggers cross the line*. Retrieved from http://www.nytimes.com/2007/02/07/us/politics/07edwards.html?scp=2&sq=Amanda+Marcotte&st=nyt

Brooksbank, D. (2000). Self employment and small firms. In Carter, S., & Jones-Evans, D. (Eds.), *Enterprise and Small Business: Principles, Policy and Practice*. London: Prentice Hall.

Brophy, D. J. (1989). Financial Women-Owned Entrepreneurial Firms. In Hagan, O., & Rivchun, C. S. D. (Eds.), *Women-Owned Businesses* (pp. 55–75). New York: Praeger.

Brosnan, M. J. (2006). Gender and Diffusion of Email: An Organizational Perspective. In Barrett, M., & Davidson, M. J. (Eds.), *Gender and Communication at Work* (pp. 260–269). Aldershot: Ashgate.

Brown, B., & Butler, J. E. (1995). Competitors as allies: a study of entrepreneurial networks in the U.S. wine industry. *Journal of Small Business Management, 33*(3), 57–66.

Browne, S., & Farrell, L., Harris & Sessions, J. (2006). Risk Preference and Employment Contract Type. *Journal of the Royal Statistical Society. Series A (General), 169*(4), 849–863.

Brunel, F. F., & Nelson, M. R. (2003). Message order effects and gender differences in advertising persuasion. *Journal of Advertising Research, 43*(3), 330–341. doi:10.1017/S0021849903030320

Bruni, A., Gherardi, S., & Poggio, B. (2004). Doing Gender, Doing Entrepreneurship: An Ethnographic Account of Intertwined Practices. *Gender, Work and Organization, 11*(4), 406–429. doi:10.1111/j.1468-0432.2004.00240.x

Brunn, P., & Jensen, M., & Skovgaard. (2002). E-marketplaces: Crafting a winning strategy. *European Management Journal, 20*(3), 286–298. doi:10.1016/S0263-2373(02)00045-2

Brush, C., & Carter, N. M. (2004). *Clearing the Hurdles: Women Building High-Growth Businesses*. Upper Saddle River, NJ: Financial Times Prentice Hall.

Brush, C. G., & Hisrich, R. (1999). Women owned businesses: Why do they matter? In Acs, Z. (Ed.), *Are Small Firms Important? Their Role and Impact*. Boston, MA: Kluwer Academic Publishers.

Brush, C. Carter, N M., Gatewood, E.J., Greene, P.G., Hart, M., (2004). *Women Entrepreneurs, Growth, and Implications for the Classroom*. United States Association for Small Business and Entrepreneurship.

Brush, C. G. (1997). Women's entrepreneurship. In *Proceedings of the OECD Conference on Women Entrepreneurs in Small and Medium Enterprises*. Paris: OECD.

Buhai, S., & van der Leij, M. (2006). *A Social Network Analysis of Occupational Segregation*. Tinbergen Institute Discussion Paper 016/1 Tinbergen Institute.

Buss, D. (1989). Sex differences in human mate preferences: Evolutionary hypotheses in 37 cultures. *The Behavioral and Brain Sciences, 12*, 1–49.

Buss, D. (1987). Sex differences in human mate selection criteria: An evolutionary perspective. In: C. Crawford, D. Crebs, & M. Smith (Eds.), *Sociobiology and psychology: Ideas, issues, and applications* (pp. 335-352). Hillsdale, NJ: Erlbaum.

Butt, G., & Lance, A. (2005). Secondary teacher workload and job satisfaction: Do successful strategies for change exist? *Educational Management Administration & Leadership, 33*(4), 401. doi:10.1177/1741143205056304

Buttner, H., & Moore, D. (1997). Women's organizational exodus to entrepreneurship: self-reported motivations and correlates with success. *Journal of Small Business Management, 35*(1), 34–47.

Bystydzienski, J. M., & Bird, S. R. (2006). *Removing barriers: Women in academic science, technology, engineering, and mathematics*. Bloomington: Indiana University Press.

Cabral, R., Kwong, H., & Tang, W. (2007). Managing ICT resources for the improvement of health quality in China. *International Journal of Healthcare Technology and Management, 8*(1-2), 5. doi:10.1504/IJHTM.2007.012107

Cabrera, S. F., & Thomas-Hunt, M. (2007). Street Cred and the executive woman: The effects of gender differences in social networks on career advancement. In Correll, S. J. (Ed.), *The Social Psychology of Gender* (pp. 123–147). Elsevier Science Press. doi:10.1016/S0882-6145(07)24006-8

Camp, T. (1997). The incredible shrinking pipeline. *Communications of the ACM, 40*(10), 103–110. doi:10.1145/262793.262813

Caoili, E. (2008). IMVU reaches 20 million registered users, largest virtual goods catalog. *Worldsinmotion.biz*. Retrieved on April 24, 2008 from http://www.worldsinmotion.biz/2008/06/imvu_reaches_20_million_regist.php

Cardinali, R., & Gordon, Z. (2001). Cliff jumping: Empowering actions for disabled women. *Equal Opportunities International, 20*(8), 17–24. doi:10.1108/02610150110786651

Carli, L. L. (2001). Gender and Social Influence. *The Journal of Social Issues, 57*(4), 725–741. doi:10.1111/0022-4537.00238

Carli, L. L. (2006). Gender Issues in Workplace Groups: Effects of Gender and Communication Style on Social Influence. In Barrett, M., & Davidson, M. J. (Eds.), *Gender and Communication at Work* (pp. 69–83). Aldershot: Ashgate.

Caro, D. H. J. (2005). The axis and nexus of e-health alliances in 2020. *Canadian Journal of Public Health, 96*(4), 1–3.

Carrington, C. (2006). Small business financing profiles: Women entrepreneurs. *Journal of Small Business and Entrepreneurship, 19*(2), 83.

Carroll, J., Howard, S., Vetere, F., Peck, J., & Murphy, J. (2002). *Just what do the youth of today want? Technology appropriation by young people.* Paper presented at the The 35th Hawaii International Conference on System Sciences. Hawaii.

Carter, S. (2000). Improving the numbers and performance of women-owned businesses: some implications for training and advisory services. *Education + Training, 42*(45), 326–334. doi:10.1108/00400910010373732

Carter, S., & Rosa, P. (1998). The financing of male and female owned businesses. *Entrepreneurship and Regional Development, 10*(3), 225–241. doi:10.1080/08985629800000013

Carter, S., & Shaw, E. (2006). *Women's Business Ownership- Recent Research and Policy Development.* Report to the Small Business Service.

Carter, S. (2000). Improving the numbers and performance of women-owned businesses: some implications for training and advisory services. *Education + Training, 42*(4/5), 326–333. doi:10.1108/00400910010373732

Carter, S., & Cannon, T. (1992). *Women as entrepreneurs.* London: Academic Press.

Carter, S., & Rosa, P. (1998). The financing of male and female owned businesses. *Entrepreneurship and Regional Development, 10*(3), 225–241. doi:10.1080/08985629800000013

Carter, N., & Kolvereid, L. (1997). *Women starting new businesses: The experience in Norway and the US".* Paper presented at the OECD conference on Women Entrepreneurs in SME's, Paris.

Carter, S., & Bennett, D. (2006). *Gender & Entrepreneurship", Enterprise and Small Business-Principles, Practice and Policy* (2nd ed., pp. 176-192). Prentice Hall Financial Times.

Catalan, J. (2004). Internet medicine sales and the need for homogeneous regulation. *International Journal of Medical Marketing, 4*(4), 342–349. doi:10.1057/palgrave.jmm.5040185

Cecez-Kecmanovic, D., & Webb, C. (2008). Towards a communicative model of collaborative web-mediated learning. *Australian Journal of Educational Technology, 16*(1), 73–85. Retrieved from http://cleo.murdoch.edu.au/ajet/ajet16/cecez-kecmanovic.html.

Cejka, M. A., & Eagly, A. H. (1999). Gender-Stereotypic Images of Occupations Correspond to the Sex Segregation of Employment. *Personality and Social Psychology Bulletin, 25*, 413–423. doi:10.1177/0146167299025004002

Center for Women's Business Research (2008). *Key Facts about Women-Owned Businesses* (2008 Update).

Chambers, R., & Campbell, I. (1996). Gender differences in general practitioners at work. *The British Journal of General Practice, 46*, 291–293.

Chang, J., & Samuel, N. (2004). Internet shopper demographics and buying behavior in Australia. *Journal of American Academy of Business, 5*(1/2), 171–176.

Chapman, P., James-Moore, M., Szczygiel, M., & Thompson, D. (2000). Building internet capabilities in SMEs. *Journal of Enterprise Information Management, 13*(6), 353–361.

Chawla, D. (2006). Subjectivity and the "Native" ethnographer: Researcher eligibility in an ethnographic study of urban Indian women in Hindu arranged marriages. *International Journal of Qualitative Methods, 5*(4), 1–13.

Chell, E., & Baines, S. (1998). Goes gender affect business performance? A study of micro business services in the UK. *Entrepreneurship and Regional Development*, *10*(2), 117–135. doi:10.1080/08985629800000007

Chen, W., & Lee, C. (2005). The impact of web site image and consumer personality on consumer behavior. *International Journal of Management*, *22*(3), 484–496.

Cherry, J. C., Moffatt, T. P., Rodriguez, C., & Dryden, K. (2002). Diabetes disease management program for an indigent population empowered by telemedicine technology. *Diabetes Technology & Therapeutics*, *11*(6), 783–791. doi:10.1089/152091502321118801

Cheung, C. M. K., Lee, M. K. O., & Chen, Z. (2002). Using the Internet as a learning medium: An exploration of gender difference in the adoption of FabWeb. In *Proceedings of the 35th Hawaii International Conference on System Sciences*.

Chowdhury, S. K. (2006). Investments in ICT-capital and economic performance of small and medium scale enterprises in East Africa. *Journal of International Development*, *18*(4), 533. doi:10.1002/jid.1250

Christensen, M. C., & Remler, D. (2007). Information and Communications technology in Chronic Disease Care. *Medical Care Research and Review*, *64*(2), 123–147. doi:10.1177/1077558706298288

Christofides, E., Islam, T., & Desmarais, S. (2009). Gender stereotyping over instant messenger: The effects of gender and context. *Computers in Human Behavior*, *25*(4), 897–901. doi:10.1016/j.chb.2009.03.004

CIA. (2010a). The world factbook: Germany. Retrieved April 4, 2010 from https://www.cia.gov/library/publications/the-world-factbook/geos/gm.html

CIA. (2010b). The world factbook: Greece. Retrieved April 4, 2010 from https://www.cia.gov/library/publications/the-world-factbook/geos/gr.html

Close, A., & Zinkhan, G. (2003). Romance and the Internet: The emergence of e-dating. In: B. Kahn & M. Luce (Eds.), *Advances in consumer research* (vol. 31, pp. 153-157). Valdosta, GA: Association for Consumer Research.

Coates, G. (2001). Disembodied cyber co-presence: The art of being there while being somewhere else. In Watson, N., & Cunningham, S. (Eds.), *Reframing the Body*. New York: Palgrave.

Cockburn, C. (1983). *Brothers: Male Dominance and Technological Change*. London: Pluto.

Cockburn, C. (1985). *Machinery of Dominance. Women, Men and Technical know-how*. London: Pluto Press.

Cockburn, C. (1999). Caught in the Wheels: The High Cost of Being a Female Cog in the Male Machinery of Engineering. In MacKenzie, D., & Wajcman, J. (Eds.), *The Social Shaping of Technology* (pp. 55–66). Buckingham: Open University Press.

Cohen, M. S., & Ellis, T. J. (2008). *The Asynchronous Learning Environment (ALN) as a Gender-Neutral Communication Environment. 38th ASEE/IEEE*. NY: Saratoga Springs.

Cohen, D. (2007, August 30). *Building social capital through online communities: The strategy of Ned Lamont's 2006 Senate Campaign.* Paper presented at the Annual Meeting of the American Political Science Association, Chicago, IL.

Cohen, M. S., & Ellis, T. J. (2008). The Asynchronous Learning Environment (ALN) as a Gender-Neutral Communication Environment. In *Proceedings of the 28th American Society for Engineering Education/IEEE*, Saratoga Springs, NY.

Collins, N. (1980). Working models of attachment: Implications for explanation, emotion and behavior. *Journal of Personality and Social Psychology*, *71*, 810–832. doi:10.1037/0022-3514.71.4.810

Collinson, D., Knights, D., & Collinson, M. (1990). *Managing to discriminate*. London: Routledge.

Comber, C., Colley, A., Hargreaves, D. J., & Dorn, L. (1997). The effects of age, gender and computer experience upon computer attitudes. *Educational Research*, *39*(2), 123–133. doi:10.1080/0013188970390201

Compeau, D. R., & Higgins, C. A. (1995, June). Computer Self-Efficacy: Development of a Measure and Initial Test. *Management Information Systems Quarterly*, *19*(2), 189–211. doi:10.2307/249688

Compton, S., Lang, E., Richardson, T. M., & Hess, E. (2007). Knowledge translation consensus conference: Research methods. *Academic Emergency Medicine*, *14*(11), 991.

Consumer Search. (2005). Online dating sites best rated dating sites, services. Retrieved May 26, 2006, from http://www.consumersearch.com/www/Internet/online-dating/fullstory.html.

Cooper, A., & Sportolari, L. (1997). Romance in cyberspace: Understanding online attraction. *Journal of Sex Education and Therapy*, *22*(1), 7–14.

Cosmides, L., & Tooby, J. (1987). From evolution to behavior: Evolutionary psychology as the missing link. In: J. Dupre (Ed.). *The latest and the best: Essays on evolution and optimality* (pp. 227-306). Cambridge, MA: MIT Press.

Cosmides, L., Tooby, J., & Barkow, J. (1992). Introduction: Evolutionary psychology and conceptual integration. In: J. Barlow, L. Cosmides, & J. Tooby (Eds.), *The adapted mind: Evolutionary psychology and the generation of culture* (pp. 3-15). New York, NY: Oxford University Press.

Coviello, N. E., & Munro, H. J. (1995). Growing the Entrepreneurial Firm: Networking for International Market Development. *European Journal of Marketing*, *29*(7), 49–61. doi:10.1108/03090569510095008

Coyle, H. E., & Flannery, D. D. (2005). *Gendered Contexts of Learning Female Entrepreneurs in Male-Dominated Industries within the United States*. Summer Institute of the National Center for Curriculum Transformation Resources on Women. Turkey.

Crepeau, R. G., Crook, C. W., Goslar, M. D., & McMurtey, M. E. (1992). Career anchors of systems personnel. *Journal of Management Information Systems*, *9*, 145–160.

Crocker, J., Major, B., & Steele, C. (1998). Social stigma. In: S. Fiske, D. Gilbert, & G. Lindzey (Eds.), *Handbook of social psychology, 2*, 504-553. Boston, MA: McGraw-Hill.

Cross, J. L. M., & Lin, N. (2008). Access to Social Capital and Status Attainment in the United States: Racial/Ethnic and Gender Differences. In Lin, N., & Erickson, B. H. (Eds.), *Social Capital* (pp. 364–379). Oxford: Oxford University Press.

Crowston, K., & Kammerer, E. (1998). Communicative style and gender differences in computer-mediated communications. In Ebo, B. (Ed.), *Cyberghetto or cybertopia? Race, class, and gender on the Internet* (pp. 185–203). Westport, CT: Praeger.

CSO. (2008, March 31). *Information Society and Telecommunications, 2007* [press release]. Retrieved May 21, 2008 from http://www.cso.ie/newsevents/pr_informationsociety2007.htm

Culbreth, J. (2005). *The boomers' guide to online dating*. U.S.: Rodale Inc.

Curran, K., Walters, N., & Robinson, D. (2007). Investigating the problems faced by older adults and people with disabilities in online environments. *Behaviour & Information Technology*, *26*(6), 447–453. doi:10.1080/01449290600740868

Cyr, D., & Bonanni, C. (2005). Gender and website design in e-business. *International Journal of Electronic Business*, *3*(6), 565–582. doi:10.1504/IJEB.2005.008536

Dabholkar, P. A., & Sheng, X. (2009). The role of perceived control and gender in consumer reactions to download delays. *Journal of Business Research*, *62*(7), 756–760. doi:10.1016/j.jbusres.2008.06.001

Danaher, P. J., Mullarkey, G. W., & Essegaier, S. (2006). Factors affecting web site visit duration: a cross-domain analysis. *JMR, Journal of Marketing Research*, *43*(2), 182–194. doi:10.1509/jmkr.43.2.182

Darden, D., & Koski, P. (1988). Using personal ads: A deviant activity? *Deviant Behavior*, *9*(3), 383–400.

Davies, I., Mason, R., & Lalwani, C. (2007). Assessing the impact of ICT on UK general haulage companies. *International Journal of Production Economics*, *106*(1), 12. doi:10.1016/j.ijpe.2006.04.007

Davies, L., & Nielsen, S. (1992). An ethnographic study of configuration management and documentation practices in an information technology centre. In: K. Kendall, K. Lyytinen, & J. De Gross (Eds.), *The impact of computer supported technology on information systems development*. Amsterdam: Elsevier/North Holland.

Davis, K. (1941). Intermarriage in caste societies. *American Anthropologist*, *43*, 376–395. doi:10.1525/aa.1941.43.3.02a00030

Davis, S. (1990). Men as success objects and women as sex objects: A study of personal advertisements. *Sex Roles*, *23*, 43–50. doi:10.1007/BF00289878

Delaney, L. (2007, February). Trading tools- gain an edge by managing global shipments online. *The Entrepreneur Magazine*. Retrieved from www.entrepreneur.com

DeMartino, R., & Barbato, R. (2003). Differences between women and men MBA entrepreneurs: Exploring family flexibility and wealth creation as career motivators. *Journal of Business Venturing*, *18*, 815–832. doi:10.1016/S0883-9026(03)00003-X

Demunter, C. (2005). The digital divide in Europe. Statistics in Focus, 28/2005.

Dennis, A. R., & Garfield, M. J. (2003). The adoption and use of GSS in project teams: Toward more participative processes and outcomes. *Management Information Systems Quarterly*, *27*(2), 289–323.

Dennis, C., Harris, L., & Sandhu, B. (2002). From bricks to clicks: Understanding the e-consumer. *Qualitative Market Research: An International Journal*, *5*(4), 281–290. doi:10.1108/13522750210443236

Dennis, C. (2000). Networking for marketing advantage. *Management Decision*, *38*(4), 287–292. doi:10.1108/00251740010371757

Department of Enterprise. Trade and Employment (1999). White Paper on Human Resource Development, Government Publications, Dublin.

DeSanctis, G., & Poole, M. (1994). Capturing the complexity in advanced technology use: Adaptive structuration theory. *Organization Science*, *5*(1), 121–147. doi:10.1287/orsc.5.2.121

Desvaux, G., Devillard-Hoellinger, S., & Baumgarten, P. (2007). *Women matter: Gender diversity, a corporate performance driver*. McKinsey & Company.

DeVoss, D. (2007). From the BBS to the Web: Tracing the spaces of online romance. In: M. Whitty, A. Baker, & J. Inman (Eds.), *Online matchmaking* (pp. 17-30). Houndmills: Palgrave Macmillan.

DeWine, S., & Casbolt, D. (1983). Networking: External communication systems for female organisational members. *Journal of Business Communication*, *20*, 57–67. doi:10.1177/002194368302000205

Dholakia, R. R., & Uusitalo, O. (2002). Switching to electronic stores: Consumer characteristics and the perception of shopping benefits. *International Journal of Retail & Distribution Management*, *30*(10), 459–469. doi:10.1108/09590550210445335

Dick, T. P., & Rallis, S. F. (1991). Factors and influences on high secondary students' career choices. *Journal for Research in Mathematics Education*, *22*, 281–292. doi:10.2307/749273

Didham, R., Martin, I., Wood, R., & Harrison, K. (2004). Information technology systems in general practice medicine in New Zealand. *The New Zealand Medical Journal*, *117*, 1198.

Diehl, W. C., & Prins, E. (2008). Unintended outcomes in Second Life: Intercultural literacy and cultural identity in a virtual world. *Language and Intercultural Communication*, *8*(2), 101–118. doi:10.1080/14708470802139619

Dion, K., & Dion, K. (1996). Cultural perspectives on romantic love. *Personal Relationships*, *3*(1), 5–17. doi:10.1111/j.1475-6811.1996.tb00101.x

Dittmar, H., Long, K., & Meek, R. (2004). Buying on the Internet: Gender differences in on-line and conventional buying motivations. *Sex Roles*, *50*(5/6), 423–444. doi:10.1023/B:SERS.0000018896.35251.c7

Dobransky, K., & Hargittai, E. (2006). The disability divide in Internet access and use. *Information Communication and Society*, *9*(3), 313–334. doi:10.1080/13691180600751298

Donn, J., & Sherman, R. (2002). Attitudes and practices regarding the formation of romantic relationships on the Internet. *Cyberpsychology & Behavior*, *5*(2), 107–123. doi:10.1089/109493102753770499

Donthu, N., & Garcia, A. (1999). The internet shopper. *Journal of Advertising Research*, *39*(3), 52–58.

Dougherty, T., Dreher, G., Arunachalam, V., & Willbanks, J. (2009). The powerful mentor effect: Differential career returns for males and females. In *Academy of Management Best Paper Proceedings*. Chicago, IL: Academy of Management.

Dourish, P., & Bellotti, V. (1992). Awareness and Coordination in Shared Workspaces. *ACM CSCW*, 107-114.

Dowd, J., & Pallotta, N. (2000). The end of romance: The demystification of love in the postmodern age. *Sociological Perspectives*, *43*, 549–581.

Duranske, B. (2008). *Virtual law: Navigating the legal landscape of virtual worlds*. American Bar Association.

Durndell, A., & Haag, Z. (2002). Computer self efficacy, computer anxiety, attitudes towards the internet and reported experience with the internet, by gender, in an East European sample. *Computers in Human Behavior*, *18*(5), 521–535. doi:10.1016/S0747-5632(02)00006-7

Durndell, A., Glissov, P., & Siann, G. (1995). Gender and computing: Persisting differences. *Educational Research*, *37*(3), 219–227. doi:10.1080/0013188950370301

Eagly, A., & Karau, S. (2002). Role Congruity Theory of Prejudice Toward Female Leaders. *Psychological Review*, *109*, 573–598. doi:10.1037/0033-295X.109.3.573

eBusiness Strategy Group (2004). *eBusiness Strategy: Optimising usage of ICTs by Irish SME's and Microenterprises*. Department of Enterprise, Trade and Employment.

Eby, L., & McManus, S. (2004). The protégé's role in negative mentoring experiences. *Journal of Vocational Behavior Psychology*, *65*, 255–275. doi:10.1016/j.jvb.2003.07.001

Eby, L., McManus, S., Simon, S., & Russell, J. (2000). The protégé's perspective regarding negative mentoring. *Personnel Psychology*, *57*, 441–447.

Edgar, H., Jr., & Edgar, H., II. (2003). *The ultimate man's guide to Internet dating: The premier men's resource for finding, attracting, meeting and dating women online*. Aliso Viejo, CA: Purple Bus Publishing.

El Sayed, H., & Westrup, C. (2003). Egypt and ICTs: How ICTs bring national initiatives, global organizations and local companies together. *Information Technology & People*, *16*(1), 76–92. doi:10.1108/09593840310463041

Elbeltagi, I. (2007). E-commerce and globalization: An exploratory study of Egypt. *Cross Cultural Management*, *14*(3), 196. doi:10.1108/13527600710775748

Ellis, B., & Symons, D. (1990). Sex differences in sexual fantasy. *Journal of Sex Research*, *27*(4), 527–555.

Elsammami, Z., Scown, O., & Hackney, R. (2001). A case study of the impact of a diffusion agent on SMEs adoption of a Web presence. In *Proceedings of the 6th UKAIS conference* (pp. 562-576). University of Wales Institute, Cardiff 26-28 April 2000.

Encyclopedia of associations. Dialog Online. Proquest. Retrieved on May 14, 2009.

Engeser, S., Limbert, N., & Kehr, H. (2008). *Abschlussbericht zur Untersuchung Studienwahl Informatik*. Retrieved October 29, 2009, from http://www.psy.wi.tum.de/Docs/Studienwahl_Informatik-Abschlussbericht.pdf

Eriksson-Zetterquist, U. (2007). Editorial: Gender and new technologies. *Gender, Work and Organization*, *14*(4), 305–311. doi:10.1111/j.1468-0432.2007.00345.x

European Commission. (2006). *Women in Science and Technology. Creating sustainable careers*. Retrieved September 1, 2009, from http://ec.europa.eu/research/science-society/document_library/pdf_06/wist2_sustainable-careers-report_en.pdf

European Commission. (2007) *Key Competencies for Lifelong Learning – A European Framework. (2007)*. Annex of the Recommendation of the European Parliament and of the Council of 18 December 2006 on key competencies for lifelong learning (2006). Official Journal of the European Union. Retrieved January 29, 2010, from http://ec.europa.eu/dgs/education_culture/publ/pdf/ll-learning/keycomp_en.pdf

European Commission. (2008). *Progress towards the Lisbon objectives 2010 in education and training*. Retrieved December 2, 2009, from http://ec.europa.eu/dgs/education_culture/publ/pdf/educ2010/indicatorsleaflet_en.pdf?aid=14505&d=2007-11

Eurostat. (2009a) *Internet-Zugangsdichte – Haushalte. Prozent der Privathaushalte mit Internet-Zugang*. Retrieved April 5, 2010, from http://epp.eurostat.ec.europa.eu/tgm/table.do?tab=table&init=1&plugin=1&language=de&pcode=tsiir040

Eurostat. (2009b). *Versorgungsgrad mit Breitbandanschlüssen. Anzahl der Breitbandanschlüsse je 100 Einwohner*. Retrieved April 5, 2010, from http://epp.eurostat.ec.europa.eu/tgm/table.do?tab=table&init=1&plugin=1&language=de&pcode=tsiir150

Eurydice (n.d.). *Key Data on Education in Europe 2009*. Retrieved January, 29th, 2010, from http://eacea.ec.europa.eu/education/eurydice/documents/key_data_series/105EN.pdf

Fallows, D. (2005). *How men and women use the internet*. Retrieved from http://www.pewinternet.org/Reports/2005/How-Women-and-Men-Use-the-Internet/06-Activities-and-Trends/02-Men-and-women-are-equally-likely-to-do-many-online-activities.aspx?r=1

Fang, X., & Yen, D. C. (2006). Demographics and behavior of internet users in China. *Technology in Society*, *28*(3), 363–387. doi:10.1016/j.techsoc.2006.06.005

Fang, Y., & Lee, L. (2009). A review and synthesis of recent research in Second Life. *Interactive Technology and Smart Education*, *6*(4), 261–267. doi:10.1108/17415650911009236

Faulkner, W. (2007). 'Nuts and bolts and people': Gender-troubled engineering identities. *Social Studies of Science*, *37*, 331–356. doi:10.1177/0306312706072175

Faulkner, W., & Lie, M. (2007). Gender in the Information Society: Strategies of inclusion. *Gender, Technology and Development*, *11*(2), 157–177. doi:10.1177/097185240701100202

Faulstich-Wieland, H., & Nyssen, E. (1998). Geschlechterverhältnisse im Bildungssystem - Eine Zwischenbilanz. In Rolff, H.-G., Bauer, K.-O., Klemm, K., & Pfeiffer, H. (Eds.), *Jahrbuch der Schulentwicklung*. Weinheim: Juventa.

Faylor, C. (2008). 1 in 5 casual gamers have disability, survey says. *Shacknews.com*. Retrieved on March 19, 2009 from http://www.shacknews.com/onearticle.x/53088

Feinberg, R., & Kadam, R. (2002). E-CRM web service attributes as determinants of customer satisfaction with retail web sites. *International Journal of Service Industry Management*, *13*(5), 432–451. doi:10.1108/09564230210447922

Feld, S. L. (1982). Social Structural Determinants of Similarity among Associates. *American Sociological Review*, *47*(6), 797–801. doi:10.2307/2095216

Ferratt, T., Enns, H., & Prasad, J. (2006). Employment arrangements, need profiles, and gender. In Trauth, E. M. (Ed.), *Encyclopedia of Gender and Information Technology* (pp. 242–248). Hershey, PA: Idea Group Publishing. doi:10.4018/978-1-59140-815-4.ch038

Fertala, N. (2005). *Do Birds of a Feather Flock Together and Perform Economically Better? A Study of the Homophily Paradox Among Immigrant Entrepreneurs in Germany*. The Twenty-Fifth Annual Research Conference, Babson College, Wellesely, MA.

Feuer, E., Messnarz, R., & Sanchez, N. (2002). Best Practices in E-Commerce: Strategies, Skills, and Processes. In *Proceedings of the eBusiness and eWork 2002 Conference proceedings, Challenges and Achievements in E-Business and E-Work* (Part 1, pp. 109-116). Amsterdam: IOS Press.

Fielden, S. L., Davidson, M. J., Dawe, A. J., & Makin, P. J. (2003). Factors inhibiting the economic growth of female-owned small businesses in North West England. *Journal of Small Business and Enterprise Development*, *10*(2), 152–166. doi:10.1108/14626000310473184

Fiore, A. (2007). Online dating research at Berkley. http://people.ischool.berkeley.edu/~atf/dating.

Fiore, A., & Donath, J. S. (2004). *Online personals: An overview*. Paper presented at the CHI. Vienna, Austria.

Fischer, E., Reuber, R., & Dyke, L. (1993). A theoretical overview and extension of research on sex, gender and entrepreneurship. *Journal of Business Venturing*, *8*(4), 151–168. doi:10.1016/0883-9026(93)90017-Y

Fisher, A. (2004). *Why Women Rule The latest numbers show that they are starting more new businesses than men and growing them faster. What's going on?* Fortune Small Business.

Fiske, S. T., & Taylor, S. E. (1984). *Social cognition*. New York: Random House.

Fiske, A., Kitayama, S., Markus, H., & Nisbett, D. (1998). The cultural matrix of social psychology. In: D. Gilbert, S. Fiske, & G. Lindzey (Eds.), *Handbook of social psychology* (4th ed., pp. 915-981). New York, NY: McGraw Hill.

FitzGerald, M., & Arnott, D. (1996). Understanding demographic effects on marketing communications in services. *International Journal of Service Industry Management, 7*(3), 31–45. doi:10.1108/09564239610122947

Fitzsimons, P., & O'Gorman, C. (2007). *Entrepreneurship in Ireland. The Global Entrepreneurship Monitor 2007: The Irish Report*. Dublin, Ireland: Enterprise Ireland.

Fitzsimons, P., & O'Gorman, C. (2008). *Entrepreneurship in Ireland. The Global Entrepreneurship Monitor 2008: The Irish Report*. Dublin, Ireland: Enterprise Ireland.

Flick, U. (1998). *An Introduction to Qualitative Research*. Thousand Oaks, CA: Sage.

Florin, J., & Lubatkin, M. (2003). A Social Capital Model of High-Growth Ventures. *Academy of Management Journal, 46*(3), 374–384. doi:10.2307/30040630

Foa, U., & Foa, E. (1974). *Societal structures of the mind*. Springfield, IL: Charles C. Thomas.

Forfas (2007). *Towards developing a policy for Entrepreneurship in Ireland*. Dublin.

Forfás. (2005). *Broadband Benchmarking* [online], Retrieved May 22, 2008 from http://www.forfas.ie/publications/forfas_annrpt05/platforms/broadband.html

Fors, M., & Moreno, A. (2002). The benefits and obstacles of implementing ICTs strategies for development from a bottom-up approach. *Aslib Proceedings, 54*(3), 198–206. doi:10.1108/00012530210441746

Forson, C., & Ozbilgin, M. (2003). Dot-com women entrepreneurs in the UK. *Entrepreneurship and Innovation*, February, 13-24.

Fraley, R., & Shaver, P. (2000). Adult romantic attachment: Theoretical developments, emerging controversies, and unanswered questions. *Review of General Psychology, 2*(2), 132–154. doi:10.1037/1089-2680.4.2.132

Frauen geben Technik neue Impulse e.V. (Ed.) (2004). *Studiengänge im Wettbewerb. Hochschulranking nach Studienanfängerinnen in Naturwissenschaft und Technik*. Retrieved March 30, 2009, from http://www.ranking-kompetenzz.de/daten/images/Ranking%20Broschuere.pdf

Friedman, E. J. (2005). The reality of virtual reality: The internet and gender equality advocacy in Latin America. *Latin American Politics and Society, 47*(3), 1–34. doi:10.1353/lap.2005.0034

Fuchs, C. (2008). The implications of new information and communication technologies for sustainability. *Environment, Development and Sustainability, 10*(3), 291–309. doi:10.1007/s10668-006-9065-0

Galanes, G. J., & Adams, K. (2007). *Effective group discussion: theory and practice* (12th ed.). New York: McGraw-Hill.

Gallagher, M. (1998). *Evolution of facilities management in the healthcare sector Construction paper no. 86, The Chartered Institute of Builders* (pp. 1–8). Ascot.

Gani, A., & Clemes, M. D. (2006). Information and communications technology: A non-income influence on economic well being. *Information and Communications Technology, 33*(9), 649–663.

Garbarino, E., & Strahilevitz, M. (2004). Gender differences in the perceived risk of buying online and the effects of receiving a site recommendation. *Journal of Business Research, 57*(7), 768–775. doi:10.1016/S0148-2963(02)00363-6

Gefen, D., & Straub, D. W. (1997). Gender differences in the perception and use of e-mail: An extension to the Technology Acceptance Model. *Management Information Systems Quarterly, 21*(4), 389–400. doi:10.2307/249720

Gefen, D., Straub, D. W., & Boudreau, M. C. (2000). Structural equation modeling and regression: Guidelines for research practice. *Communications of the AIS, 4*(7), 1–78.

Gender Equality Unit. (2000). *Women and Men in Ireland as Entrepreneurs and Business Managers*. Dublin, Ireland: Department of Justice, Equality and Law Reform.

General Secretariat of the National Statistical Service of Greece. (n.d.). *Latest Statistical Data* Retrieved January, 29th, 2010, from http://www.statistics.gr/portal/page/portal/ESYE

German, L. (2006). Theories of Patriarchy. *International socialism*.

Gesellschaft für Informatik (GI) e.V. (1999) Informatische Bildung und Medienerziehung. Empfehlung der Gesellschaft für Informatik e.V.. Beilage zu *LOG IN 19*(6).

Giddens, A. (1984). *The constitution of society: Outline of the theory of structuration*. Cambridge: Polity press.

Givens, D. (1978). The non-verbal basis of attraction: Flirtation, courtship, and seduction. *Psychiatry, 41*, 346–359.

Glancey, K. (1998). Determinants so Growth and Profitability in Small Entrepreneurial Firms. *International Journal of Entrepreneurial Behaviour and Research, 4*(1), 18–27. doi:10.1108/13552559810203948

Glaser, B. G., & Strauss, A. L. (1967). *The discovery of grounded theory*. New York, NY: Aldine Transaction.

Glick, P., & Fiske, S. T. (1999). Sexism and Other 'Isms': Interdependence, Status, and the Ambivalent Content of Stereotypes. In Swann, J. W. B., Langlois, J. H., & Gilbert, L. A. (Eds.), *Sexism and Stereotypes in Modern Society* (pp. 193–221). Washington, DC: American Psychological Association. doi:10.1037/10277-008

Goffee, R., & Skase, R. (1985). *Women in charge: The experience of female entrepreneurs*. London: Allen & Unwin.

Goffman, E. (1986). *Stigma: Notes on the management of spoiled identity*. New York: Simon & Schuster.

Goldberg, A. A., & Larson, C. E. (1975). *Group communication: Discussion processes and applications*. Englewood Cliffs, NJ: Prentice-Hall.

Gonzales, M., & Meyers, S. (1993). "Your mother would like me": Self-presentation in the personal ads of heterosexual and homosexual men and women. *Personality and Social Psychology Bulletin, 19*, 131–142. doi:10.1177/0146167293192001

Goode, W. (1996). Gender and courtship entitlement: Responses to personal ads. *Sex Roles, 34*, 141–170. doi:10.1007/BF01544293

Granovetter, M. (1973). The Strength of Weak Ties. *American Journal of Sociology, 78*(6), 1360–1380. doi:10.1086/225469

Granovetter, M. (1985). Economic Action and Social Structure: The Problem of Embeddedness. *American Journal of Sociology, 91*, 481–510. doi:10.1086/228311

Green, E., & Cohen, L. (1995). 'Women's Business': Are Women Entrepreneurs Breaking New Ground or Simply Balancing the Demands of 'Women's Work' in a New Way? *Journal of Gender Studies, 4*(3), 297–314. doi:10.1080/09589236.1995.9960615

Greenwald, R. (2003). *Find a husband after 35: Using what I learned at Harvard Business School*. Random House Publishing Book.

Gregory, A., & Windebank, J. (2000). *Women's Work in Britain and France: Practice, Theory and Policy*. Macmillan. doi:10.1057/9780230598515

Grimson, J., Grimson, W., & Hasselbring, W. (2000). The IS challenge in health care. *Communications of the ACM, 43*(6), 49–55. doi:10.1145/336460.336474

Group, S. I. A. (2001). *Policy and Planning on Developing Women in Enterprise*. Dublin: Report to Enterprise Ireland.

Gubernick, D. (1994). Bi-parental care and male female relations in mammals. In: S. Parmigiani & F. Vom Saal (Eds.), *Infanticide and parental care* (pp. 427-463). Chur, Switzerland: Harwood.

Guiller, J., & Durndell, A. (2007). Students' linguistic behaviour in online discussion groups: Does gender matter? *Computers in Human Behavior, 23*(5), 2240–2255. doi:10.1016/j.chb.2006.03.004

Gupta, Namrata and Sharma, A.K. (2003). Gender Inequality in the Work Environment at Institutes of Higher Learning in Science and Technology in India. *Work, Employment and Society, 17*(4), 597-616.

Gutek, B. A. (2006). Book Reviews ['Doing IT: Women working in Information Technology' by K. Scott-Dixon, Sumach Press, 2004.]. *Gender, Work and Organization, 13*(6), 621–623. doi:10.1111/j.1468-0432.2006.00325_1.x

Guzman, I., Stam, K., & Stanton, J. (2008). The occupational culture of IS/IT personnel within organizations. *Database, 39*(1), 33–50.

Gwinnell, E. (1998). *Online seductions: Falling in love with strangers on the Internet*. New York, NY: Kodansha International.

Hacker, S. L. (1981). The Culture of Engineering: Woman, Workplace and Machine. *Women's Studies International Quarterly*, *4*(3), 341–353. doi:10.1016/S0148-0685(81)96559-3

Hackman, R. (1990). *Groups that work (and those that don't): Creating conditions for effective teamwork*. San Francisco: Jossey-Bass.

Hagglund, M., Scandurra, I., & Koch, S. (2007). Using scenarios to capture work processes in shared home care. *Studies in Health Technology and Informatics*, *130*, 233–239.

Hahn, S., & Litwin, A. H. (1995). Women and men: Understanding and respecting gender differences in the workplace. In *Managing in the age of change* (pp. 188–198). New York: Irwin.

Hall, E., & Hall, M. (1990), *Understanding cultural differences*. Yarmouth, ME: Intercultural Press.

Halpern, D. F. (2005). How time-flexible work policies can reduce stress, improve health and save money. *Stress and Health*, *21*, 157–168. doi:10.1002/smi.1049

Hampton, A., Cooper, S., & McGowan, P. (2007). *Female Entrepreneurial Networks and Networking in Technology-Based Sectors*. Institute for Small Business & Entrepreneurship, 7-9 November 2007, Glasgow, Scotland.

Hancock, J., Toma, C., & Ellison, N. (2007). The truth about lying in online dating profiles. *Proceedings of the ACM Conference on Human Factors in Computing Systems (CHI 2007)*, (pp. 449-452).

Hansen, E. L. (1995). Entrepreneurial Networks and New Organization Growth. *Entrepreneurship. Theory into Practice*, ***, 7–19.

Hansen, E. L. (2000). Resource Acquisition as a Startup Process: Initial Stocks of Social Capital and Organizational Foundings. In *Proceedings of the Twentieth Annual Entrepreneurship Research Conference*. Babson College.

Hanson, S., & Blake, M. (2008). Gender and Entrepreneurial Networks. *Regional Studies*, *43*(1), 135–149. doi:10.1080/00343400802251452

Hargittai, E., & Shafer, S. (2006). Differences in actual and perceived online skills: The role of gender. *Social Science Quarterly*, *87*(2), 432–448. doi:10.1111/j.1540-6237.2006.00389.x

Harrigan, P. O., Boyd, M. M., Ramsey, E., Ibbotson, P., & Bright, M. (2008). The development of e-procurement within the ICT manufacturing industry in Ireland. *Management Decision*, *46*(3), 481. doi:10.1108/00251740810863906

Harrison, A., & Saeed, L. (1977). Let's make a deal: An analysis of revelations and stipulations in lonely hearts advertisements. *Journal of Personality and Social Psychology*, *35*(4), 257–264. doi:10.1037/0022-3514.35.4.257

Hartmann, H., & Bridges, A. (1981). The unhappy marriage of Marxism and feminism: towards a more progressive union. In Sargent, L. (Ed.), *Women and revolution: A Discussion of the Unhappy Marriage of Marxism and Feminism*. Boston: South End Press.

Hasan, B. (in press). Exploring gender differences in online shopping attitude. *Computers in Human Behavior*.

Hattie, J., & Fitzgerald, D. (1987). Sex differences in attitudes, achievement and use of computers. *Australian Journal of Education*, *31*(10), 3–26.

Hazan, C., & Shaver, P. (1987). Romantic love conceptualized as an attachment process. *Journal of Personality and Social Psychology*, *52*, 511–524. doi:10.1037/0022-3514.52.3.511

Headd, B. (2003). Redefining business success: Distinguishing between closure and failure. *Small Business Economics*, *21*(1), 51–61. doi:10.1023/A:1024433630958

Heer, D. (1974). The prevalence of black-white marriage in the United States, 1960 and 1970. *Journal of Marriage and the Family*, *36*, 246–258. doi:10.2307/351151

Heilman, M. E., & Block, C. J. (1989). Has Anything Changed? Current Characterizations of Men, Women, and Managers. *The Journal of Applied Psychology*, *74*(6), 935–942. doi:10.1037/0021-9010.74.6.935

Heilman, M. E., & Chen, J. (2003). Entrepreneurship as a Solution: the Allure of Self-Employment for Women and Minorities. *Human Resource Management Review*, *13*, 347–364.

Heimrath, R., & Goulding, A. (2001). Internet perception and use: A gender perspective. *Program: electronic library and information systems, 35*(2), 119-134.

Helling, K., & Ertl, B. (2009). *PREDIL - The National Context of Germany*. Munich, Heraklion: PREDIL Project Consortium.

Henry, C., & Johnston, K. (2003). *State of the Art of Women's Entrepreneurship in Ireland: Access to Financing and Financing Strategies*. Ireland: Centre for Entrepreneurship Research, Dundalk Institute of Technology.

Henry, C., & Kennedy, S. (2002). *Search of a New Celtic Tiger-Female Entrepreneurship in Ireland*. Dundalk Institute of Technology.

Hermawati, W., & Luhulima, A. S. (2000). Women in science, engineering and technology (SET): A report on the Indonesian experience. *Gender, Technology and Development, 4*(1), 87–100. doi:10.1177/097185240000400104

Herrick, J. W. (1999). And Then She Said: Office Stories and What They Tell Us about Gender in the Workplace. *Journal of Business and Technical Communication, 13*(3), 274–296. doi:10.1177/105065199901300303

Herring, S. (1996). Posting in a Different Voice: Gender and Ethics in Computer-Mediated Communication. In *Philosophical Perspectives on Computer Mediated Communication*. Albany, NY: State University of New York Press.

Herring, S. (1996). Posting in a Different Voice: Gender and Ethics in Computer-Mediated Communication. In Ess, C. (Ed.), *Philosophical Perspectives on Computer Mediated Communication* (pp. 115–143). Albany, NY: State University of New York Press.

Herring, S. C. (2003). Gender and power in online communication. In Holmes, J., & Meyerhoff, M. (Eds.), *The handbook of language and gender* (pp. 202–228). Oxford, UK: Blackwell. doi:10.1002/9780470756942.ch9

Herring, S. C., & Paolillo, J. C. (2006). Gender and genre variation in weblogs. *Journal of Sociolinguistics*.

Herring, S. C., Scheidt, L. A., Bonus, S., & Wright, E. (2004). Bridging the gap: A genre analysis of weblogs. In *Proceedings of the Thirty-Seventh Hawaii International Conference on System Sciences* (p. 11). Washington, DC: IEEE Computer Society.

Herschel, R. T. (1994). The Impact of Varying Gender Composition on Group Brainstorming Performance in a GSS Environment. *Computers in Human Behavior, 10*(2), 209–222. doi:10.1016/0747-5632(94)90004-3

Heydary, J. (2006, September, 26). Regulation of online dating services sparks controversy. *Wall Street Journal*, p. A5.

Higgins, E. (1987). Self-discrepancy theory. *Psychological Review, 94*, 1120–1134. doi:10.1037/0033-295X.94.3.319

Hiltz, S. R., Turoff, M., & Johnson, K. (1991). Group Decision Support: The Effect of Designated Leader and Statistical Feedback in Computerized Conferences. *Journal of Management Information Systems, 8*(2), 81–108.

Hindman, M., Tsioutsiouliklis, K., & Johnson, J. A. (2003, March 31). *"Googlearchy": How a few heavily-linked sites dominate politics on the web*. Paper presented at the Annual Meeting of the Midwest Political Science Association.

Hinton, A. (2006). Clues to the future. *Inkblurt.com*. Retrieved 24 April 2008 from http://www.inkblurt.com/2006/01/23/ia-summit-2006-clues-to-the-future/

Hirschman, E. (1987). People as products: Analysis of a complex marketing exchange. *Journal of Marketing, 51*(January), 98–108. doi:10.2307/1251147

Hisrich, R., & Brush, C. G. (1986). *The woman entrepreneur: Starting, financing and managing a successful new business*. Lexington, MA: Lexington Books.

Hitsch, J., Hortacsu, A., & Ariely, D. (2005). *What makes you click*. Paper presented at the AEA Meeting, Choice Symposium, Northwestern University. Estes Park.

Ho, K., Lauscher, H. N., Best, A., Walsh, G., Jarvis-Selinger, S., Fedeles, M., & Chockalingam, A. (2004). Dissecting technology-enabled knowledge translation: Essential challenges, unprecedented opportunities. *Clinical and Investigative Medicine. Medecine Clinique et Experimentale, 27*(2), 70–78.

Hofstede, G. (1991). *Culture and organizations: Software of the mind*. London: McGraw Hill.

Hofstede, G. (2001). *Culture's consequences: Comparing values, behaviors, institutions, and organizations across nations*. Thousand Oaks, CA: Sage.

Holden, L., & Holden, A. C. (1998). Woman to women: social marketing and idea to the new world. *Psychology and Marketing*, *15*(2), 175–193. doi:10.1002/(SICI)1520-6793(199803)15:2<175::AID-MAR5>3.0.CO;2-9

Holden, L. (2007, August 1). Gaining the confidence to go it alone. *The Irish Times*.

Hollis, E. (2003). ITAA: Fewer women and minorities entering IT workforce, June 2, 2003, Certification Magazine. Retrieved March 25, 2010, from http://www.certmag.com/read.php?in=265

Holzapfel, N. (2006). *IT Branche – Gehälter im Aufwind*. Retrieved, March 9, 2009, from http://www.sueddeutsche.de/jobkarriere/309/300307/text/

Hornig, S. (1992). Gender differences in responses to news anout science and technology. *Science, Technology & Human Values*, *17*(4), 532–542. doi:10.1177/016224399201700406

Horrigan, J. B. (2007). A typology of information and communication technology users. *PEW Internet & American Life Project Report*. Retrieved from http://www.pewinternet.org

Houran, J., Lange, R., Rentfrow, P., & Bruckner, K. (2004). Do online matchmaking tests work? An Assessment of preliminary evidence for a publicized 'predictive model of marital success'. *North American Journal of Psychology*, *6*, 507–526.

Howard, J. A. (1989). *Consumer behavior in marketing strategy*. Englewood Cliffs, NJ: Prentice Hall.

Hsu, J., Huang, J., Kinsman, J., Fireman, B., Miller, R., Selby, J., & Ortiz, E. (2005). Use of E-Health Services Between 1999 and 2002: A Growing Digital Divide. *Journal of the American Medical Informatics Association*, *12*(2), 164–171. doi:10.1197/jamia.M1672

Hui, T. K., & Wan, D. (2007). Factors affecting internet shopping behaviour in Singapore: Gender and educational issues. *International Journal of Consumer Studies*, *31*(3), 310–316. doi:10.1111/j.1470-6431.2006.00554.x

Humphreys, P., & Berger, J. (1981). Theoretical consequences of the status characteristics formulation. *American Journal of Sociology*, *86*(5), 953–983. doi:10.1086/227350

Hundley, G. (2001). Why Women Earn Less than Men in Self-Employment. *Journal of Labor Research*, *22*(4), 818–828. doi:10.1007/s12122-001-1054-3

Ibarra, H. (1992). Homophily and Differential Returns: Sex Differenecs in Network Structure and Access in an Advertising Firm. *Administrative Science Quarterly*, *37*(3), 422–447. doi:10.2307/2393451

Igbaria, M., & Baroudi, J. J. (1995). The impact of job performance evaluations on career advancement prospects: An examination of gender differences in the IS workplace. *Management Information Systems Quarterly*, *19*, 107–123. doi:10.2307/249713

IGM (Ed.). (2008). *Entgelt in der ITK-Branche 2008. Eine Erhebung in der Informations- und Telekomunikationsbranche. 10. Erhebung. Frankfurt a. M.* Bund.

Imhof, M., Vollmeyer, R., & Beierlein, C. (2007). Computer use and the gender gap: The issue of access, use, motivation, and performance. *Computers in Human Behavior*, *23*, 2823–2837. doi:10.1016/j.chb.2006.05.007

Ingham, T. (2008). 20% of casual gamers are disabled. *CasualGamong.biz*. Retrieved February 7, 2009 from http://www.casualgaming.biz/news/27527/20-of-casual-gamers-are-disabled

Initiative D21. (2008). *(N)ONLINER ATLAS 2008. Eine Topographie des digitalen Grabens durch Deutschland*. Retrieved January 29, 2010, from http://www.initiatived21.de/wp-content/uploads/2009/06/NONLINER2009.pdf

Inkeles, A., & Levinson, D. (1969). National character: The study of modal personality and socio-cultural systems. In: L. Gardner & E. Aronson (Eds.), *The handbook of social psychology* (vol. 4, pp. 418-516). Reading, MA: Addison Wesley.

Insight (2007). *Greece, Last revised: October 2007*, Agapi Vavouraki, Hellenic Pedagogical Institute, Retrieved January, 29, 2010, from http://insight.eun.org/ww/en/pub/insight/misc/country_report.cfm?

Irish Chamber of Commerce, (2002). *SME eBusiness Survey*.

Ituma, A. (2006). The internal career: An explorative study of the career anchors of information technology workers in Nigeria. In *Proceedings of the ACM SIGMIS Conference on Computer Personnel Research* (pp. 205-212). Claremont, CA: ACM Press.

Jackson, L. A., Ervin, K. S., Gardner, P. D., & Schmitt, N. (2001). Gender and the Internet: Women communicating and men searching. *Sex Roles*, *44*, 363–379. doi:10.1023/A:1010937901821

Jacobs, G., & Dowsland, W. (2000). The Dot-Com Economy in Wales: A Long Road Ahead. In *Proceedings of the 5th UKAIS Conference, University of Wales Institute* (pp. 590-596). Cardiff 26-28 April.

Janis, I. (1972). *Victims of Groupthink*. Boston: Houghton Mifflin.

Jansson, M., Mörtberg, C., & Berg, E. (2007). Old dreams, new means: An exploration of visions and situated knowledge in information technology. *Gender, Work and Organization*, *14*(4), 371–387. doi:10.1111/j.1468-0432.2007.00349.x

Jason, L., Moritsugu, J., & DePalma, D. (1992). Advertisements as a strategy for meeting people. *Psychological Reports*, *71*, 1311–1314. doi:10.2466/PR0.71.8.1311-1314

Jayawardhena, C., Wright, L. T., & Masterson, R. (2003). An investigation of online consumer purchasing. *Qualitative Market Research: An International Journal*, *6*(1), 58–65. doi:10.1108/13522750310457384

Johannisson, B., Ramirez-Pasillas, M., & Karlsson, G. (2002, August). Theoretical and Methodological Challenges Bridging Firm Strategies and Contextual Networking. *Entrepreneurship and Innovation*, 165-174.

Johansson, U. (2003). Regional Development in Sweden: October 2003, Svenska Kommunförbundet. Retrieved from http://www.lf.svekom.se/tru/RSO/ Regional_development_in_Sweden.pdf

Johnson, R. (1980). *Religious assortative marriage in the United States*. New York, NY: Academic Press.

Joiner, R., Gavin, J., Duffield, J., Brosnan, M., & Crook, C. (2005). Gender, internet identification, and internet anxiety: Correlates of Internet use. *Cyberpsychology & Behavior*, *8*(4), 371–378. doi:10.1089/cpb.2005.8.371

Jones, P., Beynon-Davies, P., & Muir, E. (2003). eBusiness Barriers to Growth within the SME Sector. *Journal of Systems & Information Technology*, *7*(1).

Joy, L., Carter, N. M., & Wagner, H. (2007). The bottom line: Corporate performance and women's representation on boards. *Catalyst*. Retrieved from http://www.catalyst.org/publication/200/the-bottom-line-corporate-performance-and-womensrepresentation-on-boards

Kacen, J. J. (2000). Girrrl power and boyyy nature: The past, present, and paradisal future of consumer gender identity. *Marketing Intelligence & Planning*, *18*(6/7), 345–355. doi:10.1108/02634500010348932

Kale, S. (1991). Culture-specific marketing communications: An analytical approach. *International Marketing Review*, *8*(2), 18–30. doi:10.1108/02651339110004078

Kale, S., & Barnes, J. (1992). Understanding the domain of cross-national buyer-seller interactions. *Journal of International Business Studies*, *23*(1), 101–132. doi:10.1057/palgrave.jibs.8490261

Kalmijn, M. (1991). Shifting boundaries: Trends in religious and educational homogamy. *American Sociological Review*, *56*, 786–800. doi:10.2307/2096256

Kalmijn, M. (1993). Trends in black/white intermarriage. *Social Forces*, *72*, 119–146. doi:10.2307/2580162

Kalmijn, M. (1998). Intermarriage and homogamy: Causes, patterns, trends. *Annual Review of Sociology*, *24*, 395–421. doi:10.1146/annurev.soc.24.1.395

Kalyanam, K., & McIntyre, S. (2002). The e-marketing mix: A contribution of the e-tailing wars. *Journal of the Academy of Marketing Science*, *30*(4), 483–495. doi:10.1177/009207002236924

Keddie, Z., & Jones, R. (2005). Information Communications Technology in General Practice: A Cross-Sectional Survey in London Informatics. *Primary Care*, *13*(2), 113–123.

Kelly, E. P., & Young, A. O. (1993). Sex Stereotyping in the Workplace: A Manager's Guide. *Business Horizons*, *36*(2), 23–29. doi:10.1016/S0007-6813(05)80034-5

Kennedy, R. (1952). Single or triple melting pot? Intermarriage in New Haven, 1870-1950. *American Journal of Sociology*, *58*, 56–59. doi:10.1086/221073

Kennedy, T. L. M., Robinson, J. S., & Trammell, K. (2005, October 5-9). *Does gender matter? Examining conversations in the blogosphere*. Paper presented at Internet Research 6.0: Internet Generations, Chicago, IL.

Kenrick, D., Sadalla, E., Groth, G., & Trost, M. (1990). Evolution, traits, and the stages of human courtship: Qualifying the parental investment model. *Journal of Personality, 58*, 97–116. doi:10.1111/j.1467-6494.1990.tb00909.x

Keogh, C., Moore, K., Tattersall, A., Griffiths, M., & Richardson, H. (2006). Managing Diversity or Valuing Diversity in Gender and the IT Labour Market. In Neiderman, F., & Ferratt, T. (Eds.), *IT Workers: Human Capital Issues in a Knowledge-Based Environment*. Hershey, PA, USA: Information Science Publishing.

Key US Disability Organizations. *MIUSA*. Mobility International USA. Retrieved on May 14, 2009 from http://www.miusa.org/idd/IDDresourcecenter/intldevelopment/keyorganizations/.

Kikis-Papadakis, K., Papanastasiou, R., & Margetousaki, A. (2009). *PREDIL - The National Context of Greece*. Heraklion: PREDIL Project Consortium.

Kim, E. Y., & Kim, Y. K. (2004). Predicting online purchase intentions for clothing products. *European Journal of Marketing, 38*(7), 883–897. doi:10.1108/03090560410539302

Kim, P. H., & Aldrich, H. E. (2005). *Social Capital and Entrepreneurship*. Hanover, MA: Now.

Kimmel, J., & Amuendo-Dorantes, C. (2004). The effects of family leave on wages, employment and the family wage gap: Distributional implications. *Journal of Law and Policy, 15*, 115–142.

King, N. (1998). Template Analysis. In Symon, G., & Cassell, C. (Eds.), *Qualitative Methods and Analysis in Organizational Research: A Practical Guide*. Thousand Oaks, CA: Sage.

Kirigia, J. M., Seddoh, A., Gatwiri, D., Muthuri, L. H. K., & Seddoh, J. (2005). E-Health: Determinants, Opportunities, Challenges and the Way Forward. *WHO African Region BMC Public Health, 5*, 1–11.

Kirkpatrick, L., & Davis, K. (1994). Attachment style, gender and relationship stability: A longitudinal analysis. *Journal of Personality and Social Psychology, 66*, 502–512. doi:10.1037/0022-3514.66.3.502

Kjeldsen, J., & Nielsen, K. (2000). *Women Entrepreneurs Now and in the Future*. Danish Agency for Trade and Industry.

Klein, H. K., & Huynh, M. Q. (1999). *The potential of the language action perspective in ethnographic analysis*. Binghamton, NY: School of Management, SUNY Binghamton.

Kling, R. (2000). Learning about information technologies and social change: The contribution of social informatics. *The Information Society, 16*(3), 217–232. doi:10.1080/01972240050133661

Kluckhohn, F., & Strodtbeck, F. (1961). *Variations in value orientations*. Westport, CT: Greenwood Press.

Knouse, S. B., & Webb, S. C. (2001). Virtual Networking for Women and Minorities. *Career Development International, 6*(4), 226–228. doi:10.1108/13620430110397541

Knupfer, N. N. (1998). Gender divisions across technology advertisements and the www: Implications for educational equity. *Theory into Practice, 37*(1), 54–63. doi:10.1080/00405849809543786

Koestner, R., & Wheeler, L. (1988). Self-presentation in personal advertisements: The influence of implicit notions of attraction and role expectations. *Journal of Social and Personal Relationships, 5*(1), 149–160. doi:10.1177/026540758800500202

Kolsaker, A., & Payne, C. (2002). Engendering trust in e-commerce: A study of gender-based concerns. *Marketing Intelligence & Planning, 20*(4/5), 206–214. doi:10.1108/02634500210431595

Kolsaker, A., & Payne, C. (2002). Engendering trust in e-commerce: a study of gender-based concerns. *Marketing Intelligence & Planning, 20*(4/5), 206–214. doi:10.1108/02634500210431595

Komeptenzzentrum Technik Diversity Chancengleichheit, V. (2009a). *Charta für Talente der Zukunft*. Retrieved March 11, 2009, from http://www.kompetenzz.de/Features/Charta

Compilation of References

Komeptenzzentrum Technik Diversity Chancengleichheit, V. (2009b). *Mehr Frauen in IT-Führungspositionen! Pressemitteilung.* Retrieved March 11, 2009, from http://www.komm-mach-mint.de/Startseite/Service/Presse/Pressemitteilungen/Mehr-Frauen-in-IT-Fuehrungspositionen!

Kraus, L. E., Stoddard, S., & Gilmartin, D. (1996). *Chartbook on disability in the United States: An info use report.* Washington, DC: U.S. National Institute on Disability and Rehabilitation Research.

Kraut, R., Mukhopadhyay, T., Szczypula, J., Kiesler, S., & Scherlis, B. (1999). Information and communication: Alternative uses of the internet in households. *Information Systems Research, 10*(4), 287–303. doi:10.1287/isre.10.4.287

Kray, L., Galinsky, A., & Thompson, L. (2002). Reversing the gender gap in negotiations: an exploration of stereotype regeneration. *Organizational Behavior and Human Decision Processes, 87*, 386–409. doi:10.1006/obhd.2001.2979

Krishnamurthy, S. (2006). Introducing e-markplan: A practical methodology to plan e-marketing activities. *Business Horizons, 49*(1), 51–60. doi:10.1016/j.bushor.2005.05.008

Kuhn, K., & Joshi, K. (2009). The reported and revealed importance of job attributes to aspiring information technology professionals: A policy-capturing study of gender differences. *Database, 40*(3), 40–60.

Kuruvilla, S., Dzenowagis, J., Pleasant, A., & Dwivedi, R. (2004). Digital Bridges Need Concrete Foundations: Lessons from the Health InterNetwork India. *British Medical Journal, 328*(7449), 1193. doi:10.1136/bmj.328.7449.1193

Kusku, F., Ozbilgin, M. F., & Ozkale, L. (2007). Against the tide: gendered prejudice and disadvantage in engineering study from a comparative perspective. *Gender, Work and Organization, 14*(2), 109–129.

Kvasny, L., Payton, F., Mbarika, V., Amadi, A., & Meso, P. (2008). Gendered Perspectives on IT Education and Workforce Participation in Kenya. *IEEE Transactions on Education, 51*(2), 256–261. doi:10.1109/TE.2007.909360

Kvasny, L., & Richardson, H. (2006). Critical Research in Information Systems: Looking Forward, Looking Back. *Information Technology & People, 19*(3), 196–202. doi:10.1108/09593840610689813

Kwon, H. J., Joshi, P., & Jackson, V. P. (2007). The effect of consumer demographic characteristics on the perception of fashion web site attributes in Korea. *Journal of Fashion Marketing and Management, 11*(4), 529–538. doi:10.1108/13612020710824580

Lakshman, N. (2006). Here come the bride sites. *Business Week*, 42.

Latour, B. (1993). *We Have Never Been Modern.* London: Harvester Wheatsheaf.

Lavallee, A. (2007, June 13). Firms tidy up clients' bad online reputations. *Wall Street Journal,* p. B1.

LeCompte, M., & Goetz, J. (1982). Problems of reliability and validity in ethnographic research. *Review of Educational Research, 52*(1), 31–60.

Lee, J. D. (2002). More than Ability: Gender and Personal Relationships Influence Science and Technology Involvement. *Sociology of Education, 75*(4), 349–373. doi:10.2307/3090283

Lee, J., Cain, C., Chockley, N., & Burstin, H. (2005). The Adoption Gap: Health Information Technology in Small Physician Practices. *Health Affairs, 24*(5), 1364–1366. doi:10.1377/hlthaff.24.5.1364

Lemons, M. A., & Parzinger, M. (2007). Gender schemas: A cognitive explanation of discrimination of women in technology. *Journal of Business and Psychology, 22*(1), 91–98. doi:10.1007/s10869-007-9050-0

Lenhart, A., & Madden, M. (2007). Social networking Web sites and teens: An overview. *Pew Internet & American Life Project.* Retrieved August 28, 2007, from http://www.pewInternet.org/pdfs/PIP_SNS_Data_Memo_Jan_2007.pdf.

Leon, D., Rotunda, R., Sutton, M., & Schlossman, C. (2003). Internet forewarning effects on ratings of attraction. *Computers in Human Behavior, 19*, 39–57. doi:10.1016/S0747-5632(02)00017-1

Leung, D. (2006). The male/female earnings gap and female self employment. *Journal of Socio-Economics*, *35*(5), 759. doi:10.1016/j.socec.2005.11.034

Leung, G. M., Yu, P. L. H., Wong, I. O. L., Johnston, J. M., & Tin, K. Y. K. (2003). Incentives and Barriers that Influence Clinical Computerisation in Hong Kong: A Population Based Physician Survey. *Journal of the American Medical Informatics Association*, *10*(2), 201–212. doi:10.1197/jamia.M1202

Lewins, A., & Silver, C. (2007). *Using Software in Qualitative Analysis: A Step-by-step Guide*. Los Angeles, CA: Sage.

Lewis, R., & Cockril, A. (2002). Going global - remaining local: the impact of ecommerce on small retail firms in Wales. *International Journal of Information Management*, *22*, 195–209. doi:10.1016/S0268-4012(02)00005-1

Li, N., & Kirkup, G. (2007). Gender and cultural differences in internet use: A study of China and the UK. *Computers & Education*, *48*(2), 301–317. doi:10.1016/j.compedu.2005.01.007

Liao, J., & Welsch, H. P. (2001). *Social Capital and Growth Intention: The Role of Entrepreneurial Networks in Technology-Based New Ventures*. MA: Wellesely.

Lieberson, S., & Waters, M. (1988). *From many strands: Ethnic and racial groups in contemporary America*. New York, NY: Russell Sage.

Lievens, F., & Jordanova, M. (2004). Is There a Contradiction between Telemedicine and Business? *Journal of Telemedicine and Telecare*, *10*(1), 71–74. doi:10.1258/1357633042614393

Liff, S., & Ward, K. (2001). Distorted views through the glass ceiling: the construction of women's understandings of promotion and senior management positions. *Gender, Work and Organization*, *8*(1), 19–36. doi:10.1111/1468-0432.00120

Lin, C. H., & Yu, S. F. (2008). Adolescent internet usage in Taiwan: Exploring gender differences. *Adolescence*, *43*(170), 317–331.

Liñán, F., & Santos, F. J. (2007). Does Social Capital Affect Entrepreneurial Intentions. *International Advances in Economic Research*, *13*(4), 443–453. doi:10.1007/s11294-007-9109-8

Lind, M. R. (1999). The gender impact of temporary virtual work groups. *IEEE Transactions on Professional Communication*, *42*(4), 276–285. doi:10.1109/47.807966

Lind, M. R. (2001). An Exploration of Communication Channel Usage by Gender. *Work Study*, *50*(6), 234–240. doi:10.1108/00438020110403338

Linden Labs. (2008). Economic statistics: Raw data files. *Secondlife.com*. Retrieved April 17, 2008 from http://secondlife.com/statistics/economy-data.php

Lipman-Blumen, J. (1976). Toward a Homosocial Theory of Sex Roles: An Explanation of the Sex Segregation of Social Institutions. In Blaxall, M., & Reagan, B. (Eds.), *Women and the Workplace: The Implications of Occupational Segregation* (pp. 15–32). Chicago: The University of Chicago Press.

Little, J. C. (1999). *The role of women in the history of computing* (pp. 202–205).

Ljunggren, E., & Kolvereid, L. (1996). New Business Formation: Does Gender Make a Difference? *Women in Management Review*, *11*(4), 3–12. doi:10.1108/09649429610122096

Loebbecke, C., & Wareham, J. (2003). The impact of e-business and the information society on 'strategy' and 'strategic planning': An assessment of new concepts and challenges. *Information Technology Management*, *4*, 165–182. doi:10.1023/A:1022946127615

Lopez, A. D., & Manson, P. D. (1997). A study of individual computer self-efficacy and perceived usefulness of the empowered desktop information system. *The Cal Poly Pomona Journal of Interdisciplinary Studies*, *10*, 83–92.

Loscocco, K. A., Robinson, J., Hall, R. H., & Allen, J. K. (1991). Gender and Small Business Success: An Inquiry into Women's Relative Disadvantage. *Social Forces*, *70*, 65–87. doi:10.2307/2580062

Lougheed, T. (2004). Wireless Points the Way in Africa. *Appropriate Technology*, *31*(3), 50.

Lovenduski, J. (2005). *Feminizing Politics*. Oxford: Polity Press.

Loviglio, J. (2007, June 16). Two sex convictions in online dating case. *Wall Street Journal*, p. B5.

Lowrey, Y. (2005). *US Sole Propriertorships: A Gender Comparison 1985-2000*. Washington, D.C.: Small Business Administration Office of Advocacy.

Loyd, B., & Gressard, C. (1984). The effects of sex, age and computer experience on computer attitudes. *AEDS Journal*, *18*(2), 67–77.

Lucas, H. (2008). Information and communications technology for future health systems in developing countries. *Social Science & Medicine*, *66*(10), 21–22. doi:10.1016/j.socscimed.2008.01.033

MacGregor, R. C., Harvie, C., Hyland, P. N., & Lee, B. C. (2007). Benefits Derived from ICT Adoption in Regional Medical Practices: Perceptual Differences between Male and Female General Practitioners. *International Journal of Healthcare Information Systems and Informatics*, *2*(1), 1–13. doi:10.4018/jhisi.2007010101

MacGregor, R. C., & Vrazalic, L. (2007). *E-commerce in Regional Small to Medium Enterprises*. Hershey, PA: IGI Global. doi:10.4018/978-1-59904-123-0

Malchow-Møller, A. (2003). Internet dating: A focus group investigation of young Danes' and Frenchmen's attitudes towards the phenomenon. *Kontur*, *7*, 11–20.

Maratou-Alipranti, L., Dafna, K., Yannakopoulou, L., Kymperi, Z., & Repa, P. (2002). *WOMEN AND SCIENCE: Review of the situation in Greece*. Retrieved January, 29th, 2010, from ftp://ftp.cordis.europa.eu/pub/improving/docs/women_national_report_greece_en.pdf

Marcotte, A. (2007). *Why I had to quit the John Edwards campaign*. Retrieved from http://www.salon.com/news/feature/2007/02/16/marcotte/index.html

Mare, R. (1991). Five decades of educational assortative mating. *American Sociological Review*, *56*, 15–32. doi:10.2307/2095670

Markus, M., & Robey, D. (1988). Information technology and organizational change. *Management Science*, *34*(5), 583–598. doi:10.1287/mnsc.34.5.583

Marlow, S. (2002). Self-employed women: Apart of, or apart from, feminist theory? *Entrepreneurship and Innovation*, *2*(2), 83–91. doi:10.5367/000000002101299088

Marlow, S., & Carter, S. (2004). Accounting for change: professional status, gender disadvantage and self-employment. *Women in Management Review*, *19*(1), 5–17. doi:10.1108/09649420410518395

Marlow, S., Carter, S., & Shaw, E. (2008). Constructing female entrepreneurship policy in the UK: is the US a relevant benchmark? *Environment and Planning. C, Government & Policy*, *26*, 335–351. doi:10.1068/c0732r

Marlow, S., & Carter, S. (2008). Constructing Female Entrepreneurship Policy in the UK: Is the USA a Relevant Benchmark? *Environment and Planning. C, Government & Policy*, *26*(2), 335–351. doi:10.1068/c0732r

Marlow, S., & Patton, D. (2005). The financing of small businesses- female experiences and Strategies. In Davies, M., & Fielden, S. (Eds.), *International Handbook on women and small business entrepreneurship*. Cheltenham, UK: Edward Elgar.

Marlow, S., & Strange, A. (1994). Female Entrepreneur - Success By Whose Standards? In Tanton, M. (Ed.), *Women in Management: A Developing Presence* (pp. 172–184). London: Routledge.

Martin, L. M., & Wright, L. T. (2005). No Gender in Cyberspace? Empowering Entrepreneurship and Innovation in Female-run ICT Small Firms. *International Journal of Entrepreneurial Behaviour & Research*, *11*(2), 162–178. doi:10.1108/13552550510590563

Matthews, P. (2007). ICT assimilation and SME expansion. *Journal of International Development*, *19*(6), 817–827. doi:10.1002/jid.1401

Maupin, R. (1990). Sex Role Identity and Career Success of Certified Public Accountants. *Advances in Public Interest Accounting*, 97-105.

Mazumdar, T., & Papatla, P. (1995). Gender difference in price and promotion response. *Pricing Strategy & Practice.*, *3*(1), 21–33.

McClelland, E. (2003). Following the pathway of female entrepreneurs. *International Journal of Entrepreneurial Behaviour & Research*, *11*(2), 84–107. doi:10.1108/13552550510590527

McGowan, P., & Hampton, A. (2007). An Exploration of Networking Practices of Female Entrepreneurs. In Carter, N. M., Henry, C., Cinneide, B. O., & Johnston, K. (Eds.), *Female Entrepreneurship: Implications for Education, Training and Policy* (pp. 110–134). London: Routledge.

McGowan, P., & Henry, C. (2004). Special Issue: Female Entrepreneurship-A Research Agenda. *International Journal of Entrepreneurial Behaviour and Research*.

McGregor, J., & Tweed, D. (2001). Gender and Managerial Competence: Support for Theories of Androgyny. *Women in Management Review*, *16*(6), 279–286. doi:10.1108/09649420110401540

McKenna, L., & Pole, A. (2008). What do bloggers do: An average day on an average political blog. *Public Choice*, *134*(1), 97–108. doi:10.1007/s11127-007-9203-8

McKenna, K., Green, A., & Gleason, M. (2002). Relationship formation on the Internet: What's the big attraction? *The Journal of Social Issues*, *58*, 9–31. doi:10.1111/1540-4560.00246

McKenna, L., & Pole, A. (2004). *Do blogs matter? Weblogs in American politics*. Paper presented at the Annual Meeting of the American Political Science Association, Chicago, IL.

McKinney, V., Wilson, D., Brooks, N., O'Leary-Kelly, A., & Hardgrave, B. (2008). Women and men in the IT profession. *Communications of the ACM*, *51*(2), 81–84. doi:10.1145/1314215.1340919

McKnight, D. H., & Cummings, L. L. (1998). Initial Trust Formation in New Organizational Relationships. *Academy of Management Review*, *23*(3), 473–490.

McPherson, J. M., & Smith-Lovin, L. (2001). Birds of a Feather: Homophily in Social Networks. *Annual Review of Sociology*, *27*, 415–444. doi:10.1146/annurev.soc.27.1.415

McPherson, M., & Smith-Lovin, L. (1987). Homophily in Voluntary Organizations: Status Distance and the Composition of Face-to-Face Groups. *American Sociological Review*, *52*(3), 370–379. doi:10.2307/2095356

Meaney, M., Devillard-Hoellinger, S., & Denari, A. (2008). *Room at the top: Women and success in UK business*. McKinsey&Company.

Mehra, A., & Kilduff, M. (1998). At the Margins: A Distinctiveness Approach to the Social Identity and Social Networks of Underrepresented Groups. *Academy of Management Journal*, *41*(4), 441–452. doi:10.2307/257083

Meijer, W. J., & Ragetlie, P. L. (2007). Empowering the patient with ICT tools: The unfulfilled promise. *Studies in Health Technology and Informatics*, *127*, 199–218.

Mellot, D. W. (1983). *Fundamentals of consumer bahaviour*. Oklahoma: Pen Well Publishing Company.

Melymuka, K. (2005, April 18). What IT women want, a virtual roundtable of high achievers talks about what today's women bring to IT and what they expect in return. *ComputerWorld*. Retrieved from http://www.computerworld.com/s/article/101088/What_IT_Women_Want

Meri, T. (2008) *Women in science and technology, Statistics in focus, Science and Technology, 10/2008*. Retrieved January 29, 2010, from http://bookshop.europa.eu/eubookshop/download.action?fileName=KSSF08010ENC_002.pdf&eubphfUid=554230&catalogNbr=KS-SF-08-010-EN-C

Merton, R. (1941). Intermarriage and the social structure: Fact and theory. *Psychiatry*, *4*, 361–374.

Meyers-Levy, J., & Maheswaran, D. (1991). Exploring differences in males' and females' processing strategies. *The Journal of Consumer Research*, *18*(1), 63–70. doi:10.1086/209241

Michie, S., & Nelson, D. L. (2006). Barriers women face in information technology careers: Self-efficacy, passion and gender biases. *Women in Management Review*, *21*(1). doi:10.1108/09649420610643385

Miller, N. J., & Besser, T. L. et al. (2006/7). Do Strategic Business Networks Benefit Male- and Female-Owned Small Community Businesses. *Journal of Small Business Strategy*, *17*(2), 53-74.

Minniti, M., & Bygrave, W. D. (2003). *National Entrepreneurship Assessment: United States of America Executive Report*. Kansas City, MO: Kauffman Foundation.

Mirchandani, K. (1999). Feminist Insight on Gendered Work: New Directions in Research on Women and Entrepreneurship. *Gender, Work and Organization*, *6*(4), 224–235. doi:10.1111/1468-0432.00085

Mitchell, V. W., & Walsh, G. (2004). Gender differences in German consumer decision-making styles. *Journal of Consumer Behaviour*, *3*(4), 331–346. doi:10.1002/cb.146

Mitchell, S. (2007). *Access to technology: Race, gender, class bias.* Retrieved from http://bloggingfeminism.blogspot.com/

Mitter, S., & Rowbotham, S. (Eds.). (1995). *Women encounter technology: Changing Patterns of Employment in the Third World*. London, New York: Routledge. doi:10.4324/9780203208618

Mitter, S. (1994). On organising women in casualised work: a global overview. In Rowbotham, S., & Mitter, S. (Eds.), *Dignity and Daily Bread: New Forms of Economic Organising among Poor Women in the Third World and the First*. London: Routledge. doi:10.4324/9780203422946_chapter_1

Mohr, G., & Wolfram, H.-J. (2008). Leadership and Effectiveness in the Context of Gender: The Role of Leaders' Verbal Behaviour. *British Journal of Management*, *19*(1), 4. doi:10.1111/j.1467-8551.2007.00521.x

Montgomery, J. D. (1991). Social Networks and Labour-Market Outcomes: Toward an Economic Analysis. *The American Economic Review*, *81*(5), 1408–1418.

Moore, M. (1985). Non-verbal courtship patterns in women: Context and consequences. *Ethology and Sociobiology*, *6*, 237–247. doi:10.1016/0162-3095(85)90016-0

Moore, M. (1998). Non-verbal courtship patterns in women: Rejection signalling: An empirical investigation. *Semiotica*, *118*, 201–214.

Moraski, M. (2007, June 16). Beware of digital Don Juans. *Wall Street Journal*, p. A5.

Morrison, A. M., & von Glinow, M. A. (1990). Women and Minorities in Management. *The American Psychologist*, *45*(2), 200–208. doi:10.1037/0003-066X.45.2.200

Moussi, C., & Davey, B. (2000). Internet-based electronic commerce: perceived benefits and inhibitors. In *Proceedings of the Australian Conference on Information Systems, Information Systems Management Research Centre*. Queensland, University of Technology, Brisbane, December 6-8.

MPFS - Medienpädagogischer Forschungsverbund Südwest (LFK/ LMK). (2008). *JIM-Studie 2008. Jugend, Information, (Multi-) Media. Basisuntersuchungen zum Medienumgang 12- bis 19-Jähriger*. Retrieved January 29, 2010, from http://www.mpfs.de/fileadmin/JIM-pdf08/JIM-Studie_2008.pdf

Mujtaba, B. G. (2007). *Workforce diversity management: Challenges, competencies, and strategies*. Llumina Press.

Mulrine, A. (2003, September 29). Love.com: For better or for worse, the Internet is radically changing the dating scene in America. *U.S. News and World Report*.

Munch, A., & McPherson, M. A. (1997). Gender, Children, and Social Contact: The Effects of Childrearing for Men and Women. *American Sociological Review*, *62*(4), 509–520. doi:10.2307/2657423

Munk, B. (2007). LogoGo - An approach to the design of girl-specific educational software. In Zorn, I., Maas, S., Rommes, E., Schirmer, C., & Schelhowe, H. (Eds.), *Gender Designs IT. Construction and Deconstruction of Information Society Technology*. Wiesbaden: VS Verlag für Sozialwissenschaften.

Myers, G. (1994). *Targeting the new professional woman: How to market and sell to today's 57 million working women*. Chicago: Probus Publishing Company.

Myers, M. (1999). Investigating information systems with ethnographic research. *Communications of the Association for Information Systems*, *2*(23), 1–20.

Myers, M., & Young, L. (1997). Hidden agendas, power, and managerial assumptions in information systems development: An ethnographic study. *Information Technology & People*, *10*(3), 224–240. doi:10.1108/09593849710178225

Nath, R., Akmnakigil, M., Hjelm, K., Sakaguchi, T., & Schultz, M. (1998). Electronic Commerce and the Internet: Issues, Problems and Perspectives. *International Journal of Information Management*, *18*(2), 91–101. doi:10.1016/S0268-4012(97)00051-0

National Gay and Lesbian Task Force. Retrieved on May 14, 2009 from http://www.thetaskforce.org/.

Ndubisi, N. O. (2006). Effect of gender on customer loyalty: A relationship marketing approach. *Marketing Intelligence & Planning*, *24*(1), 48–61. doi:10.1108/02634500610641552

Ndubisi, N. O., & Kahraman, C. (2005). Malaysian Women Entrepreneurs: Understanding the ICT Usage Behaviours and Drivers. *Journal of Enterprise Information Management*, *18*(6), 721–739. doi:10.1108/17410390510628418

Neely, W. (1940). Family attitudes of denominational college and university students, 1929 and 1936. *American Sociological Review*, *4*, 512–522. doi:10.2307/2084426

Nelson, M. L., & Alexander, K. (2002). The emergence of supply chain management as a strategic facilities management tool. In K. Alexander (Ed.), *Proceedings of the Euro FM Research Symposium in Facilities Management*. The University of Salford

Nemeth, C. J. (1986). Differential Contributions of Majority and Minority Influence. *Psychological Review*, *93*(1), 23–32. doi:10.1037/0033-295X.93.1.23

Nevid, J. (1984). Sex differences in factors of romantic attraction. *Sex Roles*, *11*, 401–411. doi:10.1007/BF00287468

Nillson, P. (1997). Business Counselling Services Directed Towards Female Entrepreneurs – Some Legitimacy Dilemmas. *Entrepreneurship and Regional Development*, *9*(3), 239–258. doi:10.1080/08985629700000014

Nohria, N. (1992). Information and Search in the Creation of New Business Ventures: the Case of the 128 Venture Group. In Nitin, N., & Eccles, R. G. (Eds.), *Networks and Organizations* (pp. 240–261). Boston, MA: Harvard Business School Press.

Nolan, C. (2007). *Women's voices are louder online*. Retrieved from http://bloggingfeminism.blogspot.com/

Norris, J. (2009). The Growth and Direction of Healthcare Support Groups in Virtual Worlds. *Journal of Virtual Worlds Research*, *2*(2). Retrieved on March 19, 2010 from https://journals.tdl.org/jvwr/article/view/658/500.

Noveck, B. S. (2004). Introduction: The state of play. *New York Law School Law Review. New York Law School*, *49*(1), 1–18.

Nowak, K. L., & Rauh, C. (2005). The influence of the avatar on online perceptions of anthropomorphism, androgyny, credibility, homophily, and attraction. *Journal of Computer-Mediated Communication*, *11*(1), 153–178. doi:10.1111/j.1083-6101.2006.tb00308.x

Nunamaker, J. F., Dennis, A. R., Valacich, J. S., Vogel, D. R., & George, J. F. (1991). Electronic meeting systems to support group work. *Communications of the ACM*, *34*(7), 40–61. doi:10.1145/105783.105793

O'Donnell, A., Gilmore, A., Cummins, D., & Carson, D. (2001). The Network Construct in Entrepreneurship Research: A Review and Critique. *Management Decision*, *39*(9), 749–760. doi:10.1108/EUM0000000006220

O'Dowd, T. C., McNamara, K., Kelly, A., & O'Kelly, F. (2006). Out-of-hours co-operatives: general practice satisfaction with governance and working arrangements. *The European Journal of General Practice*, *12*(1), 15–18. doi:10.1080/13814780600757195

Observatory for the Greek IS. (2008). Παρουσίαση αποτελεσμάτων έρευνας για τη χρήση των νέων τεχνολογιών από τα παιδιά. Retrieved January 29, 2010, from http://www.observatory.gr/Files/Meletes/Y8EEUR081015DOCEL_Π3%20%20Έρευνα%20Παιδιά.pdf

Observatory for the Greek IS. (2009). Measurement of eEurope/ i2010. Indicators for Greece. 2008 Findings. Retrieved April 5, 2010, from http://www.observatory.gr/files/meletes/Booklet%20eEurope%202008%20en.pdf

Odell, P. M., Korgen, K. O., Schumacher, P., & Delucchi, M. (2000). Internet use among female and male college students. *Cyberpsychology & Behavior*, *3*, 855–862. doi:10.1089/10949310050191836

OECD. (1999). *The economic and social impact of electronic commerce* (pp. 1–168). Preliminary Findings and Research Agenda.

OECD. (2004). *Promoting entrepreneurship and Innovative SME's in a global economy: Towards a more responsible and inclusive globalisation*. Paper presented at 2nd OECD Conference of Ministers responsible for Small and Medium-sized enterprises (SME's), Istanbul, Turkey, 3-5 June 2004.

Compilation of References

OECD. (2005). *Are Students Ready for a Technology-Rich World? What PISA Studies Tell Us.* OECD. Retrieved January 29, 2010, from http://www.oecd.org/document/31/0,3343,en_32252351_32236173_35995743_1_1_1_1,00.html

OECD. (2009). *Information Technology Outlook 2008.* OECD. Retrieved April 9, 2010, from http://browse.oecdbookshop.org/oecd/pdfs/browseit/9308041E.PDF

Olsen, D., Berlin, E., Olsen, E., McLean, J., & Sussman, M. (2009). *State of the blogosphere.* Retrieved from http://technorati.com/blogging/feature/state-of-the-blogosphere-2009/

Ondrejka, C. (2004). A piece of place: Modeling the digital on the real in Second Life. *SSRN.com.* Retrieved on April 28, 2008 from http://ssrn.com/abstract=555883.

Ono, H., & Zavodny, M. (2003). Gender and the internet. *Social Science Quarterly*, *84*(1), 111–121. doi:10.1111/1540-6237.t01-1-8401007

Open University. (2006). Making your teaching inclusive. *The Open University, Walton Hall Milton Keynes, UK.* Retrieved January 23, 2010 from http://www.open.ac.uk/inclusiveteaching/pages/understanding-and-awareness/models-of-disability.php.

Oppedisano, J. (2004). Giving back: women's entrepreneurial philanthropy. *Women in Management Review*, *19*(3), 174–177. doi:10.1108/09649420410529889

Orhan, M., & Scott, D. (2001). Why women enter into entrepreneurship: an explorative model. *Women in Management Review*, *16*(5), 232–247. doi:10.1108/09649420110395719

Orlikowski, W. (1991). Integrated information environment or matrix of control? The contradictory implications of information technology. *Accounting. Management and Information Technologies*, *1*(1), 9–42. doi:10.1016/0959-8022(91)90011-3

Orlikowski, W. (1992a). CASE tools as organizational change: Investigating incremental and radical changes in systems development. *MIS Quarterly*, *17*(3), 309–340. doi:10.2307/249774

Orlikowski, W. (1992b). The duality of technology: Rethinking the concept of technology in organizations. *Organization Science*, *3*(3), 398–427. doi:10.1287/orsc.3.3.398

Orlikowski, W., & Robey, D. (1991). Information technology and the structuring of organizations. *Information Systems Research*, *2*(2), 143–169. doi:10.1287/isre.2.2.143

Orr, A. (2004). *Meeting, mating and cheating: Sex, love, and the new world of online dating.* New Jersey: Reuters.

Orser, B. J., Riding, A. L., & Manley, K. (2006). Women Entrepreneurs and Financial Capital. *Entrepreneurship Theory and Practice*, *30*(5), 643. doi:10.1111/j.1540-6520.2006.00140.x

Orth, U. R., & Holancova, D. (2004). Men's and women's responses to sex role portrayals in advertisements. *International Journal of Research in Marketing*, *21*(1), 77–88. doi:10.1016/j.ijresmar.2003.05.003

Orviska, M., & Hudson, J. (2009). Dividing or uniting Europe? Internet usage in the EU. *Information Economics and Policy*, *21*(4), 279–290. doi:10.1016/j.infoecopol.2009.06.002

Osell, T. (2007). *Where are the women? Pseudonymity and the public sphere, then and now.* Retrieved from http://www.barnard.edu/sfonline/blogs/osell_01.htm

Otnes, C., & McGrath, M. A. (2001). Perception and realities of male shopping behavior. *Journal of Retailing*, *77*(1), 111–137. doi:10.1016/S0022-4359(00)00047-6

Oudshoorn, N., Rommes, E., & Stienstra, M. (2004). Configuring the user as everybody: Gender and design cultures in information and communication technologies. *Science, Technology & Human Values*, *29*(1), 30–63. doi:10.1177/0162243903259190

Owen, C. L., & Todor, W. D. (1993). Attitudes Toward Women as Managers: Still the Same. *Business Horizons*, *36*(2), 12–16. doi:10.1016/S0007-6813(05)80032-1

Ozgen, E., & Baron, R. A. (2007). Social sources of information in opportunity recognition: Effects of mentors, industry networks, and professional forums. *Journal of Business Venturing*, *22*(2), 174–192. doi:10.1016/j.jbusvent.2005.12.001

Pan, Z. X. T., & Pokharel, S. (2007). Logistics in Hospitals: A Case Study of some Singapore Hospitals. *Leadership in Health Services*, *20*(3), 195–207. doi:10.1108/17511870710764041

Parekh, R., & Beresin, E. (2001). Looking for love? Take a cross-cultural walk through the personals. *Academic Psychiatry*, *25*, 223–233. doi:10.1176/appi.ap.25.4.223

Park, S. (2009). Concentration of internet usage and its relation to exposure to negative content: Does the gender gap differ among adults and adolescents? *Women's Studies International Forum*, *32*(2), 98–107. doi:10.1016/j.wsif.2009.03.009

Park, B., & Hastie, R. (1987). Perception of variability in category development: Instance versus abstraction-based stereotypes. *Journal of Personality and Social Psychology*, *53*(4), 621–635. doi:10.1037/0022-3514.53.4.621

Parks, M., & Floyd, K. (1996). Making friends in cyberspace. *The Journal of Communication*, *46*, 80–97. doi:10.1111/j.1460-2466.1996.tb01462.x

Pasha, S. (2005, August 18). Online dating feeling less attractive. *CNN/Money*. Retrieved April 13, 2006, from http://money.cnn.com/2005/08/18/technology/online_dating/index.htm

Patton, M. Q. (2002). *Qualitative research & evaluation methods*. Thousand Oaks, CA: Sage.

Pedersen, S., & Macafee, C. (2007). Gender differences in British blogging. *Journal of Computer-Mediated Communication*, *12*(4). doi:10.1111/j.1083-6101.2007.00382.x

Pelletier-Fleury, N., Lanoe, J. L., Philippe, C., Gagnadoux, F., Rakotonanahary, D., & Fleury, B. (1999). Economic Studies and Technical Evaluation of Telemedicine: The Case of Telemonitored Polysomnography. *Health Policy (Amsterdam)*, *49*, 179–194. doi:10.1016/S0168-8510(99)00054-8

Peng, H. Y., Tsai, C. C., & Wu, Y. T. (2006). University students' self-efficacy and their attitudes toward the internet: the role of students' perceptions of the Internet. *Educational Studies*, *32*(1), 73–86. doi:10.1080/03055690500416025

Peres, Y., & Meivar, H. (1986). Self-presentation during courtship: A content analysis of classified advertisements in Israel. *Journal of Comparative Family Studies*, *17*(1), 19–31.

Perez, M. P., Carnicer, M. P. L., & Sanchez, A. M. (2002). Differential Effects of Gender Perceptions of Teleworking by Human Resources Managers. *Women in Management Review*, *17*(6), 262–275. doi:10.1108/09649420210441914

Peris, R., Gimeno, M., Pinazo, D., Ortet, G., Carrero, V., Sanchiz, M., & Ibanez, I. (2002). Online chat rooms: Virtual spaces of interaction for socially-oriented people. *Cyberpsychology & Behavior*, *5*(1), 43–51. doi:10.1089/109493102753685872

Perlmutter, D. (2008). *Blog wars*. New York, NY: Oxford University Press.

Perner, P., & Fiss, G. (2002). Intelligent e-marketing with web mining, personalization and user-adapted interfaces. In Perner, P. (Ed.), *Data Mining in E-Commerce, Medicine, and Knowledge Management* (pp. 37–52). Springer Verlag.

Perry, E. L., Davis-Blake, A., & Kulik, C. T. (1994). Explaining gender-based selection decisions: A synthesis of contextual and cognitive approaches. *Academy of Management Review*, *19*(4), 786–820.

Philipsen, R.L. & Kemp, R.G.M. (2003). *Capabilities for Growth An Exploratory Study on Medium-Sized Firms in Dutch ICT Services and Life Sciences*. Scales Paper N200313 EIM Business and Policy Research.

Phillip, M. V., & Suri, R. (2004). Impact of gender differences on the evaluation of promotional emails. *Journal of Advertising Research*, *44*(4), 360–368.

Pietromonaco, P., & Carnelley, K. (1994). Gender and working models of attachment: Consequences for perceptions of self and romantic relationships. *Personal Relationships*, *1*, 63–82. doi:10.1111/j.1475-6811.1994.tb00055.x

Pistole, C. (1995). College students ended love relationships: Attachment style and emotion. *Journal of College Student Development*, *1*, 53–60.

Pole, A. (2010). *Blogging the political: Politics and participation in a networked society*. New York, NY: Routledge.

Pole, A. (2009, April). *Clinton and Obama: Blogging the 2008 presidential primaries*. Paper presented at the Midwestern Political Science Association. Chicago, IL.

Popielarz, P. A. (1999). (In) Voluntary Association: A Multilevel Analysis of Gender Segregation in Voluntary Organizations. *Gender & Society*, *13*(2), 234–250. doi:10.1177/089124399013002005

Postmes, T., Spears, R., & Lea, M. (1998). Breaching or building social boundaries? SIDE effects of computer-mediated communication. *Communication Research*, *25*, 689–715. doi:10.1177/009365098025006006

Pouliakas, K., & Livanos, I. (2008). *The gender wage gap as a function of educational degree choices in Greece.* Retrieved January 29, 2010, from http://mpra.ub.uni-muenchen.de/14168, MPRA paper No. 14168, posted 24 March 2009

Powell, A., Piccoli, G., & Ives, B. (2004). Virtual teams: a review of current literature and directions for future research. *SIGMIS Database*, *35*(1), 6–36. doi:10.1145/968464.968467

Powell, G. N. (1993). *Women and Men in Management* (2nd ed.). Newbury Park, CA: Sage Publications.

Prakash, V. (1992). Sex roles and advertising preferences. *Journal of Advertising Research*, *32*(3), 43–52.

Protovoulia (n. d.). *Σύνοψη Μελέτης για την Απασχόληση των Πτυχιούχων Τριτοβάθμιας Εκπαίδευσης στην Ελλάδα.* Retrieved January 29, 2010, from http://studies.protovoulia.org/files/synopsi_meletis2.pdf

Pullen, S., Atkinson, D., & Tucker, S. (2000). Improvements in Benchmarking the Asset Management of Medical Facilities. In *Proceedings of the International Symposium on Facilities Management and Maintenance Brisbane* (pp 265–271).

Putnam, R. (2000). *Bowling alone: The collapse and revival of American community.* New York, NY: Simon and Schuster.

Putrevu, S. (2004). Communicating with the sex: Male and female responses to print advertising. *Journal of Advertising*, *33*(3), 51–62.

Qavi, T., Corley, L., & Kay, S. (2001). Nursing Staff Requirements for Telemedicine in the Neonatal Intensive Care Unit. *Journal of End User Computing*, *13*(3), 5–13. doi:10.4018/joeuc.2001070101

Qian, Z. (1997). Breaking racial barriers: Variations in interracial marriage between 1980 and 1990. *Demography*, *34*, 263–276. doi:10.2307/2061704

Quesenberry, J., & Trauth, E. (2007). *What do women want? An investigation of career anchors among women in the IT workforce.* In the SIGMIS Proceedings on Computer Personnel Research (pp. 122–127). St. Louis, MO: Association for Computing Machinery.

Radhakrishnan, S. (2008). Examining the "Global" Indian Middle Class: Gender and Culture in the Silicon Valley/Bangalore Circuit. *Journal of Intercultural Studies (Melbourne, Vic.)*, *29*(1), 7–20. doi:10.1080/07256860701759915

Rainer, R. K., Laosethakul, K., & Astone, M. K. (2003). Are gender perceptions of computing changing over time? *Journal of Computer Information Systems*, *43*(4), 108–114.

Rajadhyaksha, U., & Smita, S. (2004). Tracing a timeline for work and family research in India. *Economic and Political Weekly*, *39*(17), 1674–1680.

Rajecki, D., Bledsoe, S., & Rasmussen, J. (1991). Successful personal ads: Gender differences and similarities in offers, stipulations, and outcomes. *Basic and Applied Psychology*, *12*, 457–469. doi:10.1207/s15324834basp1204_6

Randall, D., Hughes, J., O'Brien, J., Rodden, T., Rouncefield, M., & Sommerville, I. (1999). Banking on the old technology: Understanding the organizational context of 'Legacy' issues. *Communications of the Association for Information Systems*, *2*(8), 1–27.

Rao, V., & Rao, V. (1982). *Marriage, the family and women in Indi.* Delhi, India: Heritage Publications.

Ratliff, C. (2007). *Attracting readers: Sex and audience in the blogosphere.* Retrieved from http://www.barnard.edu/sfonline/blogs/ratliff_01.htm

Ray, S., & Mukherjee, A. (2007). Development of a Framework Towards Successful Implementation of E-governance Initiatives in Health Sector in India. *International Journal of Health Care*, *20*(6), 464–483. doi:10.1108/09526860710819413

Rees, D. (1998). Management Structures of Facility Management in the National Health Service in England: A review of Trends 1995–1997. *Facilities*, *15*(3 /4), 254–261. doi:10.1108/02632779810229075

Renzulli, L. A., & Aldrich, H. (2000). Family Matters: Gender, Networks and Entrepreneurial Outcomes. *Social Forces*, *79*(2), 523–546. doi:10.2307/2675508

Report of the Small Business Forum. (2006). *Small Business is Big Business. The Small Business Forum.* Forfás Dublin.

Reskin, B. F. (1993). Sex Segregation in the Workplace. *Annual Review of Sociology*, *19*, 241–270. doi:10.1146/annurev.so.19.080193.001325

Reviews, D. S. com. (2006). Retrieved November, 21, 2007, from http://www.datingsitesreviews.com/staticpages/index.php?page=2010000100-FriendFinder

Reynolds, W., Savage, W., & Williams, A. (1994). *Your own business: A Practical guide to success*. ITP.

Ribeiro, J. S. (2008). Gendering Migration Flows. *Physicians and Nurses in Portugal Equal Opportunities International*, *27*(1), 77–87. doi:10.1108/02610150810844956

Rice, R., & Rogers, E. (1980). Reinvention in the innovation process. *Science Communication*, *1*(4), 499–514. doi:10.1177/107554708000100402

Richardson, I., & Hynes, B. (2007). Women in engineering and technological entrepreneurship: exploring initiatives to overcome the obstacles. In Carter, N.M., Henry, C., Cinnedie, B. Ó., & Johnston, K. (Eds.), *Female entrepreneurship implications for education, training and policy.* Routledge.

Ridgeway, C. L., & Correll, S. J. (2004). Unpacking the Gender System: A Theoretical Perspective on Gender Beliefs and Social Relations. *Gender & Society*, *18*(4), 510–531. doi:10.1177/0891243204265269

Rigby, M. (2006). Evaluation – The Cinderella Science of ICT in Health. *Yearbook of Medical Informatics*, 114–120.

Robb, A. M. (2002). Entrepreneurial Pefromance by Women and Minorities: The Case of New Firms. *Journal of Developmental Entrepreneurship*, *7*(4), 384–397.

Roberts, L. D., & Parks, M. R. (2001). The social geography of gender-switching in the virtual environments on the Internet. In Green, E., & Adam, A. (Eds.), *Virtual Gender*. London: Routledge.

Robertson, O., Hewitt, J., & Scardamalia, M. (2003). *Gender Participation Patterns in Knowledge Forum: an Analysis of Two Graduate-Level Classes.* Paper presented at the IKIT Summer Institute, Toronto, ON, Canada.

Rodgers, S., & Harris, M. A. (2003). Gender and e-commerce: An exploratory study. *Journal of Advertising Research*, *43*(3), 322–329. doi:10.1017/S0021849903030307

Rodgers, S., & Harris, M. A. (2003). Gender and e-commerce: An exploratory study. *Journal of Advertising Research*, *43*(3), 322. doi:10.1017/S0021849903030307

Rogers, C. (1951). *Client-centered therapy*. Boston, MA: Houghton-Mifflin.

Roomi, M. A. (2007). *Role of Human and Social Capital in the Growth of Women-owned Enterprises*. Glasgow: Institute for Small Business and Entrepreneurship.

Rose, D. (1999). *Internet soul mates: Finding the love of your life through the Internet*. Phoenix, AZ: Productiones Deanna, LLC.

Rosenfeld, M. (2005). A critique of exchange theory in mate selection. *American Journal of Sociology*, *110*(5), 1284–2027. doi:10.1086/428441

Rowbotham, S. (2006). Feminist approaches to technology. In Grewal, I., & Kaplan, C. (Eds.), *An Introduction to Women's studies gender in a transnational world.* McGraw-Hill.

Rudman, L. A. (1998). Self-Promotion as a Risk Factor for Women: The Costs and Benefits of Counterstereotypical Impression Management. *Journal of Personality and Social Psychology*, *74*(3), 629–345. doi:10.1037/0022-3514.74.3.629

Ryan, A. B. (2006). Post-Positivist Approaches to Research. In M. Antonesa, H. Fallon, A.B. Ryan, A. Ryan, T. Walsh, & L. Borys (Eds.), *Researching and Writing your Thesis: a guide for postgraduate students* (pp. 12-26). MACE: Maynooth Adult and Community Education.

Compilation of References

Sadalla, E., Kenrick, D., & Venshure, B. (1987). Dominance and heterosexual attraction. *Journal of Personality and Social Psychology*, *52*, 730–738. doi:10.1037/0022-3514.52.4.730

Sakai, K. (2002). *Global Industrial Restructuring: Implications for Small Firms*. STI Working Papers 2002/4, OECD, Paris (www.oecd.org/sti/working-papers.)

Samp, J. A., Wittenberg, E. M., & Gillett, D. L. (2003). Presenting and monitoring a gender-defined self on the Internet. *Communication Research Reports*, *20*(1), 1–12.

Sandberg, K. W. (2003). An Exploratory Study of Women in Micro Enterprises: Gender Related Difficulties. *Journal of Small Business and Enterprise Development*, *10*(4), 408–417. doi:10.1108/14626000310504710

Sappleton, N., & Takruri-Rizk, H. (2008). The gender subtext of science, engineering, and technology (SET) organizations: A review and critique. *Women's Studies*, *37*, 284–316. doi:10.1080/00497870801917242

Sappleton, N. (2009). Women Non-traditional Entrepreneurs and Social Capital. *International Journal of Gender and Entrepreneurship*, *1*(3), 192–218. doi:10.1108/17566260910990892

Sarton, G. (1927-48). *Introduction to the History of Science* (3 v. in 5). Carnegie Institution of Washington Publication no. 376. Baltimore: Williams and Wilkins, Co.

Scharlott, B., & Christ, W. (1995). Overcoming relationship-initiation barriers: The impact of a computer-dating system on sex role, shyness, and appearance inhibitions. *Computers in Human Behavior*, *11*, 191–204. doi:10.1016/0747-5632(94)00028-G

Schein, E. H. (1990). *Career Anchors: Discovering Your Real Values*. San Francisco, CA: Jossey-Bass.

Schein, V. E. (2001). A Global Look at Psychological Barriers to Women's Progress in Management. *The Journal of Social Issues*, *57*(4), 675–688. doi:10.1111/0022-4537.00235

Schein, V. E. (1994). Managerial Sex Typing: A Persistent and Pervasive Barrier to Women's Opportunities. In Davidson, M. J., & Burke, R. J. (Eds.), *Women in Management: Current Research Issues* (pp. 41–52). London: Paul Chapman Publishing.

Schnirch, A., & Welzel, M. (2004). Nutzung neuer Medien im Bereich des naturwissenschaftlichen Unterrichts der Realschule. Eine Studie unter Genderperspektive. In Buchen, S., Helfferich, C., & Maier, M. S. (Eds.), *Gender methodologisch. Empirische Forschung in der Informationsgesellschaft vor neuen Herausforderungen*. Wiesbaden: VS Verlag für Sozialwissenschaften.

Schoon, P., & Cafolla, R. (2002). World Wide Web Hypertext Linkage Patterns. *Journal of Educational Multimedia and Hypermedia*, *11*, 117–139.

Schultze, U., & Leidner, D. (2002). Studying knowledge management in information systems research: Discourses and theoretical assumptions. *MIS Quarterly*, *26*(3), 213–242. doi:10.2307/4132331

Schwartz, S. (1994). Beyond individualism/collectivism: new cultural dimensions of value. In: U. Kim, H. Triandis, C. Kagitcibasi, S. Choi, & G. Yoon (Eds.), *Individualism and collectivism: Theory, method and applications* (pp. 85-119). Thousand Oaks, CA: Sage.

Sealy, R., Vinnicombe, S., & Singh, V. (2008). *The Female FTSE Report 2008*. Cranfield School of Management.

Seibert, H. (2007). Gender differences in the use of computers and the Internet. Statistics in focus. Population and Social Conditions, 119/2007. Retrieved April 5, 2010, from http://epp.eurostat.ec.europa.eu/cache/ITY_OFFPUB/KS-EI-08-001/EN/KS-EI-08-001-EN.PDF

Selwood, I., & Pilkington, R. (2005). Teacher Workload: Using ICT to release time to teach. *Educational Review*, *57*(2), 163. doi:10.1080/0013191042000308341

Selwyn, N. (2007). Hi-tech=Guy-tech? An exploration of undergraduate students' gendered perceptions of information and communication technologies. *Sex Roles*, *56*, 525–536. doi:10.1007/s11199-007-9191-7

Seymour, S. (1999). *Women, family and child care in India: A world in transition*. Cambridge: Cambridge University Press.

Shactman, N. (2002). *Blogs make the headlines*. Retrieved from http://www.wired.com/culture/lifestyle/news/2002/12/56978

Shaw, L. H., & Gant, L. M. (2002). Users divided? Exploring the gender gap in internet use. *Cyberpsychology & Behavior*, *5*(6), 517–527. doi:10.1089/109493102321018150

Shih, J. (2006). Circumventing discrimination: Gender and ethnic strategies in Silicon Valley. *Gender & Society*, *20*(2), 177–206. doi:10.1177/0891243205285474

Shim, S., Eastlick, M. A., Lotz, S. L., & Warrington, P. (2001). An online prepurchase intentions model: the role of intention to search. *Journal of Retailing*, *77*(3), 397–416. doi:10.1016/S0022-4359(01)00051-3

Shohet, I. M., & Lavy, S. (2004). Healthcare Facilities Management. *State of the Art Review Facilities*, *22*(7/8), 210–220. doi:10.1108/02632770410547570

Siegel, J., Dubrovsky, V. J., Kiesler, S., & McGuire, T. W. (1986). Group processes in computer-mediated communication. *Organizational Behavior and Human Decision Processes*, *3*, 157–187. doi:10.1016/0749-5978(86)90050-6

Silverstein, J., & Lasky, M. (2004). *Online dating for dummies*. Hoboken, NJ: Wiley Publishing Inc.

Simard, C., Henderson, A., Gilmartin, S., Schiebinger, L., & Whitney, T. (2008). *Climbing the technical ladder: Obstacles and solutions for mid-level women in technology*. Retrieved December 3, 2009, from http://anitaborg.org/files/Climbing_the_Technical_Ladder.pdf

Simon, S. J., & Peppas, S. C. (2005). Attitudes towards product website design: A study of the effects of gender. *Journal of Marketing Communications*, *11*(2), 129–144. doi:10.1080/1352726042000286507

Simon, S. R., Kaushal, R., Cleary, P. D., Jenter, C. A., Volk, L. A., & Poon, E. G. (2007). Correlates of Electronic Health Record Adoption in Office Practices: A Statewide Survey. *Journal of the American Medical Informatics Association*, *14*(1), 110–117. doi:10.1197/jamia.M2187

Singh, S. (2001). Gender and use of the Internet at home. *New Media & Society*, *3*(4), 395–415.

Singh, V. (2008). *Transforming Boardroom Cultures in Science, Engineering and Technology Organizations*. Research Report Series for UKRC No. 8.

Sloan, R., & Troy, P. (2008). CIS 0.5: A better approach to introductory computer science for majors. In *Proceedings of the 39th SIGCSE Technical Symposium on Computer Science Education*. Portland, OR: Association of Computing Machinery.

Small Business Service. (2003). *A Strategic Framework for Women's Enterprise*. London: DTI Small Business Service.

Smeltzer, L. R., & Fann, G. L. (1989). Gender Differences in External Networks of Small Business Owner/Managers. *Journal of Small Business Management*, *27*(2), 25–32.

Smith, J., & Webster, L. (2000). The Knowledge economy and SMEs: a survey of skills requirements. *Business Information Review*, *17*(3), 138–146. doi:10.1177/0266382004237656

Smith, R. (2004). Access to Healthcare via Telehealth: Experiences from the Pacific. *Perspectives on Global Development and Technology*, *3*(1), 197. doi:10.1163/1569150042036693

Smith, P. L., & Smits, S. J. (1992). Female Business Owners in Industries Traditionally Dominated by Males. *Sex Roles*, *26*(11/12), 485–496. doi:10.1007/BF00289870

Smith, A. (2008). *New numbers for blogging and blog readership*. Retrieved from http://www.pewinternet.org/Commentary/2008/July/New-numbers-for-blogging-and-blog-readership.aspx

Smith, K. (2009). *The use of virtual worlds among people with disabilities*. Paper presented at the California State University, Northridge, Center on Disabilities' 24th Annual International Technology and Persons with Disabilities Conference, Los Angeles, March 16-21, 2009.

Soe, L., & Yakura, E. (2008). What's wrong with the pipeline? Assumptions about gender and culture in IT work. *Women's Studies*, *37*, 176–201. doi:10.1080/00497870801917028

Solis, B. (2009). *Are blogs losing their authority to the statusphere*. Retrieved from http://www.techcrunch.com/2009/03/10/are-blogs-losing-their-authority-to-the-statusphere/

Spears, R., & Lea, M. (1992). Social influence and the influence of the "social" in computer-mediated communication. In M. Lea (Ed.), *Contexts of computer-mediated communication* (pp. 30-65). London: Harvester-Wheatsheaf.

Spotts, T. H., Bowman, M. A., & Mertz, C. (1997). Gender and use of instructional technologies: A study of university faculty. *Higher Education*, *34*(4), 421–436. doi:10.1023/A:1003035425837

Sproull, L., & Kiesler, S. (1991). *Connections: New Ways of Working in the Networked Organization*. Cambridge, MA: The MIT Press.

Srinvasan, R., Woo, C. Y., et al. (1994). Performance Determinants for Male and Female Entrepreneurs. In *Proceedings of the Fourteenth Annual Entrepreneurship Research Conference*. Babson College, MA.

Stevanovic, R., Stanic, A., & Varga, S. (2005). Information Systems in Primary Health Care. *Acta Medica Croatica*, *59*(3), 209–212.

Stevenson, L. A. (2003). Against all odds: The entrepreneurship of women. *Journal of Small Business Management*, *24*(4), 30–36.

Still, L. V., & Timms, W. (2000). Women's business: the flexible alternative workstyle for women. *Women in Management Review*, *15*(5/6), 272–283. doi:10.1108/09649420010372931

Stiller, E., & LeBlanc, C. (2003). Creating new computer science curricula for the new millennium. *Journal of Computing Sciences in Colleges*, *18*(5), 198–209.

Storey, D. J. (1994). *Understanding the Small Business Sector*. London: Routledge.

Straeder, T., & Shaw, M. (2000). Electronic Markets: Impact & Implication. In Shaw, M., Blanning, R., Straeder, T., & Whinston, A. (Eds.), *Handbook on Electronic Commerce* (pp. 77–98). London: Springer-Verlag.

Strauss, A., & Corbin, J. (1990). *Basics of Qualitative Research: Grounded Theory Procedures and Techniques*. Newbury Park, CA: Sage.

Strauss, J., & El-Ansary, A. I. (2004). Integrating the "e" in e-marketing. *Journal of Business & Economics Research*, *2*(8), 69–80.

Studien- & Berufswahl. (n.d.). *Frauen im Studium*. Retrieved March 10, 2009, from http://www.studienwahl.de/print.aspx?f=2/2_3_0_0_0_0_0_content_01.aspx&id=

Suchman, L. (1987). *Plans and situated actions: The problem of human-machine communication*. Cambridge: Cambridge University Press.

Suler, J. R. (2002). Identity management in cyberspace. *Journal of Applied Psychoanalytic Studies*, *4*(4), 455–459. doi:10.1023/A:1020392231924

Sumner, M. Yager, S., & Frankie, D. (2005). Career orientation and organizational commitment of IT personnel. In *Proceedings of the ACM SIGMIS Conference on Computer Personnel Research* (pp. 75-80). Atlanta, GA: Association of Computing Machinery.

Suseno, Y. (2008). Examining the Role of Social Capital in Female Professional's reputation Building and Opportunities Gathering: A Network Appriach. In Aaltio, I., Kyro, P., & Sundin, E. (Eds.), *Women Entrepreneurship and Social Capital: A Dialogue and Construction* (pp. 147–166). Copenhagen.

Symons, D. (1979). *The evolution of human sexuality*. New York, NY: Oxford University Press.

Takruri-Rizk, H. (2006). *Women in North West Engineering. ESF Report: University of Salford*. Informatics Research Institute.

Tatli, A., Ozbilgin, M. F., & Kusku, F. (2008). Gendered occupational outcomes: The case of professional training and work in Turkey. In J. Eccles & H. Watt (Eds.), *Explaining Gendered Occupational Outcomes*. Michigan: American Psychological Association (APA) Press.

Teo, T. S. H. (2002). Attitudes toward online shopping and the internet. *Behaviour & Information Technology*, *21*(4), 259–271. doi:10.1080/01449290210000018342

Teo, T. S. H., & Lim, V. K. G. (1997). Usage patterns and perceptions of the internet: The gender gap. *Equal Opportunities International*, *16*(6/7), 1–8. doi:10.1108/eb010696

Teo, T. S. H., & Lim, V. K. G. (2000). Gender differences in internet usage and task preferences. *Behaviour & Information Technology*, *19*(4), 283–295. doi:10.1080/01449290050086390

Tetteh, E., & Burn, J. (2001). Global Strategies for SME-business: Applying the SMALL Framework. *Logistics Information Management*, *14*(1-2), 171–180. doi:10.1108/09576050110363202

Thatchenkery, T., & Stough, R. R. (2005). *Information Communication Technology and Economic Development – Learning from the Indian Experience*. Edward Elgar Publishing Ltd.

Thomson, R. (2006a). The Effect of Topic of Dicussion on Gendered Language in Computer-Mediated Discussion. *Journal of Language and Social Psychology, 25*(2), 167–178. doi:10.1177/0261927X06286452

Thomson, R. (2006a). Gender and Electronic Discourse in the Workplace. In Barrett, M., & Davidson, M. J. (Eds.), *Gender and Communication at Work* (pp. 239–249). Aldershot, UK: Ashgate.

Thoreau, E. (2006). Ouch!: An examination of the self-representation of disabled people on the Internet. *Journal of Computer-Mediated Communication, 11*(2), 442–468. doi:10.1111/j.1083-6101.2006.00021.x

Thornhill, R., & Thornhill, N. (1983). Human rape: An evolutionary analysis. *Ethology and Sociobiology, 4*, 137–173. doi:10.1016/0162-3095(83)90027-4

Torp, S., Hanson, E., Ulstein, I., & Magnusson, I. (2008). A pilot study of how information and communication technology may contribute to health promotion among elderly spousal carers in Norway. *Health & Social Care in the Community, 16*(1), 75–85. doi:10.1111/j.1365-2524.2007.00725.x

Torres, I. M., Summers, T. A., & Belleau, B. D. (2001). Men's shopping satisfaction and store preferences. *Journal of Retailing and Consumer Services, 8*(4), 205–212. doi:10.1016/S0969-6989(00)00024-2

Townsend, J. (1993). Sexuality and partner selection: Sex differences among college students. *Ethology and Sociobiology, 14*, 305–330. doi:10.1016/0162-3095(93)90002-Y

Townsend, J., & Wasserman, T. (1997). The perception of sexual attractiveness: Sex differences in variability. *Archives of Sexual Behavior, 26*, 243–268. doi:10.1023/A:1024570814293

Trauth, E., Quesenberry, J., & Yeo, B. (2008). Environmental influences on gender in the IT workforce. *Database, 39*(1), 8–32.

Trauth, E., Quesenberry, J., & Morgan, A. J. (2004). Understanding the under-representation of women in IT: Toward a theory of individual differences. In *Proceedings of the ACM SIGMIS Conference on Computer Personnel Research (pp.* 114-119). Phoenix, AZ: Association of Computing Machinery.

Travers, C., & Pemberton, C. (1997). Women's Networking Across Boundaries: Recognizing Different Cultural Agendas. *Women in Management Review, 12*(2), 61–67. doi:10.1108/09649429710162820

Trompenaars, F., & Hampden-Turner, C. (1998). *Riding the waves of culture: Understanding cultural diversity in global business* (2nd ed.). New York, NY: McGraw-Hill.

TruDating. (2006). Online dating service directory. Retrieved May 25, 2006, from the http://www.trudating.com/.

Truman, G. E., & Baroudi, J. J. (1994). Gender differences in the information systems managerial ranks: An assessment of potential discriminatory practices. *Management Information Systems Quarterly, 18*, 129–141. doi:10.2307/249761

Tsai, C. C., & Lin, C. C. (2004). Taiwanese adolescents' perceptions and attitudes regarding the internet: Exploring gender differences. *Adolescence, 397*(156), 725–734.

Tsai, C. C., Lin, S. S. J., & Tsai, M. J. (2001). Developing an internet attitude scale for high school students. *Computers & Education, 37*(1), 41–51. doi:10.1016/S0360-1315(01)00033-1

Tsai, M. J., & Tsai, C. C. (2010). Junior high school students' internet usage and self-efficacy: A re-examination of the gender gap. *Computers & Education, 54*(4), 1182–1192. doi:10.1016/j.compedu.2009.11.004

Tyran, C. K., Dennis, A. R., Vogel, D. R., & Nunamaker, J. F. (1992). The application of electronic meeting technology to support strategic management. *Management Information Systems Quarterly, 16*(3), 313–334. doi:10.2307/249531

U.S. Gay/Lesbian Business Groups. Small Business Association. Retrieved on May 14, 2009 from http://www.smallbusinessnotes.com/interests/usgaybuslks.html.

United States Access Board. Retrieved on May 14, 2009 from http://www.access-board.gov/.

Urberg, K. (1979). Sex role conceptualization in adolescents and adults. *Developmental Psychology*, *15*, 90–92. doi:10.1037/h0078082

Usunier, J. (1996). *Marketing across cultures* (2nd ed.). Hertfordshire, UK: Prentice Hall.

Uzzi, B. (1996). The Sources and Consequences of Embeddedness for the Economic Performance of Organizations: The Network Effect. *American Sociological Review*, *61*(4), 674–698. doi:10.2307/2096399

Valacich, J. S., Jessup, L. M., Dennis, A. R., & Nunamaker, J. F. (1992). A conceptual framework of anonymity in group support systems. *Group Decision and Negotiation*, *1*, 219–241. doi:10.1007/BF00126264

Valcke, M., & De Wever, B. (2006). Information and Communication Technologies in Higher Education: Evidence-based Practices in Medical Education. *Medical Teacher*, *28*(1), 40–48. doi:10.1080/01421590500441927

Valiulis, M., Drew, E., Humbert, A., & Daverth, G. (2004). *Springboard for Women in Business Initiative Wicklow Chamber of Commerce*. Centre for Gender and Women's Studies, Trinity College Dublin.

Van Akkeren, J., & Cavaye, A. (1999). Factors affecting Entry Level Internet Adoption by SMEs: An Empirical study. In *Proceedings from the Australasian Conference in Information Systems* (pp. 1716-1728).

Van den Brink-Muinen, A., Bensing, J. M., & Kerssens, J. J. (1998). Gender and Communication Style in general practice: Differences between Women's Health care and regular health care. *Medical Care*, *36*, 100–106. doi:10.1097/00005650-199801000-00012

Van den Poel, D., & Buckinx, W. (2005). Predicting online-purchasing behavior. *European Journal of Operational Research*, *166*(2), 557–575. doi:10.1016/j.ejor.2004.04.022

Varadarajan, P. R., & Cunningham, M. (1995). Strategic Alliances: A Synthesis of Conceptual Foundations. *Journal of the Academy of Marketing Science*, *23*(4), 282–296. doi:10.1177/009207039502300408

Venkatesh, V., & Morris, M. G. (2000). Why don't men ever stop to ask for directions? Gender, social influence, and their role in technology acceptance and usage behaviour. *Management Information Systems Quarterly*, *24*(1), 115–139. doi:10.2307/3250981

Verba, S., Lehman Schlozman, K., & Brady, H. E. (1995). *Voice and equality: Civic voluntarism in American politics*. Boston, MA: Harvard University Press.

Verheul, I., Risseeuw, P., & Bartelse, G. (2002). Gender Differences in Strategy and Human Resource Management. *International Small Business Journal*, *20*(4), 443–476. doi:10.1177/0266242602204004

Verheul, I., & Caree, M. (2009). Allocation and Productivity of Time in New Ventures of Female and Male Entrepreneurs. *Small Business Economics*, *33*(3). doi:10.1007/s11187-009-9174-x

Verheul, I., & Risseeuw, P. (2002). Gender Differences in Strategy and Human Resource Management: The Case of Dutch Real Estate Brokerage. *International Small Business Journal*, *20*(4), 443–476. doi:10.1177/0266242602204004

Villanueva, J., & Pavone, C. (2007). *The Effect of Entrepreneurial Motives on Growth: A Study of Women Entrepreneurs*. Babson College.

Volman, M., van Eck, E., Heemskerk, I., & Kuiper, E. (2005). New technologies, new differences: Gender and ethnic differences in pupils' use of ICT in primary and secondary education. *Computers & Education*, *44*(1), 35–55. doi:10.1016/S0360-1315(04)00072-7

Volman, M., & Eck, E. V. (2001). Gender equity and information technology in education: The second decade. *Review of Educational Research*, *71*(4), 613–634. doi:10.3102/00346543071004613

Walby, S. (1990). *Theorising Patriarchy*. Oxford: Blackwell.

Walby, S. (1997). *Gender Transformations*. London: Routledge.

Waldstrom, C., & Madsen, H. (2007). Social relations among managers: Old boys and young women's networks. *Women in Management Review*, *22*(2), 136. doi:10.1108/09649420710732097

Walker, T. (2001). Wooing female consumers reaps rewards for health plans. *Managed Healthcare Excutive*, *11*(3), 42–45.

Walker, E. A., & Webster, B. J. (2007). Gender, age and self-employment: Some things change, some stay the same. *Women in Management Review*, *22*(2), 122. doi:10.1108/09649420710732088

Wallsten, K. (2007). Political blogs: Transmission belts, soapboxes, mobilizers, or conversation starters? *Journal of Information Technology & Politics*, *4*(3), 19–40. doi:10.1080/19331680801915033

Walther, J. (1996). Computer-mediated communication: Impersonal, interpersonal and hyperpersonal interaction. *Communication Research*, *23*, 3–43. doi:10.1177/009365096023001001

Walther, J., Slovacek, C., & Tidwell, L. (2001). Is a picture worth a thousand words? Photographic images in long-term and short-term computer-mediated communication. *Communication Research*, *28*, 105–134. doi:10.1177/009365001028001004

Walther, J. (1996). Computer-mediated communication: Impersonal, interpersonal, and hyperpersonal interaction. *Communication Research*, *23*, 3–43. doi:10.1177/009365096023001001

Wanberg, C., Welsh, E., & Hezlette, S. (2003). Mentoring research: A review and dynamic process model. *Research in Personnel & Human Resources Management*, *22*, 39–124. doi:10.1016/S0742-7301(03)22002-8

Ward, K. (2001). Crossing cyber boundaries: Where is the body located in the online community? In Watson, N., & Cunningham, S. (Eds.), *Reframing the Body*. New York: Palgrave.

Waring, T., & Wainwright, D. (2002). Enhancing Clinical and Management Discourse in ICT Implementation. *Journal of Management in Medicine*, *16*(2/3), 133–149. doi:10.1108/02689230210434880

Watkins, J., & Watkins, D. (1984). The Female Entrepreneur: Backgrounds and Determinants of Business Choice – Some British Data. *International Small Business Journal*, *2*(4), 21–31.

Watson, J. (2002). Comparing the Performance of Male- and Female-Controlled Businesses: Relating Outputs to Inputs. *Entrepreneurship: Theory and Practice*, *26*(3), 91–100.

Weber, B., Wittchen, M., & Hertel, G. (2009). Gendered Ways to Motivation Gains in Groups. *Sex Roles*, *60*, 731–774. doi:10.1007/s11199-008-9574-4

Weiler, S., & Bernasek, A. (2001). Dodging the Glass Ceiling? Networks and the New Wave of Women Entrepreneurs. *The Social Science Journal*, *38*(1), 85–103. doi:10.1016/S0362-3319(00)00111-7

Weiser, E. B. (2000). Gender differences in internet use patterns and internet: Application preferences: A two-sample comparison. *Cyberpsychology & Behavior*, *3*(2), 167–178. doi:10.1089/109493100316012

Wellman, B., & Salaaf, J. (1996). Computer Networks as Social Networks: Collaborative Work, Telework and Virtual Community. *Annual Review of Sociology*, *22*, 213–238. doi:10.1146/annurev.soc.22.1.213

Welter, F., & Trettin, L. (2006). The Spatial Embeddedness of Networks for Women Entrepreneurs. In Fritsch, M., & Schmude, J. (Eds.), *Entrepreneurship in the Region* (pp. 35–59). New York: Springer. doi:10.1007/0-387-28376-5_3

West, C., & Fenstermaker, S. (1993). Power, Inequality and the Accomplishment of Gender: An Ethnomethodological View. In England, P. (Ed.), *Theory on Gender/Feminism on Theory*. New York: Aldine.

White, D., & Winn, P. (2008). *State of the blogosphere*. Retrieved from http://technorati.com/blogging/feature/state-of-the-blogosphere-2008/

White, E. (2007, January, 1). Employers reach out to recruit with Facebook. *Wall Street Journal*, p. D3.

Whitley, E. A. (1997). In cyberspace all they see is your words. A review of the relationship between body, behavior and identity drawn from the sociology of knowledge. *Information Technology & People*, *10*(7), 147–163. doi:10.1108/09593849710174995

Whitty, M. (2002). Liar, Liar! An examination of how open, supportive and honest people are in chat rooms. *Computers in Human Behavior*, *18*(4), 343–352. doi:10.1016/S0747-5632(01)00059-0

Whitty, M. (2003). Cyber-flirting: Playing at love on the Internet. *Theory & Psychology*, *13*(3), 339–357. doi:10.1177/0959354303013003003

Whitty, M. (2004). Cyber-flirting: An examination of men's and women's flirting behaviour both off-line and on the Internet. *Behaviour Change*, *21*(2), 115–126. doi:10.1375/bech.21.2.115.55423

Compilation of References

Whitty, M. (in press). Revealing the 'real' me, searching for the 'actual' you: Presentations of self on an Internet dating site. *Computers in Human Behavior*.

Whitty, M., & Gavin, J. (2001). Age/sex/location: Uncovering the social cues in the development of online relationships. *CyberPsychology & Behaviour, 4*, 623–630. doi:10.1089/109493101753235223

Whitty, M. (2007). The art of selling one's self on an online dating site: The BAR Approach. In: M. Whitty, A. Baker, & J. Inman (Eds.), *Online matchmaking* (pp. 57-69). Houndmills: Palgrave Macmillan.

Whitty, M., & Buchanan, T. (2007, manuscript under preparation). What's in a 'screen' name? The types of screen names online daters find attractive.

Whitty, M., & Carr, A. (2006). *Cyberspace romance: The psychology of online relationships*. Basingstoke: Palgrave Macmillan.

Wildermuth, S. (2004). The effects of stigmatizing discourse on the quality of online relationships. *Cyberpsychology & Behavior, 7*(1), 73–84. doi:10.1089/109493104322820147

Wilson, M. (2004). A world of differences. *Chain Store Age, 80*(9), 126.

Wilson, F. (2003). Can compute, won't compute: Women's participation in the culture of computing. *New Technology, Work and Employment, 18*(2), 127–142. doi:10.1111/1468-005X.00115

Witt, P. (2004). Entrepreneurs' Networks and the Success of Start-Ups. *Entrepreneurship & Regional Development, 16*, 391–412. doi:10.1080/0898562042000188423

Wood, L. M. (1999). The Use of Electronic Networking by Australian Small Business Women. In *Proceedings of the Australian Community Networking Alliance, Balart, ACNA*.

Wu, Y. T., & Tsai, C. C. (2006). Developing an information commitment survey for assessing students' web information searching strategies and evaluative standards for web materials. *Journal of Educational Technology & Society, 10*(2), 120–132.

Wysocki, D. (1998). Let your fingers to do the talking: Sex on an adult chat line. *Sexualities, 1*, 425–452. doi:10.1177/136346098001004003

Wysocki, D., & Thalken, J. (2007). Whips and chains? Fact or fiction? Content analysis of sadomasochism in Internet personal advertisements. In: M. Whitty, A. Baker, & J. Inman (Eds.), *Online matchmaking* (pp. 178-196). Houndmills: Palgrave Macmillan.

Yancey, G., & Yancey, S. (1998). Interracial dating: Evidence from personal advertisements. *Journal of Family Issues, 19*(3), 334–348. doi:10.1177/019251398019003006

Yang, B., & Lester, D. (2005). Gender Differences in e-commerce. *Applied Economics, 37*, 2077–2089. doi:10.1080/00036840500293292

Ye, J. (2006). Seeking love online: A cross-cultural examination of personal advertisements on American and Chinese dating Web sites. *Global Media Journal, 5*(8).

Yee, N. (2007). *The Proteus Effect: Behavioral Modifications via Transformations of Digital-Self Representations*. Unpublished Ph.D. Thesis, Department of Communications, Stanford University.

Yin, R. (1994). *Case study research: Design and methods*. Beverly Hills, CA: Sage.

Yin, R. K. (2008). *Case study research: Design and methods*. Thousand Oaks, CA: Sage.

Zander, A. (1971). *Motives and Goals in Groups*. New York: Academic Press.

Zaucher, S., Korunka, C., Weiss, A., & Kafka-Lützow, A. (2000). Gender-related effects of information technology implementation. *Gender and Information Technology, 7*(2), 119–132.

Zhou, L., Dai, L., & Zhang, D. (2007). Online shopping acceptance model – A critical survey of consumer factors in online shopping. *Journal of Electronic Commerce Research, 8*(1), 41–62.

Zuboff, S. (1988). *In the age of the smart machine*. New York, NY: Basic Books.

About the Contributors

Celia Romm Livermore is Professor at Wayne State University, Detroit and the editor-in-chief of the International Journal of E-Politics (IJEP). She published five books: "Virtual Politicking" (1999), "Electronic Commerce: A Global Perspective" (1998), "Doing Business on the Internet" (1999), "Self Service on the Internet" (2008) and "Social Networking Communities and eDating Services" (2008) and over a hundred and fifty journal articles, book chapters, and conference papers. Her research was published in The Harvard Business Review, Communications of the ACM, Information & Management, Transactions on Information Systems and others.

* * *

Elizabeth Koh is a PhD candidate and instructor in the School of Computing, Department of Information Systems, at the National University of Singapore. Her research interests include IT and education, virtual teams, and social computing. She has published in several international peer-reviewed periodicals such as the Information Resources Management Journal, the International Conference on Human-Computer Interaction, the Pacific-Asia Conference on Information Systems, and the International Simulation And Gaming Association Conference. She has also contributed chapters to the Encyclopedia of Multimedia Technology and Networking (2nd edition), and the Handbook of Research on Social Interaction Technologies and Collaboration Software: Concepts and Trends.

Na Liu received her B.S. degree in Computing (Hons) from the Department of Information Systems, National University of Singapore in 2006. She is currently a Doctoral candidate and teaching assistant in the School of Computing, Department of Information Systems, at the National University of Singapore. Her research interests include collaborative technologies, online healthcare communities and human-computer interactions. Her works have been published in several international conferences and journal, including Hawaii International Conference on System Sciences (HICSS), Pacific Asia Conference on Information Systems (PACIS), International Conference on Human-Computer Interaction (HCII), International Federation For Information Processing Working Group 8.2 Conference (IFIP WG8.2) and IEEE Transactions on Engineering Management.

John Lim is Associate Professor in the School of Computing, Department of Information Systems, at the National University of Singapore. Dr. Lim graduated with First Class Honors in Electrical Engineering and a M.Sc. in MIS from the National University of Singapore, and a Ph.D. from the University of British Columbia. His current research interests include IT and education, collaborative technology,

About the Contributors

negotiation support, and media effects. He has published in MIS and related journals including Journal of Management Information Systems, Journal of Global Information Management, Decision Support Systems, International Journal of Human Computer Studies, Organizational Behavior and Human Decision Processes, Behaviour and Information Technology, International Journal of Web-based Learning and Teaching Technologies, Journal of Database Management, and Small Group Research.

Yvonne Costin is a Junior Lecturer in Entrepreneurship in the Department of Management and Marketing, Kemmy Business School, University of Limerick. Yvonne is responsible for teaching Enterprise Formation, Enterprise Development and Business Consulting at an undergraduate and postgraduate level. She currently pursuing a PhD study in the area of female entrepreneurship. Her main research interests include entrepreneurship education at both primary and tertiary levels, female entrepreneurship and growth achievement in entrepreneurial firms.

Ruth Guthrie (raguthrie@csupomona.edu) is chair of the CIS department at California Polytechnic University, Pomona. She has a professional background in software test and program management of large scale software programs in the aerospace industry. Before becoming chair of the CIS department, Ruth worked as the Cal Poly NSF Advance Scholar to help women faculty, in STEM disciplines, on their professional development. Her research interests are in educational technology, user interface design, and women in technology careers. She has authored several papers in a variety of areas including podcasting, accreditation maturity, and technological changes to the music industry. Ruth has a PhD in Information Science from the Claremont Graduate University, where she was the first woman graduate in this discipline in 1994.

Louise Soe (llsoe@csupomona.edu) is a Professor of Computer Information Systems at California State Polytechnic University, Pomona. She teaches web-based applications and coordinates the CIS senior projects class, which completes service-learning projects for the larger community. Her current research interests include Women in Technology and IS Curriculum. Most of her recent efforts have been on a research project on women in technology careers. She has coauthored several papers resulting from this research, as well as papers on IS education, and accreditation. She has a Ph.D. from the Anderson School at UCLA.

Elaine K. Yakura (yakura@msu.edu) teaches courses in Negotiations, Organizational Development & Change, and Organizational Behavior & Leadership at the School of Labor and Industrial Relations at Michigan State University. She received the Excellence in Education Award from the Industrial Relations Research Association in 2002 and a MSU Lilly Teaching Fellowship in 1998. Her past articles on information technology consulting include ethnographic studies of billables as well as the use of timelines. With Louise Soe & Ruth Guthrie, she is also studying work and careers in information technology organizations. Elaine has a PhD in Organization Studies from MIT, and a JD from UC Berkeley.

Sunrita Dhar-Bhattacharjee is a Research Fellow at the School of the Built Environment, University of Salford. Her research interests are in the area of comparative gender relations and systems, and women's paid work, particularly in relation to Britain and India. She is interested in cross-national comparative methodology and has a lively interest in policy and practice in this field. She is the beneficiary of the Vice

Chancellor's Early Research Career Award and is currently undertaking a doctoral research on 'Gender Segregation in science, engineering, construction and technology (SECT): An Indo-British comparison.

Haifa Takruri-Rizk is a senior lecturer in the School of Computing, Science and Engineering at the University of Salford, UK. Her current technical research is within the Computer Networking and Telecommunications Research group, focusing on the development of wireless sensor networks for varied applications. Haifa has been an advocate for the promotion of women in engineering and science, and in particular encouraging young women to study science and engineering which has recently won her two national awards. She is leading research investigating the factors contributing to the under-representation and the leaky-pipeline phenomena of women and minorities in science, engineering and technology (SET) and their lack of progression into SET careers. In the last a few years, she led a number of projects in this area including the Women Audio Video Engineers (WAVE) project, the Women in North West Engineering (WEWIN) project, Participation of BMEs in SET and the Developing Female Engineers (DFE) project.

Erkan Özdemir is currently a research assistant in marketing program of the Department of Business Administration at Uludag University in Bursa, Turkey. He received his PhD in marketing from Uludag University in 2005. Dr. Özdemir's current research interests include e-marketing, online consumer behaviors, marketing to women, and gender based marketing strategies.

Olca Sürgevil, Ph.D., is a researcher in Management and Organization at the Department of Business Administration, Dokuz Eylül University (DEU) Faculty of Economics and Administrative Sciences. Her research interests include organizational psychology and diversity issues. She received her bachelor's degree in Business Administration from DEU, and master's degree in both Management and Organization from DEU and in Social Psychology from Ege University. She received her Ph.D. degree in Business Administration from DEU with her dissertation on workforce diversity and diversity management. She has been employed at DEU as a research assistant since 2005.

Mustafa F. Özbilgin is the Director of Diversity and Equality in Careers and Employment Research (DECERe), and Professor of Human Resource Management at the University of East Anglia. His research appears in journals such as the International Journal of Human Resource Management, Journal of Vocational Behavior and Gender Work and Organization. He has also authored/edited nine research monographs, including Global Diversity Management (Palgrave 2008). He is the editor of the journal Equality, Diversity and Inclusion (Emerald Press) and British Journal of Management. Holding a doctoral degree from the University of Bristol, he has previously worked at Queen Mary (University of London) and held visiting posts at CEPS-INSTEAD (Luxembourg), Cornell University (USA), St Gallen University (Switzerland), Sorbonne University (France) and Japan Institute of Labor and Policy (Japan).

Bernhard Ertl is senior researcher at the Universität der Bundeswehr München. He has realized several research projects in the context of gender in computer and science teaching which includes projects with national and EU funding, e.g. SESTEM (Supporting Equality in Science Technology and Mathematics related choices of careers), PREDIL (Promoting Equality in Digital Literacy) and "Comparative study on gender differences in technology enhanced and computer science learning: Promoting

About the Contributors

equity". A further focus of research is on issues like video-mediated learning, Internet collaboration and online-courses with a particular focus on the support of collaborative knowledge construction by the methods of scripts and structured communication interfaces. He recently edited books about E-collaborative knowledge construction and Technologies and Practices for Constructing Knowledge in Online Environments. Bernhard Ertl earned his Diploma in computer science from the Ludwig Maximilian University Munich in 1998 and his Doctorate in education 2003. From 1999 to 2006, he was researcher at the Department Psychology of Ludwig Maximilian University of Munich and worked with Professor Heinz Mandl in DFG-funded research projects focusing on collaborative learning, e.g. "Collaborative Learning in Graphics-enhanced Tele-learning Environments" and "Collaborative Knowledge Construction in Desktop Videoconferencing".

Kathrin Helling, M.A. is research associate at the Universität der Bundeswehr München and University of Innsbruck, Department of Education. She has experience as researcher and project manager in several national projects and European projects in the frame of the Lifelong Learning Program. A focus of her research is on gender aspects in the context of computer-supported mathematics, science and informatics teaching and related career choices of women. At the Institute for Future Studies in Innsbruck she worked on the development of computer-based learning scenarios and curricula for specific target groups (e.g. people of the age group 50+, learners with low educational achievement). She also trained trainers in using learning management systems and educational technologies. Kathrin Helling has worked in a DFG-funded research project at the Ludwig Maximilian University of Munich. The focus of this research was on supporting collaborative learning processes in video conferencing by the methods of scripts and structuring the communication of learners. She has gained her magister diploma in education science in 2006 at the Ludwig Maximilian University Munich.

Kathy Kikis-Papadakis received her PhD in Educational Planning and Evaluation (area of investigation: Employability and Language Training of Minority Groups in the USA). In the past she held the post of leading the section of bilateral cooperation at the Hellenic Ministry of Research and Technology and taught at the Department of Philosophical and Social Science at the University of Crete. Since 1993 she is leading the Educational Research and Evaluation Group at the Institute of Applied and Computational Mathematics at FORTH. Her research interests are in the study of impact of technology enhanced learning, both from an effectiveness and innovation introduction perspectives, and on the socio-cultural aspects of learning. Specific areas of interest include gender and mathematics and curriculum development for teachers' professional development. All of her research work is supported from competitive grants and concern educational policy and evaluation.

Robert MacGregor is an Honorary Fellow in the School of Information Systems and Technology at the University of Wollongong in Australia. He is also the former Head of Discipline in Information Systems. His research expertise lies in the areas of information technology (IT) and electronic commerce (e-commerce) in small to medium enterprises (SMEs). He has authored over 80 journal and conference publications examining the use and adoption of IT in SMEs. Along with Lejla Vrazalic, Rob was the recipient of the 2004 Australian Prime Minister's Award for Excellence in Business Community Partnerships (NSW). Rob is also the founding Editor of the Australasian Journal of Information Systems and was Conference Chair of the Australian Conference of Information Systems in 1992. In his spare time, Rob writes music. His most recent work is the symphony 'Alba'.

Peter Hyland is an Associate Professor at the University of Wollongong, Australia where he has been a researcher for 16 years. Peter is currently the Head of the School of Information Systems and Technology. His PhD examined the usability of OnLine Analytical Processing (OLAP) tools by casual and novice users. Peter's research interests include usability and interface design and community informatics. He has worked on projects with NSW Police Service, NSW State Emergency Services, Illawarra Division of General Practitioners and the Australian Community Domain name authority. He is currently starting community informatics projects in Saharakham, Thailand and in Hoi An and Danang, Vietnam.

Charles Harvie obtained his Ph.D in Economics from the University of Warwick (UK) in 1986 and is currently an Associate Professor in the School of Economics at the University of Wollongong. He is the Director of the Centre for Small Business and Regional Research which he established in 2000. He also initiated the 'SMEs in a Global Economy' conference series. He has published over 150 papers on the economies and economic integration of East Asia and on SMEs in East Asia. He recently completed a four volume book series with Dr Boon-Chye Lee on SMEs in East Asia. His current research interests include: private sector and small business development in transition economies (particularly that in Vietnam), enhancing the export performance of Australia's small businesses, modelling economic integration in East Asia, and China's integration into the global economy (currently writing a book on this topic).

Abbe E. Forman, Ph.D. is an Assistant Professor of Teaching/Instruction in the Computer Information Science department at Temple University. She holds a Ph.D. in Information Systems from Nova Southeastern University and a Master of Science in Information Science from Penn State University. She is a digital/information ethics researcher with recent work including online identity representations regarding both disability and gender. This has led to the formation of a research group studying the use of social networking sites to better serve the educational needs of students with disabilities. Additional research areas include digital piracy and online privacy. Most recently, she has been involved with Wikygy, and open source textbook project providing online course related content at no cost for students.

Paul M.A. Baker, Ph.D., is Director of Research, Center for Advanced Communications Policy, and an Adjunct Professor with the School of Public Policy, Georgia Institute of Technology. He also holds appointments as Professor, (Courtesy Appointment), Ph.D. Program in Public Affairs, College of Health and Public Affairs, University of Central Florida. Recent projects include policy barriers to the adoption of wireless technologies by people with disabilities, teleworking and people with disabilities, online collaboration and virtual communities, collaborative policy networks and e-accessibility, social media and policy development. He is also involved in international policy research, especially as it relates to issues of technological accessibility and information technologies. Baker holds a Ph.D. in Public Policy from George Mason University, an M.P. in Urban Planning from the University of Virginia. He is chair of the Rehabilitation Engineering Society of North America's (RESNA) Government Affairs Committee and serves on the editorial boards of seven journals.

About the Contributors

Jessica Pater is a digital media research scientist within the Information Technology and Telecommunications Laboratory at the Georgia Tech Research Institute. Her research centers on the use of digital immersive environments for identity development and education purposes. She currently serves as the Associate Director for the Foundations for the Future Program, which focuses on technology-enabled learning in Georgia and beyond.

Kel Smith's publications have been cited by the Pentagon Library, Kent State's Knowledge Management Program and the Sandra Day O'Connor College of Law. He is a frequent speaker who presented at the CSUN 2009 Los Angeles Conference for Persons with Disabilities, the TechShare RNIB Conference in London, St. Joseph's University Center for Consumer Research, Temple University's College of Science, the IxDA 10 Conference in Savannah, and CSUN 2010 in San Diego. A current member of the Interaction Design Association (IxDA) and the Usability Professionals' Association (UPA), Kel serves as Vice Chair of the Philadelphia chapter of ACM/SIG-CHI for computer-human interaction. He earned his BFA in photography from the Maryland Institute College of Art and studied cognitive science as part of the MS program at Philadelphia University. More information can be found at http://anikto.com.

Natalie Sappleton is a doctoral researcher at Manchester Metropolitan University. Natalie graduated from The University of Glasgow in 2002 with First Class Honours in Economics and Politics, and began a career in Policy and Research. In 2003, she worked at the Equal Opportunities Commission on an Investigation into occupational segregation and gender stereotyping in science and engineering. Since then, she has worked at Manchester Metropolitan University and the University of Salford on research projects relating to gender and the labour market. Her doctoral research is investigating whether the networks of male and female entrepreneurs operating in gender atypical industries differs from those of business owners in gender typical sectors, and whether this has an impact on their performance of their businesses. In 2008/09, Natalie was a scholar at the Kluge Center, Washington DC.

Antoinette Pole is an Assistant Professor of Political Science & Law at Montclair State University. She studies the intersection of information technology and politics, exploring theoretical questions related to representation and political participation. Professor Pole recently authored Blogging the Political: Politics and Participation in a Networked Society (Routledge, 2010). An expert on political blogs, she has been interviewed for her work by the Chicago Tribune, Philadelphia Tribune, Boston Globe, and Rhode Island Monthly Magazine. Her work appears in peer-reviewed journals such as Public Choice, American Journal Public Health, and International Journal of Technology, Knowledge and Society. Additionally, Professor Pole writes about state politics, coauthoring New York Politics: A Tale to Two States, Second Edition (ME Sharpe, 2010).

Timothy R. Hill is the chair of the Management Information Systems department and a Linda F. Morasch Fellow for Innovative e-Learning in the College of Business at San Jose State University. He has published research on innovative applications of emerging technologies, spanning artificial intelligence, neural networks, graphic visualization and manipulation, and more recently, mobile, multi-layered, media-rich e-learning environments and Web-related phenomenon involving social perception, identity and content aging. He continues to experiment in e-Learning with a variety of techniques, technologies and pedagogies.

Leslie Jordan Albert is an Assistant Professor in the Management Information Systems department in the College of Business at San Jose State University. She received her Ph.D. in MIS from the University of Oklahoma. Her research interests include computer-mediated communication, gender differences in IS, social networks, online learning, and the ability of technology to enhance individual creativity. Her work has been published in Information Systems Journal and in numerous conference proceedings.

Shailaja Venkatsubramanyan is an Assistant Professor in the Management Information Systems department in the College of Business at San Jose State University. She has a Ph.D. in MIS from the University of Arizona and has worked at Price Waterhouse, Tulane University, and Kanisa (a Knowledge Management startup in the Silicon Valley). Her research interests include knowledge management, information retrieval, search engine valuation, web log analysis, digital personas, and aging content on the Internet. She is also involved in studies involving the use of cloud computing to streamline operations in developing countries.

Monica Whitty is lecturer in the Division of Psychology at Nottingham Trent University in the UK. She is the first author of *Cyberspace Romance: The Psychology of Online Relationships* (2006, Palgrave), and *Trust, Lies and Truth on the Internet* (2008, Routledge). In recent years, her work has focused on online dating, cyber-relationships, Internet infidelity, misrepresentation of self online, cyberstalking, cyberethics, and Internet and e-mail surveillance in the workplace.

Toni M. Somers is professor of information systems management in the School of Business Administration at Wayne State University. She received her Ph.D. from the University of Toledo and an MBA degree from Bowling Green State University. Her research interests focus on information technology adoption and implementation. She has published in *Information Systems Research, Decision Sciences, Journal of Management Information Systems, Information and Management,* and *Production and Operations Management Journal*, among other journals. She is on the editorial board of *Journal of Information Technology Case and Application Research.*

Kristina S. Setzekorn is a financial advisor with Smith Barney, Inc. in Evansville, IN. She earned her Ph.D. at Southern Illinois University-Carbondale, where she majored in MIS and minored in operations management. Her MBA is from Southern Illinois University-Edwardsville and her B.S. is from Iowa State University. In her previous academic career, Dr. Setzekorn taught MIS, global IT strategy, telecommunications, microeconomics and operations. Her research dealt with performance impacts of information and coordination, at individual, organizational and supply chain levels.

Ashley King is currently undertaking a bachelor's degree in business, specializing in finance, at Wayne State University, Michigan. She expects to graduate in May 2008. In addition to her studies, she plays Division I ice hockey at Wayne State University. Her future plans include moving back to her hometown of Winnipeg, Manitoba, where she intends to pursue a career in finance.

Sudhir H. Kalé, Ph.D. is professor of marketing and co-director of the Globalization and Development Center at Bond University in Australia. With over 100 publications to his credit, Dr. Kalé's research has been published in top scholarly journals including *Journal of Marketing, Journal of Marketing Research,*

About the Contributors

Journal of Applied Psychology, Journal of International Business Studies, and *Journal of International Marketing.* Of late, he has shifted his research focus to the study of customer relationship management (CRM), spirituality in marketing, and the marketing aspects of casino management. Dr. Kalé serves on the editorial review board of several journals in gaming, management and marketing. He has conducted over 200 executive development seminars and workshops across four continents on a range of topics such as the Enneagram, applications of the Myers-Briggs type indicator in management, corporate culture, and psychological aspects of selling.

Mark T. Spence, Ph.D., joined Bond University, Australia, in 2002, where he is associate professor of marketing. He has taught a variety of marketing courses to undergraduates, postgraduates and executives, including consumer behavior, marketing strategy, market research and entrepreneurship. He has presented executive development workshops in the United States, Macau and in Europe. He has published in top academic journals including the *Journal of Marketing Research, Journal of Consumer Research, the European Journal of Marketing, Organizational Behavior and Human Decision Processes, Business Horizons,* and *Psychology and Marketing.*

Nainika Seth is an assistant professor of information systems in the Department of Economics and Information Systems at the University of Alabama in Huntsville. She received her Ph.D. from Southern Illinois University at Carbondale in 2002. Her research interests include IT infrastructure, IT value and the role of IT in supply chains. Her research has been published in *MIS Quarterly, Journal of Management Information Systems, Communications of the ACM, Omega, Information Resource Management Journal* and presented at the *ICIS conference*.

Ravi Patnayakuni is an associate professor of information systems in the Department of Economics and Information Systems at the University of Alabama in Huntsville. His research focuses on supply chain partnerships, information technology business value and knowledge management. His research has been published in *MIS Quarterly, Journal of Management Information Systems, IEEE Transactions on Engineering Management, Communication of the ACM, Journal of the Association of Information Systems, Omega, Communications of the AIS* and *Information Systems Journal*. He received his Ph.D. from Southern Illinois University at Carbondale in 1997.

Index

A

adult attachment theory 239
advertisements, classified 273, 276, 282, 288, 292
agriculture 98, 181
all-female learning group 11
anonymity 1-4, 6, 8-9, 11-15, 60, 145, 151, 172, 174, 192, 206, 223, 287-288
arranged marriages 272-273, 275-281, 284-285, 287, 289-291
attention phase 224, 226, 232
Australia 82, 120-121, 124, 126-127, 136, 257-258, 261-262, 265-268
authoritative language 3
avatar 145, 149-152, 157, 160, 204

B

balance between and attractive and real self (BAR) 19, 26, 159, 227-228, 230, 234, 256
Big 4 consulting firms 46
blogs 183-198, 201, 203-204, 216, 218
Brahmins 283, 285, 295
BrainDates.com 269
bulletin board systems (BBs) 222-223, 233
businesses, women-led 17-18, 24, 168
businesses, women-owned 18, 20, 31, 137, 162-163, 166, 168, 174-177, 180-181
business growth 19, 21, 28-29
business mentor 21
business networking 125
business ownership 162-164, 166-167, 175
business transactions 17-18, 82

C

career values 107
caste groups 285, 295
castes 67, 252, 260, 275-276, 280-281, 283-288, 290, 294-295

Central Statistics Office (CSO) 22-24, 31
cognitive-development theory 146-147
collaborative systems 3
commercial business growth objectives 29
computer information systems 36, 38, 85
computer mediated communications (CMC) 2-3, 7, 10-11, 13-15, 158, 160, 178, 196, 203-205, 216-217, 234, 271
confucian dynamism 261
cross-case analysis 6, 9, 11
cultural convergence 272, 281, 288
cultural differences 84, 89, 146, 257-258, 269-270
customer base 26-27
cyber-love 241
cyber-sex 237, 241, 243

D

data collection process 246, 278
dating behavior 259
digital divide 93, 100-101, 114, 117, 136, 139, 146, 175
disability 44, 87, 144-149, 151-159
disability schema theory 144-147, 157
DocDates.com 269
dowry 275-277, 291
duality of structure 274

E

e-business 2, 22-24, 28, 32-33, 82, 86, 96
e-commerce 32, 75, 78, 84-85, 91, 126, 138-142, 237, 273
economic development 17-18, 51, 69, 71, 99
e-daters 236-237, 239-243, 245-252
e-dating 222, 235-237, 239-252, 254, 256-258, 268-269, 271, 290-291, 293
e-dating evolvement 236
e-dating services 235, 240, 244, 250-252, 256-257, 271, 293

e-dating theory development 236
eHarmony 224, 236, 250
electronic dating 273
entrepreneurial intentions 163, 178
entrepreneurial segregation 163, 175
entrepreneurship 18-20, 28, 31-34, 92, 95, 136-137, 139-140, 162, 164-165, 167, 174-176, 178-181
ePerception 201-202, 204-205, 214, 218
ePersona 201-206, 208, 214, 218
essentialists 90
ethnography 277-278, 289, 291
Eurostat 100-101, 108, 113, 117, 119
evolutionary psychology 259, 267, 269-270

F

Facebook 145, 157, 204, 216-217, 230, 253
face-to-face interaction 163, 172-174, 213, 258
face-to-face meeting 226, 228, 232, 276, 288
family commitments 19, 21, 26, 113
family disintermediation 272
female entrepreneurs 17-22, 25-27, 29-30, 33, 138, 140, 164, 170-171, 177, 179-180
finance 24-25, 31, 124-126, 163, 168, 182, 206
finance availability 124
financial performance 17-18

G

gender 1-4, 6, 9, 11-15, 19-20, 29, 31-35, 37, 46-62, 64, 66-77, 79-102, 105, 107-116, 118-122, 124-127, 136-142, 144-149, 151-154, 157-160, 163-169, 171-181, 184-186, 188, 192-194, 196, 198, 200-205, 207, 209-217, 223-224, 238, 240, 244-245, 247, 249-250, 253-254, 262, 264-268, 270-271, 282-283, 290
gender anonymity 1, 12
gender-role socialization 3
gender schema theory 146-147, 158
gender stereotyping 14, 89, 93, 163, 167, 174-175
general practitioners (GP) 120-121, 125-127, 131-136
Germany 66, 98-108, 110-111, 114-117, 119, 177
GP, gender differences among 126
GP, male and female 120-121, 131, 135-136
graduate business programs 38
Greece 98-119
group dynamics 1, 4-5, 9, 11-12
group productivity 5
groupthink 5, 14

H

healthcare sector 121
homophilious networks 163, 166
homophily 160, 163, 165-166, 170-172, 174, 177-179, 182
homophily, choice 165
homophily, induced 165-166
horoscopes 281, 284-285, 295
human resource management (HRM) 27

I

ICT adoption 29-30, 120-123, 126-128, 131-136, 140
ICT sector 50-51, 59, 61, 64, 68, 88, 94, 99, 107, 114
ICT usage 98-100, 102, 114-115, 140
idea gathering 163
identity 2-3, 8, 37, 53-54, 93, 140, 144-149, 151-153, 156-161, 165, 174, 179, 203, 215-216, 218, 223, 226, 232-233, 254, 278
India, arranged marriage in 272-273, 275, 289
individualistic societies 267
industry 17-20, 25-26, 29-30, 33, 35-36, 41-47, 50, 60-61, 63, 69, 83, 88, 98, 110, 139, 162-163, 166-168, 170, 174, 176, 180, 237, 246, 251, 281
information rich environment 288
interaction phase 226-227, 231-232
interpersonal communication 75, 91-92, 125, 175, 216-217

J

joint family 273, 286, 288

L

leader emergence 1, 5, 11-12, 14
long-term orientation (LTO) 261-262
love marriage 273

M

management style 124-125
massive multiplayer online role playing game (MMORPG) 269
masters degree 25
Match.com 224, 236-237, 246-248, 254, 256
matchmaking 222, 225, 233-235, 273, 276, 280-284, 286, 288
mate-seeking behavior 257

matrimonial Web sites 272-273, 275, 277-279, 281, 288-290, 294
Media education 102-103, 115
media literacy 102
medical practice 120, 122-123
member awareness 1, 5, 11-12
mentoring 21, 30, 35-42, 45, 47, 49, 89
models of courtship 222
modern technologies 272, 274
mompreneurs 17-18, 21-22, 24-25, 29-30
MUDs, object oriented (MOO) 222-223
multi-user dungeons (MUD) 222-223

N

networking 21, 30, 32, 36, 39-42, 82, 124-125, 138-139, 145, 157, 163-164, 169, 171-179, 181, 200-201, 203-218, 230, 233, 235, 251, 256, 271, 293
nuclear family 61, 273, 276-277

O

online courtship 286
online daters 224, 226-228, 230-232, 234, 241
online dating 222-227, 229-234, 237, 241, 252-253, 270
online dating, phases of 222, 225
online matrimonial services 276-279, 281-283, 285, 287, 289-290
online matrimonial sites 257, 272-273
online personal ads 257-258, 272-273, 275, 288, 290-291
online profiles 283
Online Publishers Association (OPA) 237
online virtual worlds 149
opportunity recognition 163, 180

P

paid employment 17-18, 21, 56
partial least square model 121
Perceiver 201-205, 214-215, 218
personal ad dating 259
power distance 260-261, 268

R

recognition phase 224, 226-227, 232
resolution 225-226, 229, 232
romantic attachments 239

S

Schema Theory 144-147, 157-158
science, engineering, technology (SET) 1, 9, 36, 48, 56, 61, 67, 73, 80, 94, 96, 127-129, 205, 209, 222, 224, 229-231, 237, 241, 250-251, 256, 262, 274
science, technology, engineering and mathematics (STEM) 36, 87-89, 108-109, 114-115, 119
screen name 226, 229
search, matching, and interaction (SMI) framework 48, 225, 279, 289
Second Life 144-146, 148-160
self-concepts 53-54, 107, 115
sex discrimination 68, 163
sex roles 16, 54, 83, 85, 95-97, 139, 179-180, 261, 270-271
short-term orientation 261
small and medium enterprises (SME) 22-24, 31-34, 89, 120, 124-125, 140, 176
SMI framework 279
SMI model 225
social capital 165, 170-171, 173, 176-178, 180-181, 196
social constructionists 90
social construction of technological systems 274, 291
social construction of technology theory 273
social factors 90
social identity model of deindividuation effects (SIDE) model 15, 42, 121, 203-204, 213, 218, 261, 284-285, 295
social learning theory 146
social networking 30, 157, 200-201, 203-217, 230, 233, 235, 251, 256, 271, 293
social networking activity 201, 203-205, 213-215
social networks 29, 80, 163, 165, 171, 174-177, 179, 181-182, 208, 214, 273, 276, 282, 285-290
social penetration theory 230
social progress 17-18
social taboos 283
societal divisions 284
socio-emotional content 3
start-up completion 163
stay-at-home mum 21
stereotype-based models 202
Strategies of Inclusion: Gender and the Information Society (SIGIS) 93
structural factors 90, 175
sub-caste 275-276, 283, 285
supply chains 123

Index

T

team collaboration 1
technology adoption 275
Technology-as-designed 274
technology workers 48, 88
telephony system 41
traditional model of courtship 222
trait-based models 201

U

uncertainty avoidance 260-261

V

venture capital 163, 176, 281
virtual communities 73, 163, 172, 216
virtual dating 272
virtual flirt form note 229
virtual networking 30, 163-164, 171-174, 178
virtual teams 1-6, 10, 12-15, 215

W

warranting theory 204-205, 217-218
Web-based impressions 215
Web-based instant messaging client 8
Web-based phenomena 215
Web search skills 207
western stereotype 99
work environment 28, 46, 61, 70, 88
working models 239-240, 252-253
World of Warcraft 150

Y

YahooPersonals 236